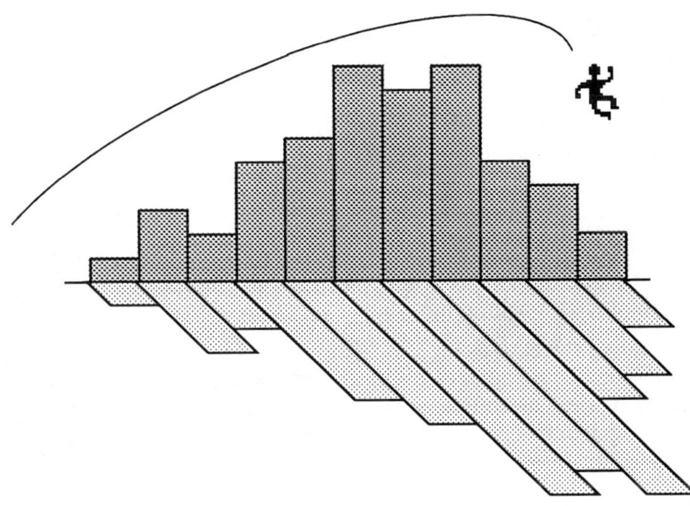

JMP IN® User's Guide

Statistical Visualization Software for the Macintosh®
Student's Edition
From SAS® Institute Inc.

Duxbury Press
An Imprint of Wadsworth Publishing Company
Belmont, California

Duxbury Press
An Imprint of Wadsworth Publishing Company
A division of Wadsworth, Inc.

1993 Duxbury Press Edition

© Copyright 1989 by SAS Institute Inc., Cary, NC, USA

ISBN 0-534-20544-5

All rights reserved. No part of this publication may be reproduced, stored in a retrieval system, or transmitted, in any form or by any means, electronic, mechanical, photocopying, or otherwise, without the prior written permission of SAS Institute Inc.

Information in this document is subject to change without notice. The software described in this document is furnished under the license agreement printed on the last page of this book, opposite the envelope that contains the software. The software may be used or copied only in accordance with the terms of the agreement. It is against the law to copy the software on any medium except as specifically allowed in the license agreement.

Printed in the United States of America

1 2 3 4 5 6 7 8 9 10—97 96 95 94 93

JMP® and JMP IN® are registered trademarks of SAS Institute Inc.
SAS® is the registered trademark of SAS Institute Inc.
Microsoft® is the registered trademark of Microsoft Corporation.
Apple®, Appletalk®, HyperCard®, ImageWriter®, LaserWriter®, Macintosh®, MacWrite®, MacWriteII®, and QuickDraw® are registered trademarks of Apple Computer, Inc.
MacDraw® and MacPaint® are registered trademarks of Claris Corporation.
PostScript® is a registered trademark of Adobe Systems, Inc.
SuperPaint® is a registered trademark of Silicon Beach Software.
Fullwrite® is a registered trademark of Ashton-Tate Corporation.
VAX® is a registered trademark of Digital Equipment Corporation.
WriteNow™ is a trademark of NeXT Inc.
Expressionist™ is a trademark of Allan Bonadio Associates.

This document was created with Microsoft Word software. Graphics and equations were produced using MacDraw, MacPaint, SuperPaint, HyperCard, and Expressionist software. The index was created with Microsoft Word and JMP software. The document was produced on an Apple LaserWriter printer and copies printed by offset lithography.

Credits

JMP was designed and developed by John Sall, Katherine Ng, and Michael Hecht. This book was written and produced by Ann Lehman, John Sall, and Mary Cole. Cover design was by Stuart Paterson/Image House. Statistical testing was done by Steve Shack. William R. Gjertsen administered the test sites.

Special thanks to Robert N. Rodriguez for statistical editorial support. Additional editorial support and testing was provided by John Boling, Philip Shelton, William Fellner, Eddie Routten, Donna Fulenwider, David Schlotzhauer, Maura Stokes, Donna Woodward, and Larry Crum. Technical support was provided by Jeffrey Perkinson. Document production assistance was provided by Eric Gjertsen. Thanks to the Statistical R&D Numerical Group for the linear algebra and probability routines we borrowed from the SAS System for JMP.

The product consists of about 60,000 lines of C code compiled in Think LightspeedC™ from Symantec, Bedford MA.

Acknowledgments

We owe special gratitude to the people that encouraged us to start JMP, to the alpha and beta testers of JMP, and to the reviewers of the documentation. In particular we thank Al Best, Harold Gugel, Stan Young, and Kris Rahn. Also, we thank Charles Shipp, David Morganstein, Mei-Mei Ma, Tom Esposito, Susan West, Jim Winters, Chris Fehily, James Mulherin, Dan Chilko, Jim Shook, William Lisowski, Bud Martin, and George Fraction. Special thanks also to R. Hocking and P. Spector for advice on effective hypotheses.

Limited Warranty

Duxbury Press warrants the diskette(s) on which the software is furnished to be free from defects in materials and workmanship under normal use for a period of ninety (90) days from the date of purchase. Your copy of the invoice is your evidence as to when your copy of the software was purchased. Your exclusive remedy shall be replacement of the diskette(s).

Contents

Chapters 9-19 contain detailed chapter contents.

Preface

We are happy to share JMP IN with you. This product lets you use the extraordinary capability of the Apple Macintosh to display and analyze data. JMP IN is software for interactive statistical graphics and includes

- a spreadsheet for viewing, editing, transforming, and manipulating data
- graphical and statistical methods for data analysis
- options to highlight and display subsets of the data
- tools for moving analysis results between applications and for printing.

JMP IN is easy to learn. You can begin immediately and see graphical and statistical results quickly by applying the techniques presented in this book. JMP IN is the student edition of JMP Version 1. It is a fully functional version of JMP but is restricted to saving JMP data tables with 4000 or fewer cells; however, JMP IN can read JMP tables of any size.

This book, the *JMP IN User's Guide*, contains complete documentation of the JMP IN software product and includes some discussion of statistical theory and methodology. It is designed to support the use of JMP IN software as a learning aid for students in beginning or advanced statistics courses. The content assumes no previous knowledge of statistics and should be used in conjunction with a statistical textbook.

The disk that contains JMP IN software, sample data, and Help files is included at the back of this book.

Happy JMPing

SAS Institute Inc.
SAS Campus Drive
Cary, NC 27513

Chapter 1.
Introducing JMP

JMP IN software is an interactive statistical visualization application for the Macintosh.

It is easy to learn and use. Statistical concepts are automatically displayed in graphs and plots and are supported with the appropriate analysis tables. The data and all its related plots and graphs are linked so that relationships between variables are quickly seen. The interactive nature of JMP IN encourages data exploration, and visualization of concepts makes learning about statistics much easier than ever before, and even fun.

This chapter gives you the *getting started* information you need. It tells you the hardware requirements and how to install JMP IN. It includes a section describing the organization of JMP data tables in memory and on disk, and defines terms used throughout the following chapters. It also takes you on a guided tour, a simple tutorial that covers the basics of using JMP IN.

Preliminaries

Hardware requirements

To run JMP, you need an Apple Macintosh with a hard disk and at least one megabyte of memory. Two megabytes are recommended but not required.

Installing JMP

Before installing JMP IN, it is best to make a copy of the disk and store the original. To install, create a new folder on your hard disk. Then insert the program disk copy and double–click on the icon, 'Install JMP IN'. A standard Macintosh dialog appears asking you to select a destination folder for JMP IN. Select your hard disk and the newly created folder. Then click the 'Extract' button in the dialog to complete the installation.

Learning About JMP

On Your Own

If you are familiar with Macintosh software, you may want to proceed on your own. After you install JMP, you can open any of the JMP files in the SAMPLE DATA folder and experiment with analysis tools. Help windows are available for most menus and options.

To access JMP help windows, select **About JMP** from the Apple menu. You can also choose **Help** from pop-up menus in JMP report windows, or you can select the help tool (**?**) from the **Tools** menu and click on items.

Reading the JMP User's Guide

If you prefer to learn JMP from documentation, read this book. It is divided into three sections.

Part I is a collection of tutorials designed to help you learn JMP strategies. Each tutorial uses a file from the SAMPLE DATA folder. By following through these step-by-step processes, you can quickly become familiar with JMP menus, options, and report windows.

Part II contains complete documentation of all JMP menus, an explanation of data table manipulation, and a description of the calculator. These chapters show you how to do common tasks such as manipulating files, selecting subsets, transforming data table columns, and cutting and pasting JMP data and analyses.

Part III documents each of the statistical platforms. This section discusses statistical methods and describes all report windows and options.

Completing Project Examples

This book includes a collection of tutorials designed to help you learn about each of the JMP analysis platforms. These tutorials range from single step procedures to complex analyses. You can read through the tutorials for reference, or you can work through them step-by-step.

What You Need to Know...

About the Macintosh

Before you begin using JMP, you should be familiar with standard Macintosh operations and terminology such as *click*, *double-click*, *shift-click*, *drag*, and *select*. You should also know how to use menu bars and scroll bars, move and resize windows, and how to manipulate files in the Finder desktop. If you are using your Macintosh for the first time, consult the reference guides that came with your system for more information.

About Statistics

You can use JMP with a minimal background in formal statistical training. All analysis platforms include graphical displays with options which help you review and interpret the results. Each platform also includes access to help windows which offer general help and some statistical details. There are also tutorials in this book which provide examples of displaying and analyzing data.

JMP Basics

To use JMP, you need to know how to

- start a JMP session
- create a data table and use a JMP spreadsheet
- select an analysis platform
- end a JMP session.

The following sections describe these activities. Follow through each of these steps to ensure that you have all the basic techniques at your fingertips.

Starting a JMP Session

Start a JMP session by double-clicking on the JMP icon. When JMP is active you see the following menu bar:

Figure 1.1 The Menu Bar

 File Edit Rows Columns Analyze Tools Window

You also see the blank untitled JMP spreadsheet shown in **Figure 1.2**. The new spreadsheet is available for you to key in your own data but is not needed for this tutorial. To close the spreadsheet, click the close box in the upper left corner of the window.

Figure 1.2 A New Untitled JMP Spreadsheet

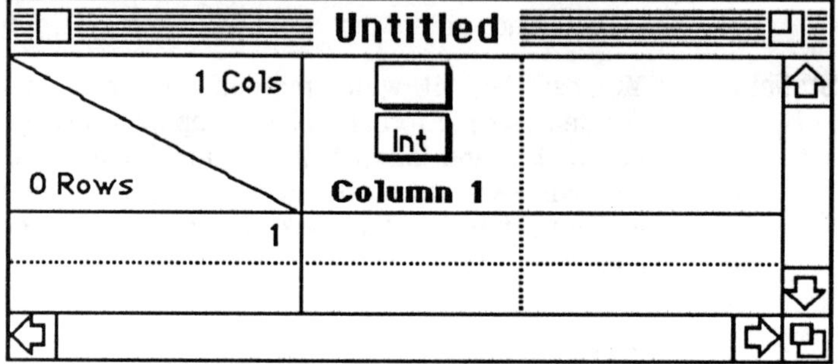

Creating a JMP Data Table

Your next action is to enter your data in the JMP system. There are three ways to do this.

- The **New** command creates an empty data table with a spreadsheet available for you to paste new data.
- The **Import** command lets you use data created by other applications.
- The **Open** command gives you the standard Macintosh file selection dialog with a list of existing JMP files.

To continue with this demonstration select **Open** from the **File** menu.

Figure 1.3 The **Open** command in the **File** menu

When the open file dialog appears, select and open the file called TYPING DATA .

Using the Spreadsheet

Opening a file creates a data table in memory that appears in spreadsheet form with rows and columns as shown in **Figure 1.4**. The counts of rows and columns appear in the upper left corner of the spreadsheet. Each row is identified by a row number and each column by a column name. The rows and columns are sometimes referred to as observations and variables in an analysis.

Figure 1.4 The TYPING DATA Spreadsheet

TYPING DATA

2 Cols / 17 Rows	Nom brand	Int speed
1	REGAL	70
2	SPEEDYTYPE	87
3	SPEEDYTYPE	79
4	REGAL	73
5	SPEEDYTYPE	77
6	REGAL	72

The JMP spreadsheet is a flexible worksheet for preparing data. Using it, you can perform a variety of data table management tasks including

- editing the value in any cell
- changing a column's width by dragging the column line
- hiding columns temporarily or deleting columns permanently
- rearranging the order of columns
- selecting a subset of rows for analysis
- sorting a table
- combining tables.

Cursor Forms

To navigate in the spreadsheet, you need to understand how the cursor works in each part of the spreadsheet. The cursor assumes different forms and performs different activities depending on its location in the spreadsheet. These forms and functions are described below.

Arrow Cursor

The cursor displays as a standard arrow when it is in the *measurement level box* or the *role assignment box* as shown in **Figure 1.5**. It is also a standard arrow when it is in the *Cols* or *Rows* triangular areas in the upper left corner of the spreadsheet.

Figure 1.5 Arrow Cursor in the Spreadsheet

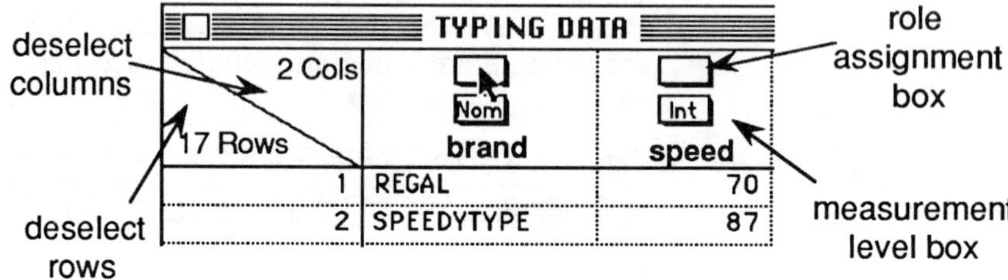

Cross Cursor

When the cursor is within a column heading or a data cell in an active spreadsheet, it becomes a large cross indicating that it is available to select text. When you click the cross cursor, the text is highlighted and can be edited. The cursor then becomes an I-beam.

Figure 1.6 Cross Cursor in the Spreadsheet

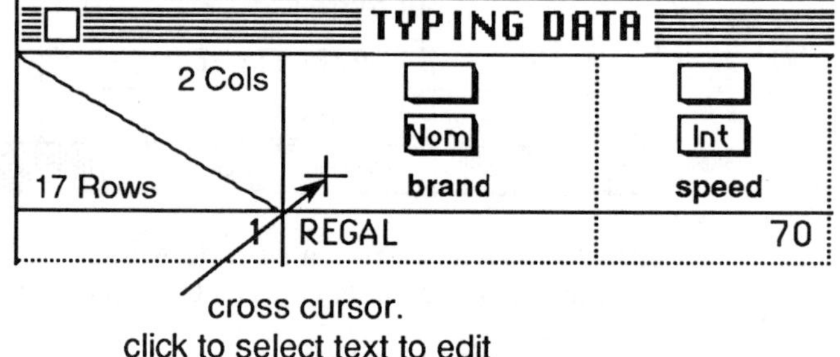

I-beam Cursor

To edit text, position the I-beam within highlighted characters. Click next to any character to mark an insertion point. The I-beam deposits a vertical blinking bar. Alternatively, drag the I-beam to select a portion of the text for replacement (by default, the entire character string is selected). Use the keyboard to make changes after positioning the I-beam.

Figure 1.7 I-beam Cursor in the Spreadsheet

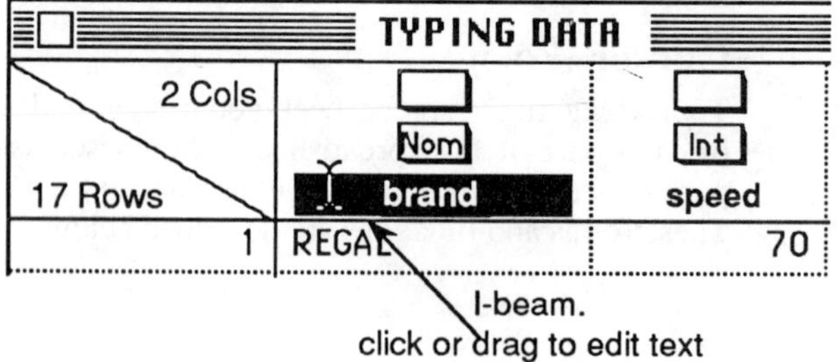

Open Cross Cursor

The cursor becomes a large open cross when you move it into a column or row selection area. Use the open cross cursor to select a single row or column. Shift-click a beginning row or column and an ending row or column to select an entire array. Use command-click to select multiple rows or columns that are not contiguous. See the next section for a more detailed explanation of selecting rows and columns.

Figure 1.8 Large Cross Cursor in the Spreadsheet

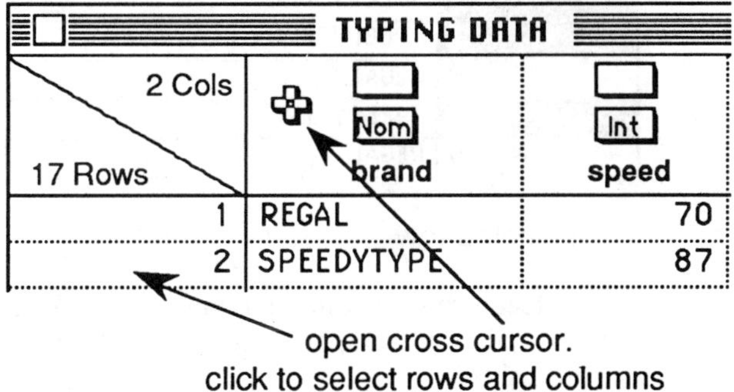

Double Arrow Cursor

The cursor changes to a double arrow cursor when it is placed on a column boundary. This boundary dragger is used to change the width of columns.

Figure 1.9 Double Arrow Cursor in the Spreadsheet

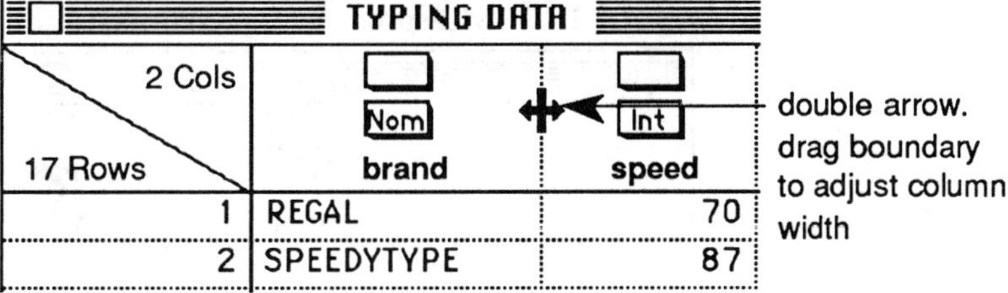

Selecting Rows and Columns

The selection of rows and columns in a JMP spreadsheet is done by highlighting them as shown in **Figure 1.10**. To highlight a row, click on the space that contains the row number. To highlight a column, click on the background area above the column name.

To extend a selection of rows or columns, drag the cursor across the array or shift-click the first and last rows or columns of the desired range. To make a discontinuous selection, command-click on the desired selections.

Figure 1.10 Select Rows and Columns

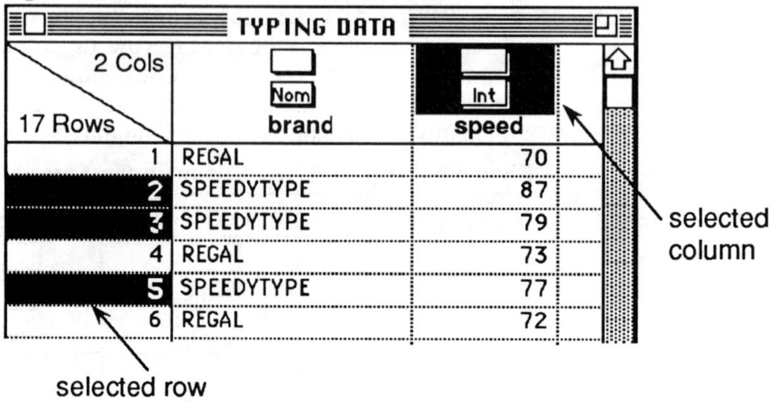

TYPING DATA

2 Cols / 17 Rows	brand (Nom)	speed (Int)
1	REGAL	70
2	SPEEDYTYPE	87
3	SPEEDYTYPE	79
4	REGAL	73
5	SPEEDYTYPE	77
6	REGAL	72

selected column

selected row

To deselect a row or column, command-click on it. To deselect all rows or columns, click in the appropriate triangular area in the upper left corner of the spreadsheet.

Specifying the Level of Measurement

The small box just above the column name is a pop-up menu used to declare the measurement level of the values in the column.

Figure 1.11 Measurement Levels

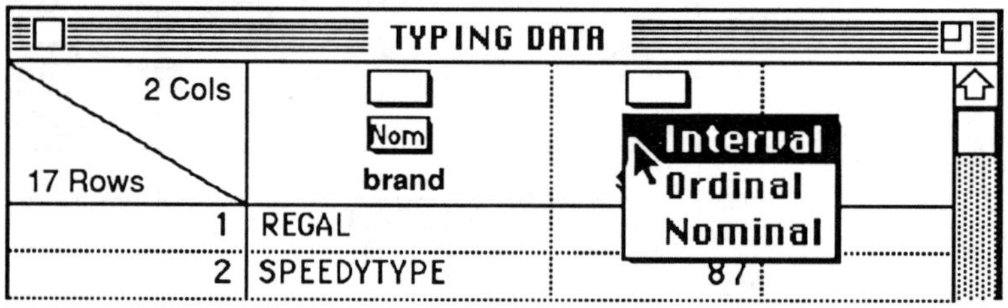

TYPING DATA

2 Cols / 17 Rows	brand (Nom)	
1	REGAL	
2	SPEEDYTYPE	87

Interval
Ordinal
Nominal

There are three levels of measurement.

- **Interval** values are numeric measurements.
- **Ordinal** values are ordered numeric or character values.
- **Nominal** values are numeric or character classifications.

The default measurement is nominal for character type values and interval for numeric values. In **Figure 1.11**, the column called brand is a nominal variable and speed is an interval variable.

You can assign each column a measurement level appropriate for the next analysis using its measurement level pop-up menu (see **Figure 1.11**).

The measurement level is important so that JMP can automatically determine how to analyze the column's values.

Choosing Variable Roles

The JMP analysis methods are like stages or platforms for variables to dramatize their values. Each analysis platform requires information about what role the variables are to play in analysis. You can specify variable roles by

- using the role assignment box at at the top of each column in the spreadsheet
- selecting a command from the **Analyze** menu and responding to a specific assign roles prompt
- using the **Assign Roles** command from the **Columns** menu.

Role assignment boxes in the spreadsheet are pop-up menus. Drag the arrow to the appropriate selection for your analysis as shown in **Figure 1.12**. You can change a variable's role at any time or drop it from analysis by selecting **None** from the role assignment pop-up menu. Each role is described below.

Figure 1.12 The Variable Role Menu

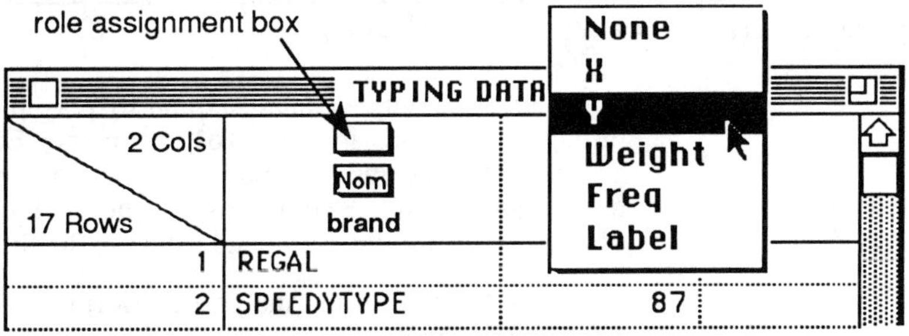

- **NONE** means the column is not included in the analysis.
- **X** identifies a column as a classification or independent variable whose values divide the rows into sample groups.
- **Y** identifies a column as a response or dependent variable whose distribution is to be studied.
- **Weight** identifies a column whose values are used to supply weights for each response.
- **Freq** identifies a column whose values assign a frequency to each row for the analysis.
- **Label** identifies a column whose values are used to specify a labelling convention for points in plots.

When you choose a variable's role using the role selection menu or the **Assign Roles** command, it appears in the role selection box as shown in **Figure 1.13**. Roles remain in effect until they are changed.

Proceed with this demonstration by identifying both the brand and speed columns as Y variables in the TYPING DATA spreadsheet.

Figure 1.13 Choose Variable Roles

TYPING DATA

	Nom brand	Int speed
REGAL		70
SPEEDYTYPE		87

TYPING DATA

	Y Nom brand	Y Int speed
REGAL		70
SPEEDYTYPE		87

Selecting an Analysis Platform

There are a variety of analysis and display platforms available in JMP through commands in the **Analyze** menu. Each command produces a graphical display and optional statistical text reports. For example, to see histograms of all Y columns in the active spreadsheet, select the **Distribution of Y's** command from the **Analyze** menu.

Figure 1.14 Select an Analysis Platform

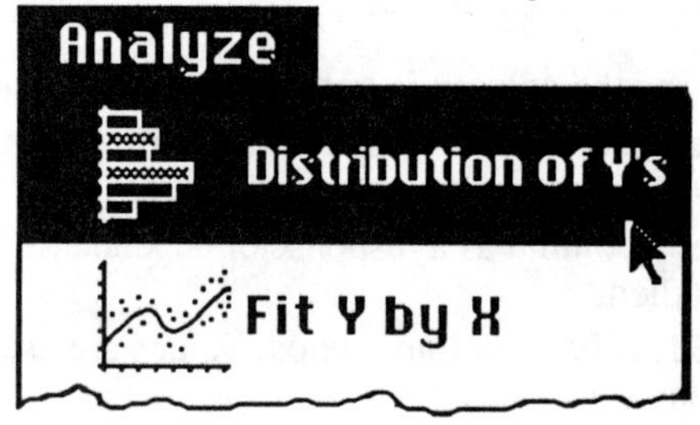

Graphical Display of Data

When you choose the **Distribution of Y's** analysis, a report window is created containing a graphical display of all Y columns in the TYPING DATA table. For the TYPING DATA example, the **Distribution of Y's** command produces the graphical displays shown in **Figure 1.15** which include

- histograms of both the brand and speed columns
- a mosaic (stacked bar) chart of the nominal variable, brand
- a quantile box plot of the interval variable, speed.

Figure 1.15 Graphical Displays from **Distribution of Y's** Platform

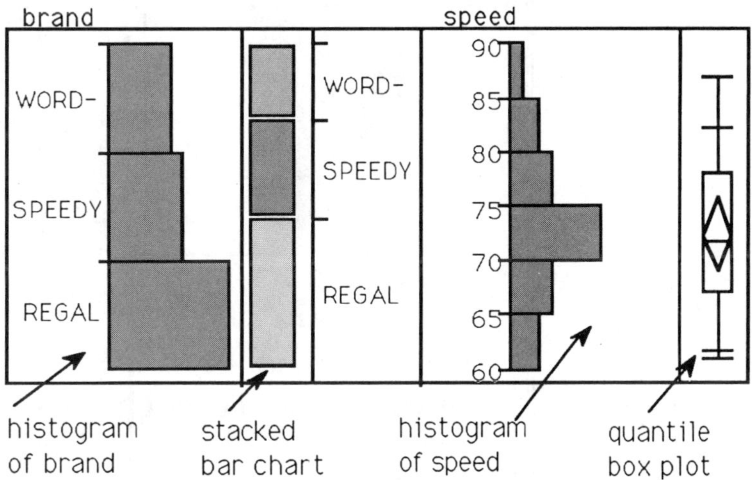

Adjust the histogram bars

To adjust the histogram bars, select the hand tool from the **Tools** menu. Use the hand to *grab* the plot you want to change. Moving the hand to the left combines intervals and shows fewer bars. Moving the hand to the right produces a finer division of each intervals showing more bars. Moving the hand up or down maintains the number of bars but changes the boundary of the bars on the axis, redistributing the data.

Figure 1.16 Adjust the Number of Bars

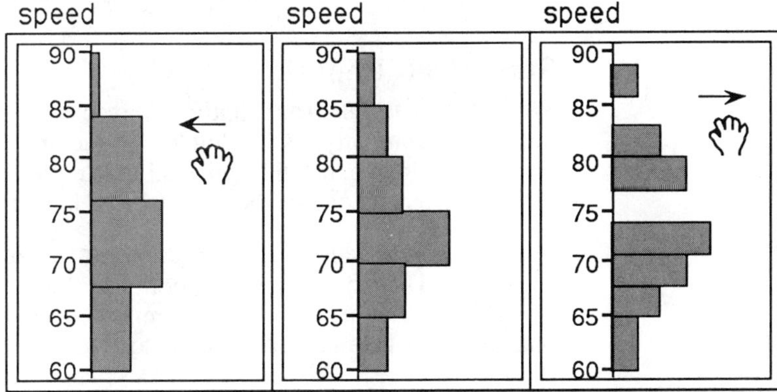

Display options

The plots are one of several optional representations of data. The check mark at the left corner of the horizontal scroll bar accesses a pop-up menu which lists report options available in the **Distribution of Y's** platform.

Figure 1.17 Report Options for **Distribution of Y's** Platform

TYPING DATA: Distribution

brand

WORD-

WORD-

SPEEDY

REGAL

✓Text Report
✓Histogram
✓Box/Mosaic Plot
Normal Quantile Plot

Horizontal Layout
Normal Curve
Count Axis
Prob Axis

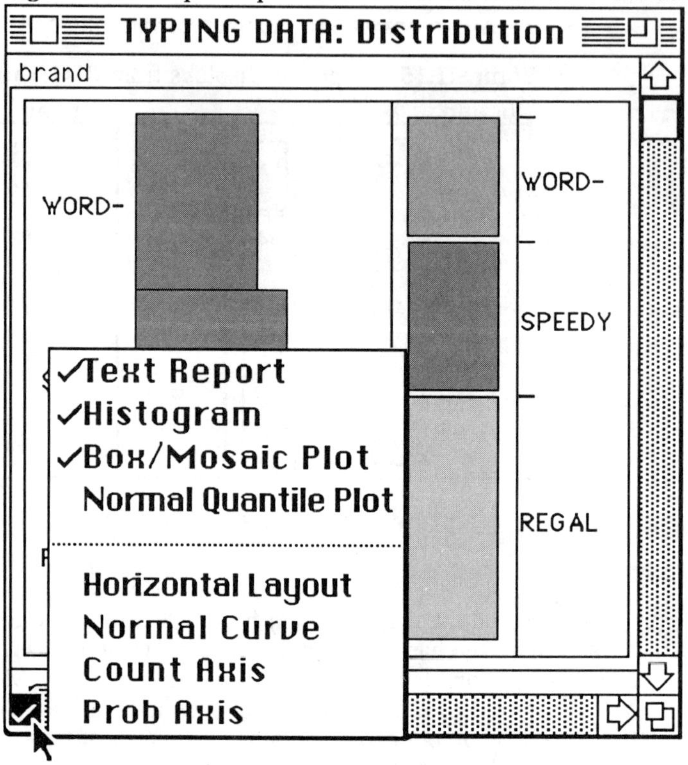

For practice, try selecting different combinations of these options and watch the effect they have on the displays and reports.

Statistical Reports

When the **Text Report** option is checked, *reveal/conceal* text report buttons accompany each graph. You can click on these buttons to alternately show and hide statistical text reports. The kinds of reports you can reveal depend on whether a variable is interval, ordinal, or nominal.

For an interval variable, there are Quantiles and Moments reports. **Figure 1.18** shows these reports for the speed variable from the TYPING DATA table. The Test report, which is not revealed, is a t-test that compares the mean of a sample to zero.

Note➔ The layout of the text report shown in **Figure 1.18** results when the **Horizontal Layout** display option is checked.

Figure 1.18 Reports for Interval Variables

Quantiles | Moments | Tests

Quantiles

100% maximum	87
99.5%	87
97.5%	87
90%	82.200
75% quartile	78
50% median	72
25% quartile	67
10%	61.800
2.5%	61
0.5%	61
0% minimum	61

Moments

N	17
Sum Wgts	17
Mean	72.4705
Std Dev	7.00105
Std Err Mean	1.69800
upper 95% Mean	75.7986
lower 95% Mean	69.1425

Tests

For nominal and ordinal variables, the Summary and Profile reports show the total sample frequency, category frequencies and associated probabilities. **Figure 1.19** shows the reports for the variable brand from the TYPING DATA table.

Figure 1.19 Reports for Nominal or Ordinal Variables

Summary | Profile

Summary

Number of Levels	3
Total Frequency	17
Total LogLikelihood	17.936727

Profile

Level	Count	Probability	Cum Prob
REGAL	8	.470588	.470588
SPEEDYTYPE	5	.294117	.764705
WORD-O-MATIC	4	.235294	1

Completing a JMP Session

After you have completed an analysis, you can use JMP tools to copy any part of a report window into the Macintosh clipboard. When you have completed the analysis, you can close the report window or the spreadsheet without quitting JMP.

- To close a report window, click the close box in the upper left corner of the window.
- To close a spreadsheet, either click the close box in the upper left corner of the spreadsheet or choose **Close** from the **File** menu while the spreadsheet is the active window. If you have made changes to the spreadsheet, a dialog asks if you want to save or discard the changes to the data table. The dialog also asks if you want to close related windows (if any are still open).

To end a JMP session, select **Quit** from the **File** Menu.

Take A Quick Spin

The following example is designed to take you on a short guided tour through a JMP session. Follow the steps to see a three-dimensional spinning plot.

Open a Data Table

1 Begin a JMP session by opening the file called COWBOY HAT. The spreadsheet shown in **Figure 1.20** appears.

Figure 1.20 The COWBOY HAT Spreadsheet

COWBOY HAT

4 Cols / 441 Rows	x (Int)	y (Int)	z (Int)	markers
1	5	5	10.71	▪
2	5	5.5	10.43	▪
3	5	6	10.12	▪
4	5	6.5	9.821	▪
5	5	7	9.563	▪
6	5	7.5	9.361	▪

This spreadsheet consists of three numeric columns and a *row state* column. **x** and **y** are x and y coordinates, and **z** is a result of the function

$$z = SIN(SQRT(x^2 + y^2)) \quad .$$

The fourth column, called **markers**, contains special row state information used to plot points.

2

The pop-up menu at the top of a row state column differs from those above numerical and character data columns. Use the row state pop-up menu at the top of the markers column to activate its row state information. The active row state information appears to the left of the row numbers as shown in **Figure 1.21.**

Figure 1.21 Active Row State

COWBOY HAT

4 Cols / 441 Rows		x (Int)	y (Int)	z (Int)	markers
▪	1	5	5	10.71	▪
▪	2	5	5.5	10.43	▪
▪	3	5	6	10.12	▪
▪	4	5	6.5	9.821	▪

Copy to RowState

active row state information

Select an Analysis Platform

3

To plot the three columns of information from the COWBOY HAT data table, choose the **Spin** platform from the **Analyze** menu.

Figure 1.22 The **Spin** Command in the **Analyze** Menu

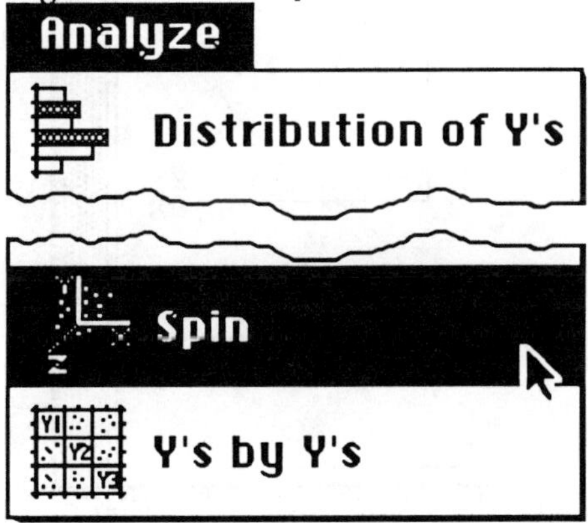

4 When you release the mouse, the dialog shown in **Figure 1.23** prompts you to select columns for the Spin platform. Select the **x**, **y**, and **z** columns from the column selector list on the left side of the dialog, and click the **Add** button. These column names now appear in the column on the right side of the dialog.

Figure 1.23 The **Spin** Column Selection Dialog

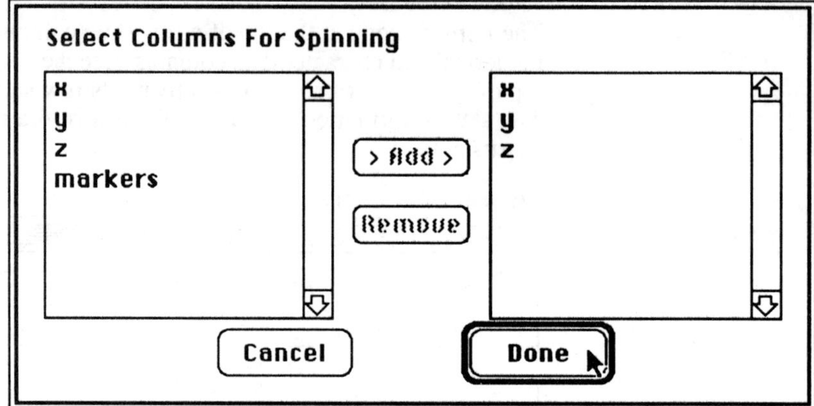

5 When you click **Done**, the spin platform appears. The points in the COWBOY HAT data table are plotted on the spin platform as shown in **Figure 1.24**. They appear as a contour plot because the Z dimension is projected onto the X, Y plane.

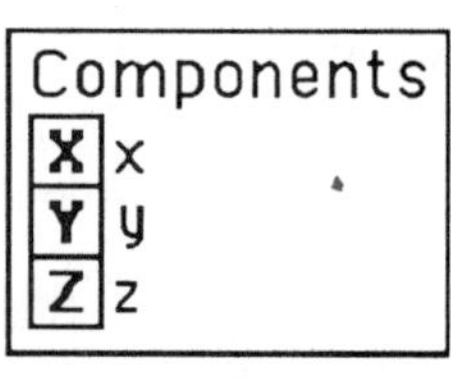

Figure 1.24 Contour Representation of the COWBOY HAT Data Table

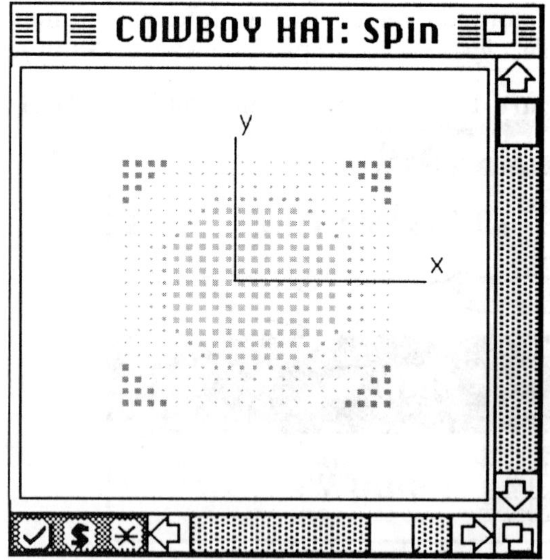

Spin the Cowboy Hat

Next, using the **Tools** menu, change the standard arrow cursor to the hand tool. Position the hand on the cowboy hat spin plot, hold down the mouse button, and move the hand about. The cowboy hat moves in three dimensions as shown in **Figure 1.25**.

The plot can also spin by itself. Hold down the shift key and give the plot a push with the hand.

To stop the spinning plot, click in any JMP window.

Figure 1.25 Three-dimensional Plot of the COWBOY HAT Data Table

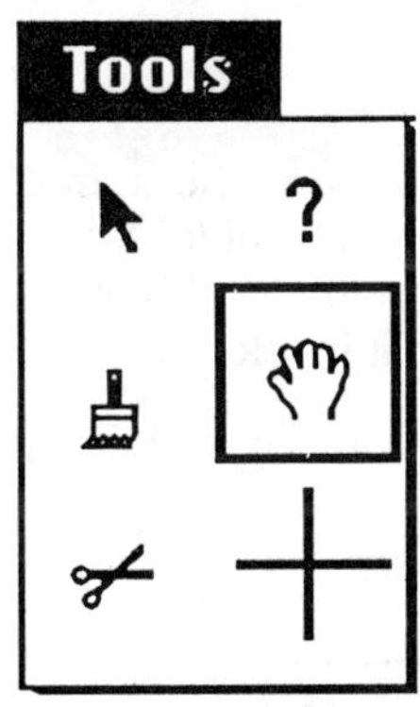

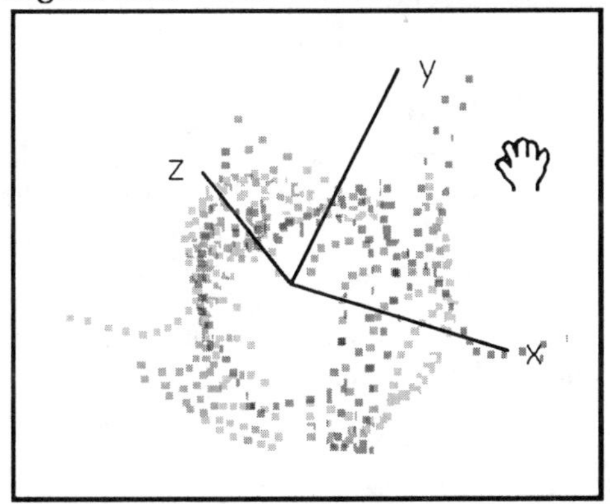

Precision Spinning

The spin control panel, shown in **Figure 1.26**, contains twelve small icons. These icons control the spin plot.

The six directional arrows are buttons which control the spin of the three dimensional image. If you click on one of these arrows, the plot spins in the direction indicated. As long as you hold down the mouse button, the spin persists. However, you can shift-click on a directional arrow to make the plot spin continuously. Clicking the mouse anywhere in the display stops the spin. The other six icons control spin rate, plot scale, and various display options.

- Clicking on one of the angle icons containing the + and – symbols increases or decreases the rate of the next spin, respectively.
- The two four-way arrows adjust the scale of the plot as indicated by their arrows.
- The triangle icon accesses a pop-up menu of display options which are described in the next section.
- The home icon resets the spin axes to their original position.

Figure 1.26 Precision Spinning

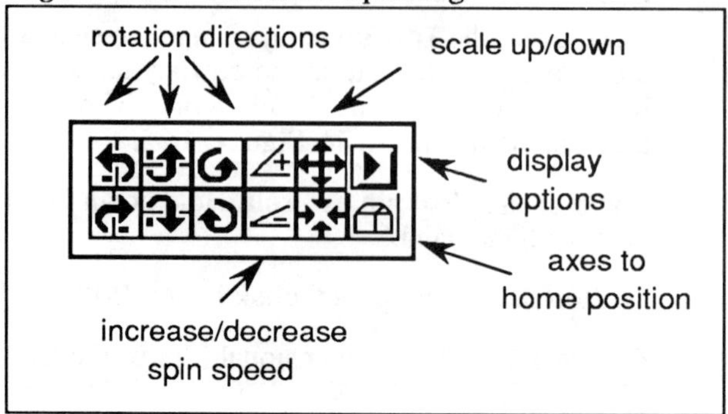

Display Options

The display options shown in **Figure 1.27** let you tailor the spinning plot. By default, the background is black and the axes are visible. You can change these conditions by selecting different options from this pop-up menu. Each option switches on or off alternately when it is selected.

Complete your experiment with the spinning cowboy hat by trying out a few of the display options.

Figure 1.27 Display Options

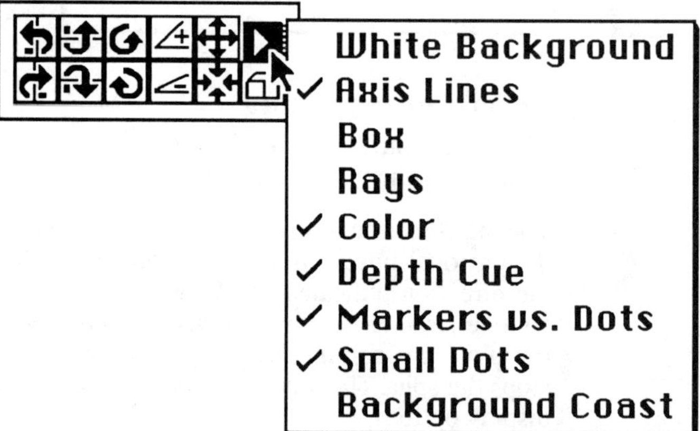

When you finish,

- close the spinning platform by clicking the close box in the upper left corner of the window.
- close the COWBOY HAT spreadsheet by clicking in its close box.

To end a JMP session, choose **Quit** from the **File** menu.

Chapter 2.
Looking at Distributions

School Health Study

The students in a local school are participating in a health study. The data analyst for the school system has been asked to summarize basic information about the students for the school system health care specialists. The data collected include age, sex, weight, and height.

The summaries are needed to document the sample of participating students, identifying any students with unusual characteristics who may need special medical attention. The job of the analyst is to produce these reports with graphs and short straightforward explanations.

You can follow the path the analyst takes to summarize demographic and health data. The data are in the JMP file called STUDENTS, found in the SAMPLE DATA folder.

Look Before You Leap

The school analyst knows that the first step is to become familiar with the data in the STUDENTS file. Looking at the information using the JMP spreadsheet helps the analyst to determine which summary charts and tables to use in the health report.

Open a JMP File

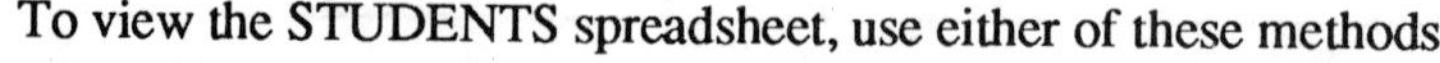

1 To view the STUDENTS spreadsheet, use either of these methods

- double-click on the JMP icon and use the **Open** command in the **File** menu to open the JMP file called STUDENTS
- double-click on the STUDENTS file icon to start the JMP application automatically and display the STUDENT spreadsheet.

You should see the spreadsheet shown in **Figure 2.1.**

Figure 2.1 The STUDENTS Spreadsheet

STUDENTS

5 Cols / 233 Rows	Ord age	Nom sex	Int height	Int weight	Nom idnum
1	11	F	56	85	874
2	11	F	57	69	31
3	11	F	54	69	613
4	11	F	62	104	814
5	11	F	51	51	628
6	11	F	62	85	895

The STUDENTS file contains the height, weight, age, sex and a student identification number for each student participating in the health study. The data table is in order by age, and sex is ordered within each age group.

Even though there are only five columns of information, these variables can be used to answer a number of questions that interest the health researchers.

- How many boys and how many girls are there?
- How old are they?
- What is the average height and weight for the group?
- Are there any students much younger or older than the average age?
- Are there any students whose height or weight might signal the need for medical attention?

Choose Variable Roles

2 The school analyst begins by examining the demographic distribution (age and sex) of the students. To identify the columns for analysis, the analyst assigns the role of Y to both the age and sex columns. The empty box at the top of each column is a pop-up role assignment menu.

Figure 2.2 Choose Variable Roles

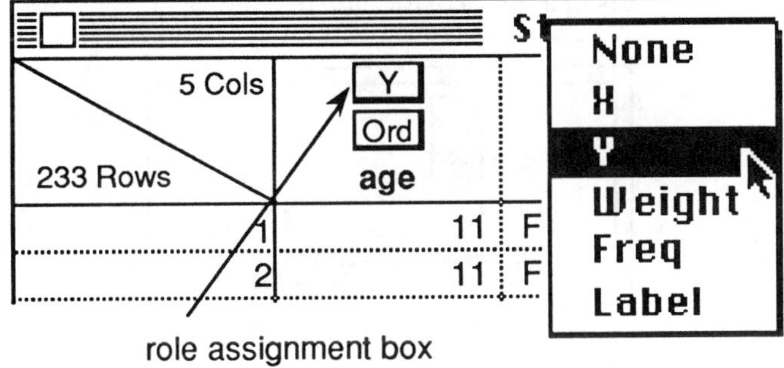

Note ➡ You can examine as many or as few columns as you like during a JMP session. When you assign the role of Y to a column, you cast it as a response variable for the analysis. As the investigation proceeds, you can change the role of a variable or select **None** from the pop-up menu to eliminate the column from further analysis.

Graphical Display of Distributions

3 Now the analyst is ready to select a command from the **Analyze** menu to summarize the information. For this problem the statistical platform called **Distribution of Y's** is appropriate. Each column identified as Y becomes a response variable for the analysis and is summarized individually.

Figure 2.3 The Analyze Menu

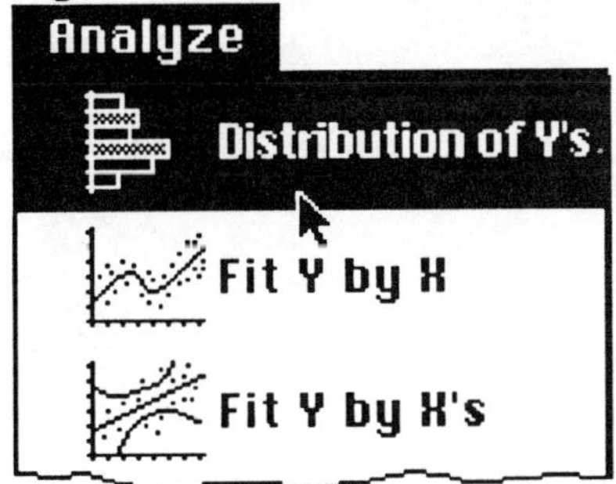

Histograms of Nominal and Ordinal Values

After selecting **Distribution of Y's**, a summary plot for each of the **Y** columns is presented. The values of age are ordinal and the values of sex are nominal so a histogram is displayed with a mosaic bar plot to the right for each of these variables. The histogram displays a bar for each level of an ordinal or nominal variable.

Figure 2.4 Histogram of Age Values

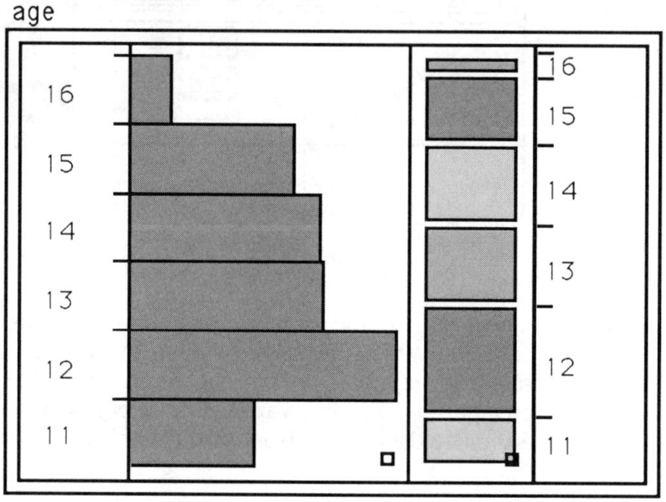

What can be said about age?

The analyst sees at a glance that the distribution of age is slightly skewed with a larger proportion of younger students and a small number of older students.

Figure 2.5 Histogram of Sex Values

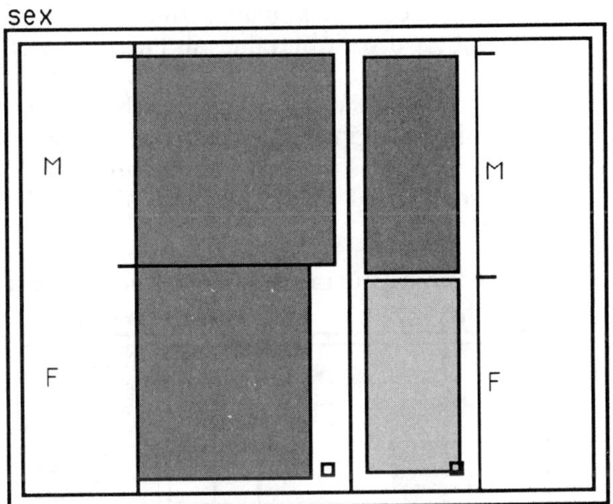

What can be said about gender?

The bars for sex appear nearly the same. The proportion of boys and girls in the school is about even, which is what you expect to see.

Mosaic Bar Plots

For ordinal or nominal variables, a mosaic bar is displayed with each frequency histogram. The mosaic bar is a way to visualize the the proportion of each ordinal or nominal level within the sample. The bar consists of a section for each level, where the size of the section is proportional to the corresponding group's size. You can think of a mosaic plot as a bar chart with its bars stacked end to end.

Figure 2.6 Horizontal Display of Mosaic Age Bar

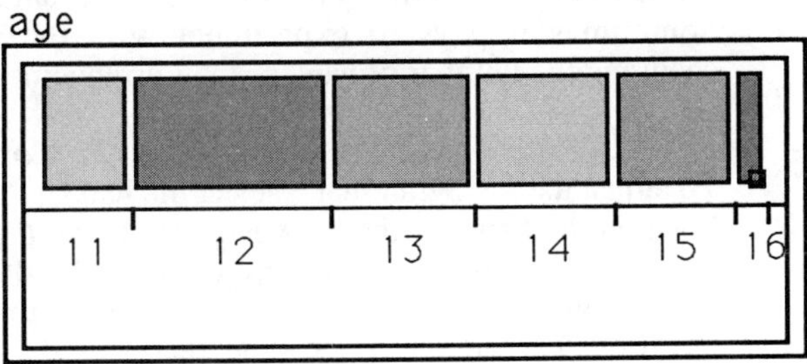

Histograms of Interval Values

4

The analyst continues summarizing the data by looking at the distribution of height and weight values in the sample of students. To follow along with the analyst, click on the STUDENTS spreadsheet to make it the active window. Using the role assignment box on the spreadsheet, designate the height and weight columns as response variables (**Y**). Designate the age and sex columns as **None** to eliminate them from the next analysis. **Distribution of Y's** displays the plots for height and weight shown in **Figure 2.7.**

Figure 2.7 Histograms of Height and Weight

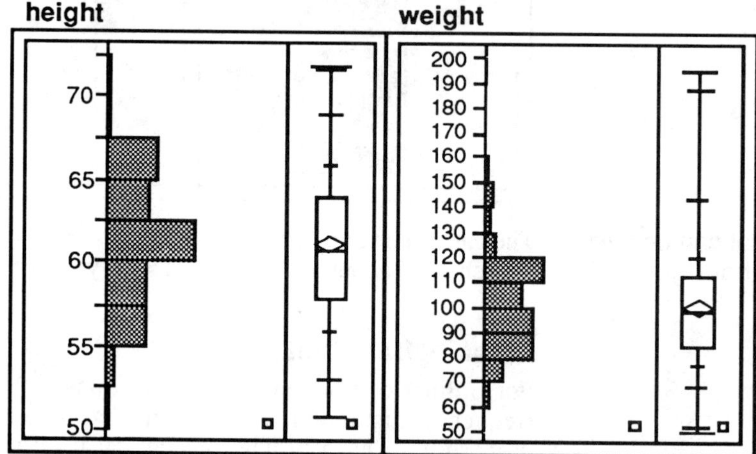

What can be said about height and weight?

Both height and weight appear to have normal (bell shaped) distributions, but the analyst notices a few extreme values and plans to examine them more closely later.

The analyst knows the value of presenting data in the best possible form. Sometimes it pays to experiment with the shape of a histogram by changing the number of bars, or their arrangement on the axis.

5

To adjust the histogram bars, select the hand, , from the **Tools** menu. Position the hand on the bars and hold down the mouse button to *grab* the plot. If you think of each bar as a bin that holds a number of observations, then moving the hand to the left increases the bin width and combines intervals. The number of bars decreases as the bar size increases. Moving the hand to the right decreases the bin width showing more bars. Moving the hand up or down changes the boundaries of the bins. The height of each bar adjusts according to the new number of observations within each bin.

Exchange the hand for the standard arrow cursor in the **Tools** menu after adjusting the bars.

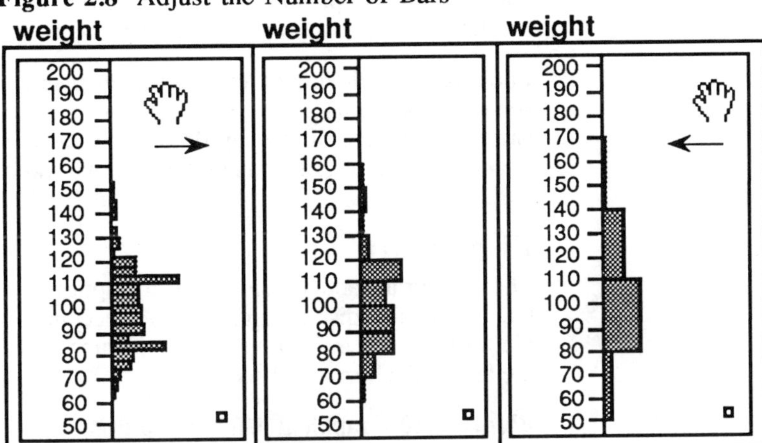

Figure 2.8 Adjust the Number of Bars

Box Plots

The box plot, also called a quantile plot, is a simple schematic for looking at the shape of a distribution. It shows the location of selected percentiles, called quantiles, on the response axis as shown in **Figure 2.9**. Display options (discussed in the next section) produce the horizontal view of the box plot and the quantiles report shown on the right.

Figure 2.9 Box Plot of Weight Distribution

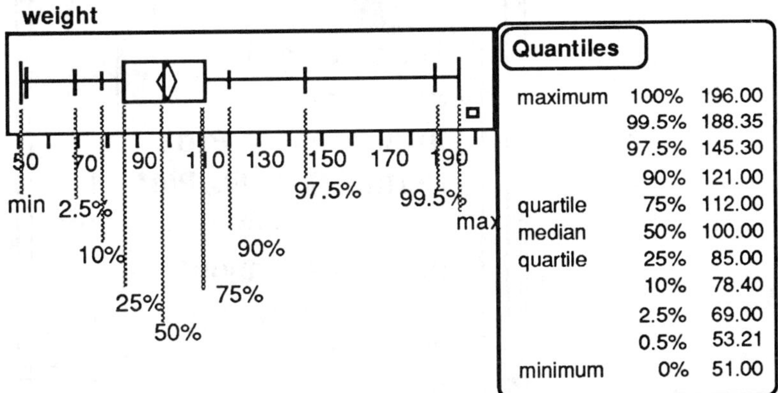

Quantiles

maximum	100%	196.00
	99.5%	188.35
	97.5%	145.30
	90%	121.00
quartile	75%	112.00
median	50%	100.00
quartile	25%	85.00
	10%	78.40
	2.5%	69.00
	0.5%	53.21
minimum	0%	51.00

The box plot shows the median as a line in the body of the box. The ends of the box locate the 25th and 75th quantiles. The number of other quantile lines depends on the available space. The accompanying report shows the data values for each of the standard quantiles.

The box also contains a means diamond. The mean of the sample is identified by drawing a line that crosses the axis and passes through the two diamond points spanning the box. The two diamond points within the box identify the 95% confidence interval of the mean.

Looking at the quantile box plot and means rectangle together helps you to see if data are distributed normally. If data are distributed normally (bell shaped), then the 50th quantile and the mean are the same and the other quantiles are arranged symmetrically above and below them.

Display Options

6

The histogram and Mosaic plot are optional representations of the data. Other graphical display options which accompany the **Distribution of Y's** platform are listed in a pop-up menu which is accessed by the check mark to the left of the horizontal scroll bar.

Figure 2.10 Graphical Display Options

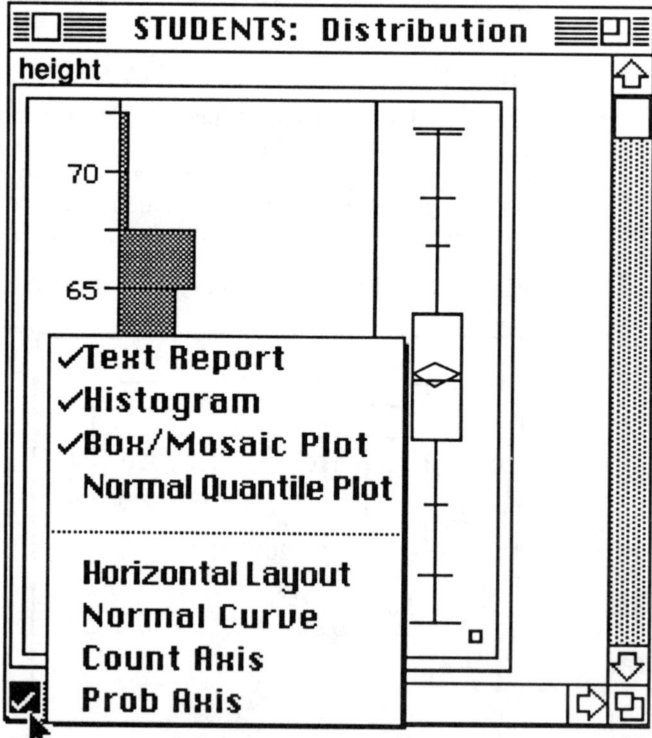

Try different combinations of these options and watch the effect they have on the graphical displays and text reports.

Text Reports

The analyst needs tables of statistical summaries to accompany the graphs. The graphical display option called **text report** is a toggle for viewing or hiding text reports buttons. When **text report** is selected, reveal/conceal buttons show as additional toggles for viewing or hiding each individual report.

Reports for Interval Values

7

The kinds of reports JMP produces depend on whether a variable is interval, ordinal, or nominal. In **Figure 2.11** you see reports for the interval variable height, produced with the **horizontal layout** display option. The reveal/conceal buttons for the Quantiles, Moments, and Tests reports are visible, and the Quantiles and Moments reports are displayed. The Tests report is appropriate only when the mean of a variable is compared to zero.

- The Quantiles report displays the maximum value, minimum value, and other values for selected quantiles.
- The Moments report displays the mean, standard deviation, and other summary statistics.
- The Tests report displays the t-test statistic that compares the sample mean to zero. It includes the probability of obtaining a greater t value by chance alone. The table displays both one-tailed and two-tailed tests.

Figure 2.11 Reports for Interval Values

Quantiles | Moments | Tests

height

Quantiles

maximum	100.0%	72.000
	99.5%	71.830
	97.5%	69.000
	90.0%	66.000
quartile	75.0%	64.000
median	50.0%	61.000
quartile	25.0%	58.000
	10.0%	56.000
	2.5%	53.000
	0.5%	51.000
minimum	0.0%	51.000

Moments

Mean	61.33047
Std Dev	3.89825
Std Err Mean	0.25538
upper 95% Mean	61.83102
lower 95% Mean	60.82992
N	233
Sum Wgts	233

Reports for Ordinal or Nominal Values

8

The **Distribution of Y's** text report option displays different reports depending on whether the response variable is nominal or ordinal. Click on the report window showing sex and age to make it active. Then click the Summary and Profile reveal/conceal buttons to see the text reports for the age variable. You should see the tables shown in **Figure 2.12.**

For nominal and ordinal variables, the reports show frequencies and probabilities associated with the frequencies.

- The Summary report shows the number of different ages (Number of Levels), the total number of rows (Total Frequency), and Total LogLikelihood. The total log likelihood is a measure of the total uncertainty in the sample.
- The Profile report lists the frequency (**Count**), proportion (**Probability**), and cumulative proportion (**Cum Prob**) for each level of the nominal or ordinal variable.

Figure 2.12 Reports for Ordinal or Nominal Values

Summary | Profile

age

Summary

Number of Levels	6
Total Frequency	233
Total LogLikelihood	393.71647

Profile

Level	Count	Probability	Cum Prob
11	30	0.12876	0.12876
12	63	0.27039	0.39914
13	46	0.19742	0.59657
14	45	0.19313	0.78970
15	39	0.16738	0.95708
16	10	0.04292	1.00000

Subsets

The health study specialists want to be on the alert for any unusual subjects such as students who have extreme height or weight values. The analyst knows that a good indicator of extreme values is the proportion of weight to height.

Add a Computed Column

 9

To examine the weight/height ratio, the analyst creates a new column called ratio, computed as weight divided by height. This is done by clicking on the STUDENTS spreadsheet to make it the active window and then selecting the **New Column** command in the **Columns** Menu.

Figure 2.13 The Columns Menu

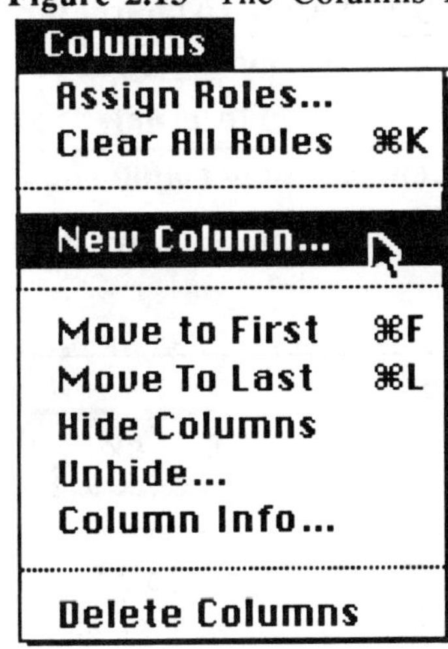

10 To create the new column of ratios, complete the new column characteristics definition panel shown in **Figure 2.14**.

- Type the new name, ratio, in the **Col Name** area.
- The default data type, **Num**, is correct as is.
- Set the spreadsheet ratio display format to **Fixed Dec** with 2 decimal places. (The **Best** format displays the values according to internal precision and may display more decimal places than needed.)
- Type any documentation for the new column into the **Notes** area.

Figure 2.14 The Variable Definition Panel

Table Name: STUDENTS
Col Name: ratio ☒ Lock
Data Type: ◉ Numeric ○ Character ○ Row State
Field Width: 10 Format: ○ Best ◉ Fixed Dec 2
Measurement Type: ◉ Interval ○ Ordinal ○ Nominal
Notes: compute new column as weight/height
☒ Use Formula
OK Cancel

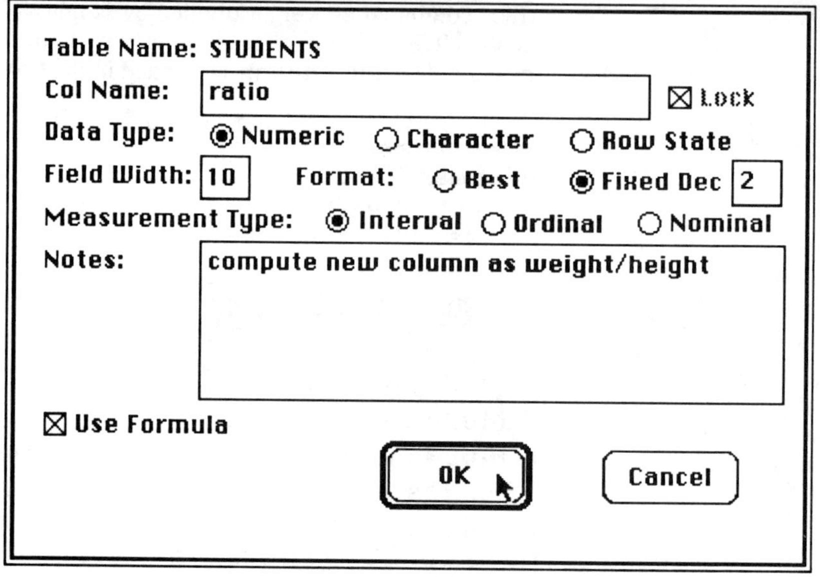

11 To complete the variable definition panel, click the **Use Formula** box to indicate that the new column is to be computed. After you click **OK,** the *calculator window* appears. To construct the formula that calculates values for the ratio column, follow these steps:

1. Select **weight** from the list of column names in the upper left corner of the calculator.
2. Press the divide (÷) key on the calculator keypad.
3. Select **height** from the list of column names.

Figure 2.15 The Calculator Window

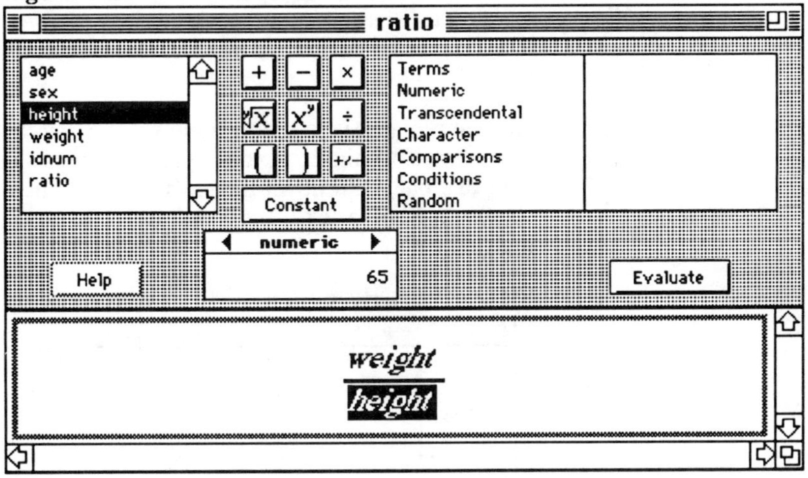

When the formula is complete, click the close box in the upper left corner of the calculator window. The new column called ratio is added to the STUDENTS data table and contains a weight/height ratio for each student.

Figure 2.16 Adding a New Column

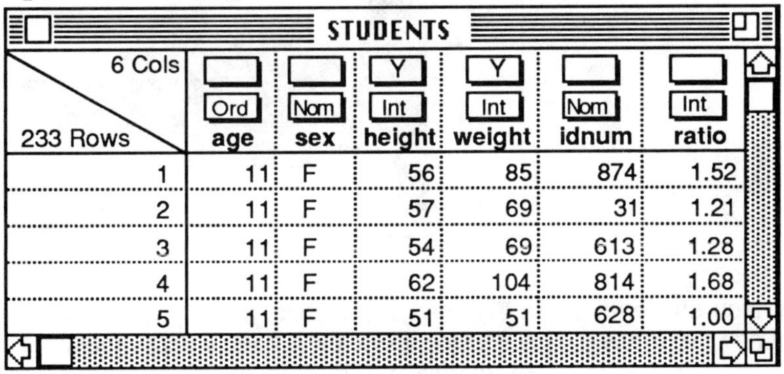

6 Cols / 233 Rows	age (Ord)	sex (Nom)	height (Y, Int)	weight (Y, Int)	idnum (Nom)	ratio (Int)
1	11	F	56	85	874	1.52
2	11	F	57	69	31	1.21
3	11	F	54	69	613	1.28
4	11	F	62	104	814	1.68
5	11	F	51	51	628	1.00

12

Next, the analyst assigns Y as the variable role for ratio and sets all other existing role assignments to **None**. The analyst uses the **Distribution of Y's** command in the **Analyze** menu to look at the distribution of the ratios.

13

All that remains is to identify the subjects having extreme values. One way to do this is to highlight the appropriate histogram bars. To highlight more than one bar, hold down the shift key and click each bar. In the histogram shown in **Figure 2.17**, the hand tool was used to reshape the chart before highlighting.

Figure 2.17 Histogram of Ratio with Extreme Values Highlighted

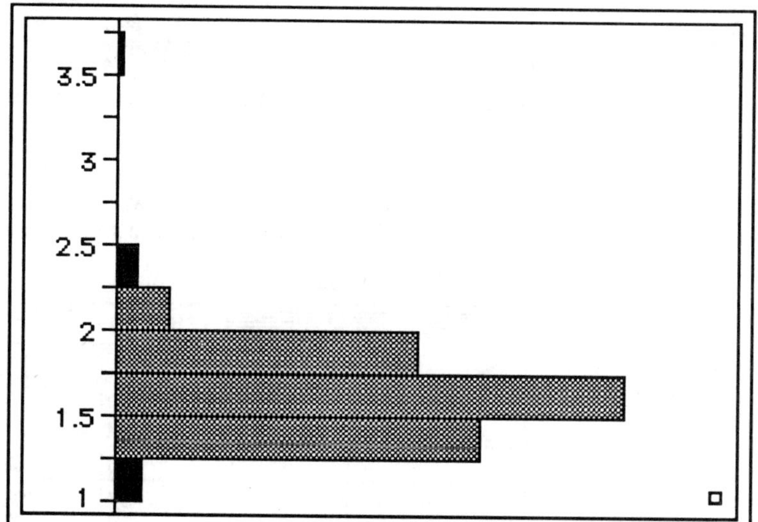

Which students have extreme values?

In **Figure 2.17**, the bars that represent a ratio either greater than or equal to 2.25 or less than 1.25 are highlighted. The corresponding values are automatically highlighted in all other reports generated from the STUDENTS data table. The corresponding rows are also automatically selected on the spreadsheet.

Figure 2.18 Selected Rows in the Data Table

STUDENTS

6 Cols / 233 Rows	age (Ord)	sex (Nom)	height (Int)	weight (Int)	idnum (Nom)	ratio (Y, Int)
146	14	F	62	143	56	2.30
147	14	F	63	103	251	1.63
148	14	F	62	95	986	1.53
149	14	F	62	99	723	1.59
150	14	F	61	112	921	1.83
151	14	F	64	149	867	2.32
152	14	F	67	118	687	1.76

Create a New View

The analyst clicks on the STUDENTS spreadsheet to examine the selected rows but finds that scrolling through a large spreadsheet can be tedious. For the final report to the health researchers, the analyst decides to include a separate list containing only the highlighted students.

14

The **Transform** command in the **File** menu contains a pop-up menu of commands that modify data tables. Select the **Subset** command from the pop-up menu to create a new data table that consists of the selected rows in the active data table.

Figure 2.19 **Subset** command in the **Transform** Pop-up Menu

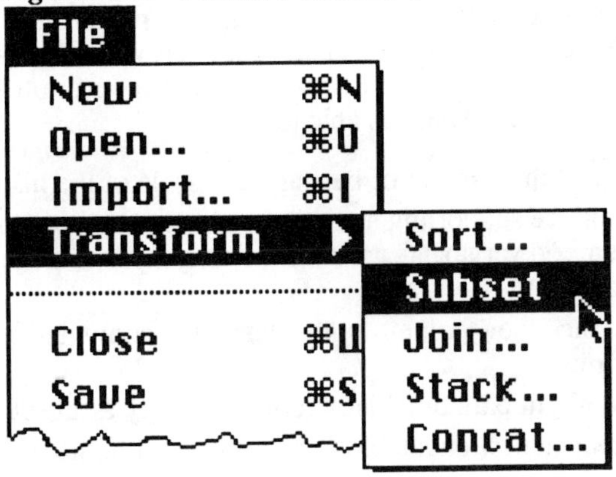

The new spreadsheet shown in **Figure 2.20** contains all the students with extreme weight/height ratios. This table can be saved as a JMP file, exported for use in another application, or printed from the JMP session.

Figure 2.20 Data Table Containing a Selected Subset

Subset of STUDENTS

6 Cols / 10 Rows	age (Ord)	sex (Nom)	height (Int)	weight (Int)	idnum (Nom)	ratio (Int)
1	11	F	57	69	31	1.21
2	11	F	51	51	628	1.00
3	11	F	53	64	238	1.21
4	12	F	52	64	804	1.23
5	13	F	56	67	635	1.20
6	13	M	62	140	262	2.26
7	14	F	62	143	56	2.31
8	14	F	64	149	867	2.33
9	14	M	54	196	999	3.63
10	15	M	66	151	710	2.29
11						

Chapter Summary

In this chapter, a school analyst summarized the demographic and vital data of students participating in a health study. The profile was completed using the **Distribution of Y's** analysis platform and data management features of the JMP spreadsheet.

The **Distribution of Y's** platform produced histograms and box plots or divided (stacked) bar charts for each variable assigned the role of response variable (Y). Using display and text report options to look more closely at data, the analyst was able to

- adjust the number of bars and the scale of the histograms
- produce supporting statistical reports showing moments and quantiles of numeric variables and frequencies and proportions of nominal and ordinal variables
- create a new column in the data table computed as a function of existing columns
- highlight bars of a histogram to identify a subset of rows in the data table
- create a new data table from a subset of highlighted rows.

The analyst can print graphical displays and text reports from JMP. They can also be copied into other applications to complete a report to the school system health care specialists.

Chapter 3.
Comparing Group Means

The Corporate Scene

In keeping with a recent corporate policy to modernize operations, all the typewriters in the typing pool are to be replaced with modern word processors. The typists are eager for this change and willingly participated in a study to help determine what kind of equipment to buy. Three brands of word processor were brought in to the company on a trial basis, and one machine was assigned to each typist. A test score in words-per-minute was recorded for each typist.

The corporate statistician is asked to find out if the typing scores are significantly better on any one brand of machine than on the others.

If statistics can show that one machine promotes significantly faster typing than the others, that machine wins the corporate contract and becomes the new company standard. If no such difference can be found, the company will buy the word processor that the employees like best.

Look Before You Leap

The first step is to become familiar with the data. The typing test scores are in a JMP file so the statistician can review the information and determine the kind of analysis needed.

Open a JMP file

1 The statistician begins a JMP session and opens the file called TYPING DATA. Alternatively, you can double-click on the TYPING DATA file to open it and start JMP running automatically. The TYPING DATA table appears in spreadsheet form as shown in **Figure 3.1**.

Figure 3.1 The Typing Data Table

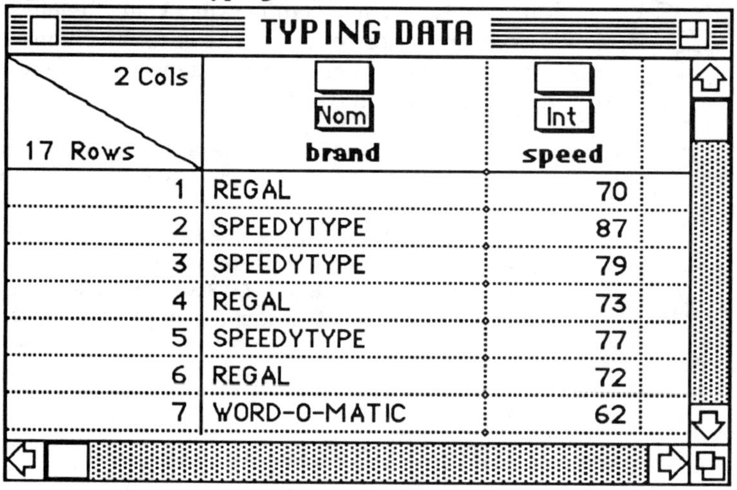

TYPING DATA

2 Cols / 17 Rows	brand (Nom)	speed (Int)
1	REGAL	70
2	SPEEDYTYPE	87
3	SPEEDYTYPE	79
4	REGAL	73
5	SPEEDYTYPE	77
6	REGAL	72
7	WORD-O-MATIC	62

The data table has columns called brand and speed. The level of measurement for each column is indicated in the box above the column name. The variable brand has *nominal* values and speed has *interval* values.

There are 17 rows representing typing scores for 17 typists. Because some of the scheduled participants did not show up for the study, the number of participants in each group differs. The statistician wonders if any other statistics for the groups differ also. In particular, the statistician asks

Is the mean (average) typing speed the same for each brand?

Does any one of the three brands of word processor stand out from the others, or does it make no difference which brand is used? A large investment will be made based on this statistical analysis so the statistician must evaluate the data carefully.

Choose Variable Roles

2 The statistician first assigns variable roles using role assignment pop-up menus as shown in **Figure 3.2**. Select **X** for brand and **Y** for speed to follow along with the statistician.

- **Y** identifies a response or dependent variable.
- **X** identifies a classification or independent variable.

JMP produces a statistical analysis appropriate for the variable roles, the measurement level of each variable, and the selected analysis platform. The next step is to choose an analysis that answers the question,

> Is there a statistical difference between the mean values of Y within the levels of X?

Figure 3.2 Select X and Y Variables

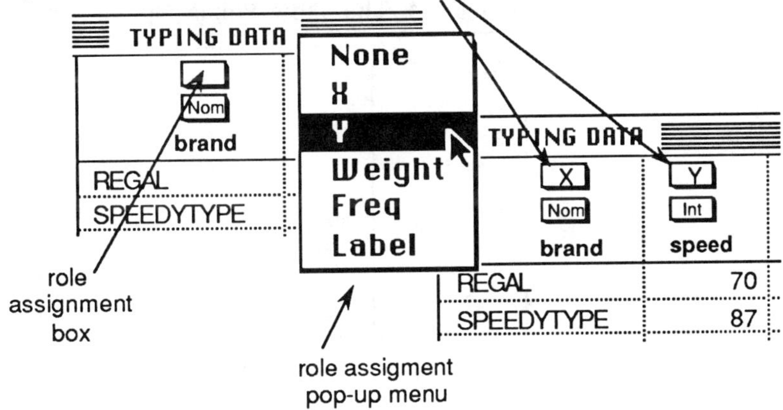

Graphical Display of Grouped Data

3 To compare the mean typing scores for each brand of word processor, the statistician selects the **Fit Y by X** platform from the **Analyze** menu.

Figure 3.3 The **Analyze** Menu

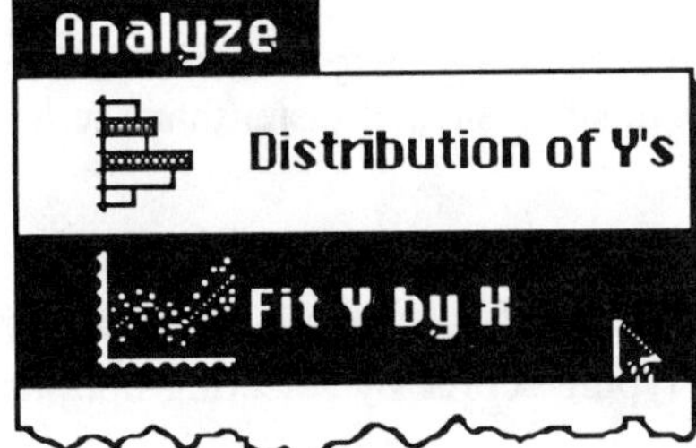

Note ➡ **Fit Y by X** can be used for four kinds of analyses:

- categorical analysis when both **X** and **Y** have nominal/ordinal values
- analysis of variance when **X** is nominal/ordinal and **Y** has interval values
- logistic regression when **X** is interval and **Y** has nominal/ordinal values
- regression analysis when both **X** and **Y** have interval values.

Show Points

After selecting **Fit Y by X** from the **Analyze** menu, the plot shown in **Figure 3.4** appears. Each of the typing test scores is plotted for each brand of word processor with the distance between tick marks on the brand axis proportional to the sample size of each group. The mean typing score for the total sample is shown as the dotted line across the plot.

Figure 3.4 The Show Points Option

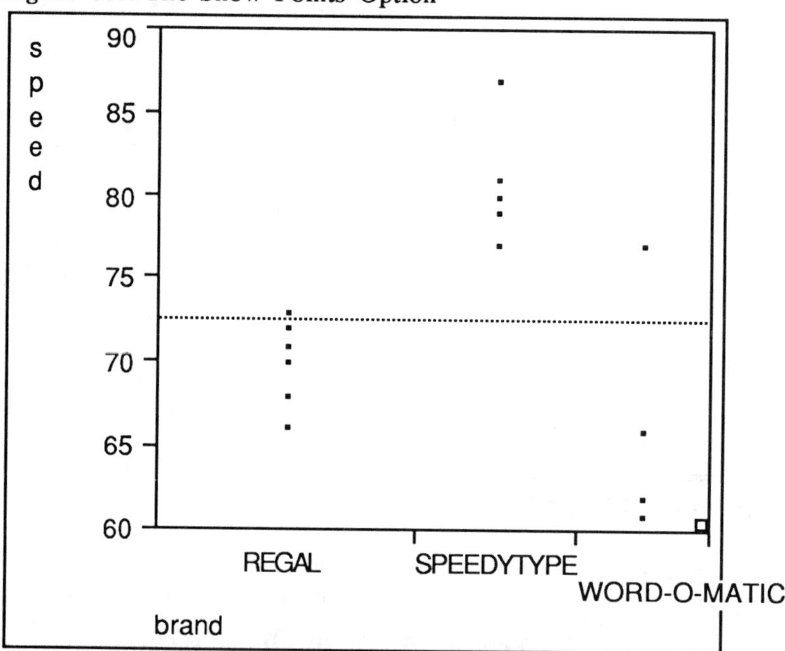

How do the data points look?

The statistician sees at a glance that most participants who typed on the SPEEDYTYPE machines typed faster than the others.

Fit Means Option

The statistician now chooses to view more graphical information about the distribution of typing scores by selecting options from the display options pop-up menu. This menu is accessed by the triangle located in the lower left corner of the plot. Initially, only the **Show Points** option is in effect.

Each of the options displays one aspect of the distribution of typing scores for each word processor. The options are *toggles* meaning you turn any option alternately on and off each time you select it.

4 To follow the path of the statistician, use the display options menu and select the **Fit Means** option.

Figure 3.5 The Plot Options Menu

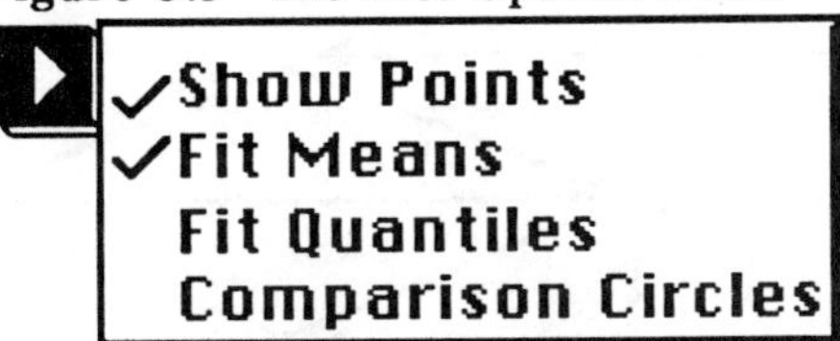

When you choose the **Fit Means** option, the plot is enhanced with a *means diamond* constructed for each group as shown in **Figure 3.6**. A line is drawn at the mean (average) value of words-per-minute for each brand of word processor. The upper and lower points of the means diamond span a 95% confidence interval computed from the sample values for each machine as illustrated in **Figure 3.6**.

Figure 3.6 The Fit Means Option

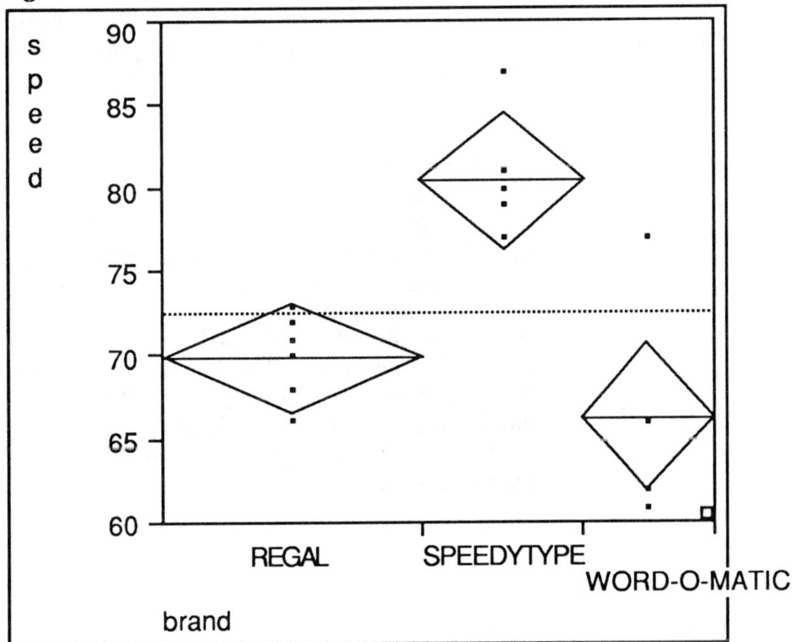

The difference in the number of group participants displays as a difference in diamond width with larger sample sizes having wider means lines. The width of each diamond spans the distance on the horizontal axis proportional to the group size.

Figure 3.7 The Means Diamond

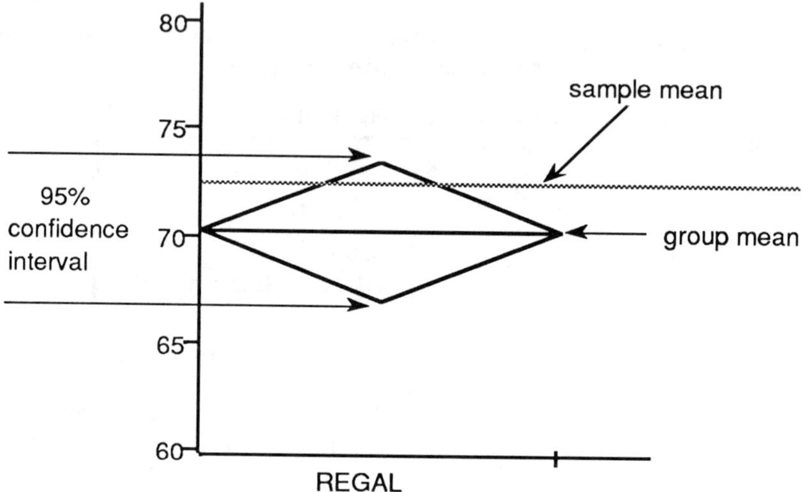

How do the means compare?

The mean scores of the REGAL and WORD-O-MATIC word processors appear to be nearly the same, but the statistician sees that the SPEEDYTYPE mean is much higher.

Fit Quantiles

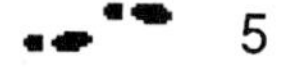 5

The statistician wants to check the distribution of points within each group to get a better idea of the spread of the values and to visualize the distance of the extreme values from the center of the data. The **Fit Quantiles** display option is useful for making this comparison.

When you choose **Fit Quantiles**, a quantile box plot is drawn for each group of typing scores. The median, or 50th quantile, is shown by a line in the body of the box. The top and bottom of the box are drawn at the 75th and 25th quantiles, also called the upper and lower quartiles. The box encompasses the *interquartile range* of the sample data. The 10th and 90th quantiles are shown with lines above and below each box.

Figure 3.8 The Fit Quantiles Option

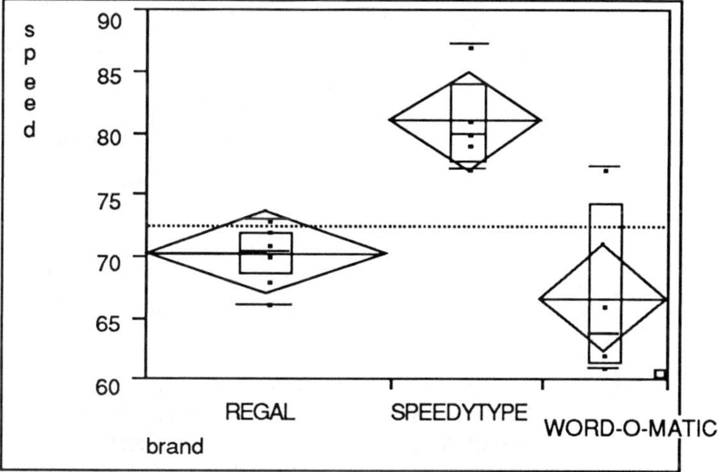

Looking at the quantile box plot and the means rectangle together helps you to see if data are distributed normally within a group. If data are normally distributed (bell shaped), then the 50th percentile and the mean are the same and the other quantiles are arranged symmetrically above and below the median.

Figure 3.9 The Quantiles Box Plot

90th
75th
sample mean
50th percentile (median)
group mean
25th
10th
WORD-O-MATIC

How are the typing scores distributed within each brand?

The quantile box plots show that there is a difference in variation of scores across the three groups. The scores in the REGAL group are clustered tightly around the mean score but the WORD-O-MATIC scores vary. One WORD-0-MATIC score is much higher than the others. However, even with this variation among the groups, the SPEEDYTYPE brand still appears to promote the best performance.

Comparison Circles

6

In order to complete the inspection of the typing data, the statistician chooses **Comparison Circles** from the display options menu. The circles displayed to the right of the plot provide a graphic test of whether or not the mean typing scores are statistically different.

The statistician examines whether the SPEEDYTYPE machine is statistically better than the other machines by clicking on the SPEEDYTYPE comparison circle. The confidence circles highlight to show the statistical magnitude of the difference between typing scores. Groups that are the same are displayed in the same color and in the same font on the axis.

The center of each circle is aligned with the mean of the group it represents. The diameter of each circle spans the 95 percent confidence interval for each group. Whenever two circles intersect, the confidence intervals of the two means overlap indicating that the means may not be significantly different. Whenever two circles do not intersect, the group means they represent are significantly different.

Figure 3.10 Comparison Circles

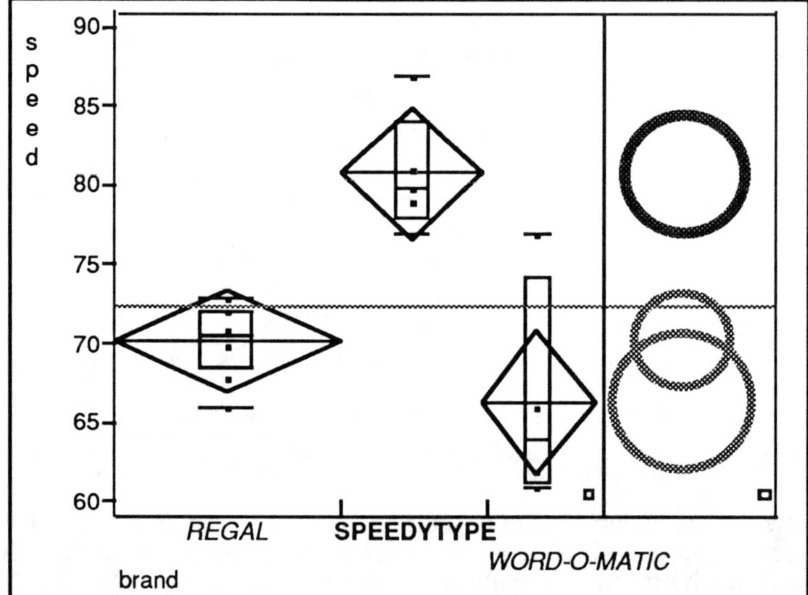

Are the mean typing scores different?

You can see that the confidence circle for the SPEEDYTYPE brand does not intersect with either of the other two. While both the REGAL and WORD-O-MATIC brands appear slower than SPEEDYTYPE, they are not statistically different from each other.

Quantify Results: Analysis of Variance

7 To see what quantitative information is available, the statistician clicks the Means button below the plot. This displays three reveal/conceal text report buttons. The first report, Summary of Fit, is automatically open.

Figure 3.11 Table Selection Menu

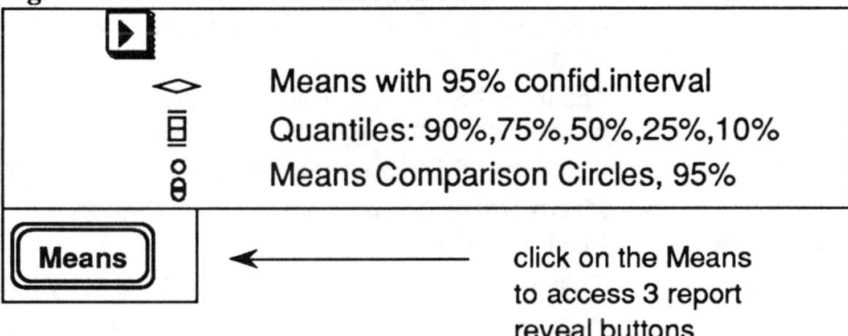

Summary of Fit

8 The Summary of Fit report, shown in **Figure 3.12**, is a numeric summary of the typing data distribution.

- Rsquare quantifies the proportion of total variation in the typing scores resulting from different word processors rather than different people.
- Root Mean Square Error (RMSE) is a measure of the variation in the typing scores that can be attributed to different people rather than different machines.
- Mean of Response is the mean (arithmetic average) of all the typing scores.
- Observations is the total number of scores recorded.

Figure 3.12 Summary of Fit Table

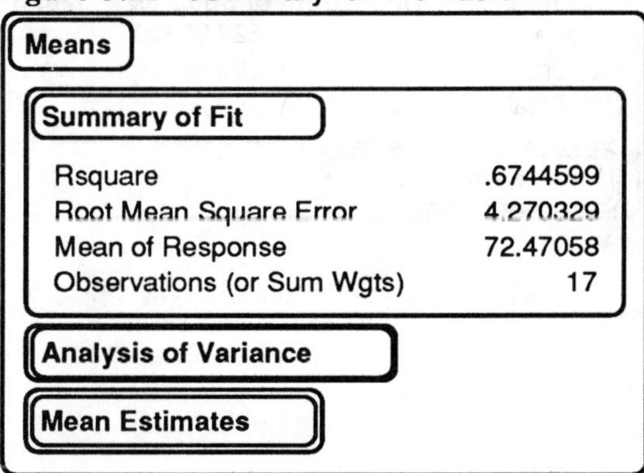

Analysis of Variance

9

Each report can be closed by clicking its reveal/conceal button again. The view shown in **Figure 3.13** is displayed when you close the Summary of Fit report and click on the Analysis of Variance reveal/conceal button.

The Analysis of Variance report is in the form of a table.

- **Sum of Squares** (SS for short) identifies the sources of variation in the typing scores. C Total is the corrected total SS. It is divided into the SS for Model and SS for Error.

  The model SS is the variation in the typing scores explained by the analysis of variance model, hypothesizing that the word processors are different.

  The error SS is the remaining or unexplained variation.
- **Mean Square** is the sum of squares divided by its associated degrees of freedom (**DF**).
- **F ratio** is the model mean square divided by the error mean square.
- **Prob > F** is the probability of a greater F value occurring if the mean typing scores for the word processors differed only because different people were typing on them rather than because the word processors differed in any way.

In this example, the significance of the F value is .0004. This implies that the differences observed in this typing trial would be expected only 4 times in 10,000 similar trials if the word processors did not really promote different typing performances.

Figure 3.13 Analysis of Variance Table

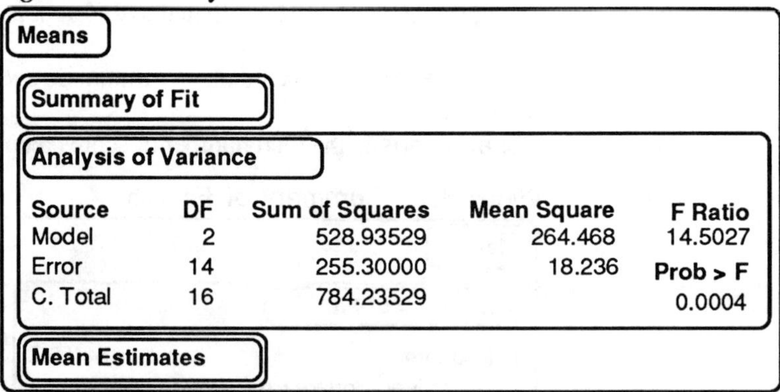

Means

Summary of Fit

Analysis of Variance

Source	DF	Sum of Squares	Mean Square	F Ratio
Model	2	528.93529	264.468	14.5027
Error	14	255.30000	18.236	Prob > F
C. Total	16	784.23529		0.0004

Mean Estimates

Mean Estimates

10

To see the list of means for each group, the statistician opens the Means Estimates report as shown in **Figure 3.14.** This produces a table summarizing the scores for each brand and tells the statistician the level of performance to expect.

- **Level** lists the name of each group.
- **Number** is the number of scores in each group.
- **Mean** is the mean of each group.
- **Std Error** is the standard error of each group mean.

Figure 3.14 The Means Estimates Table

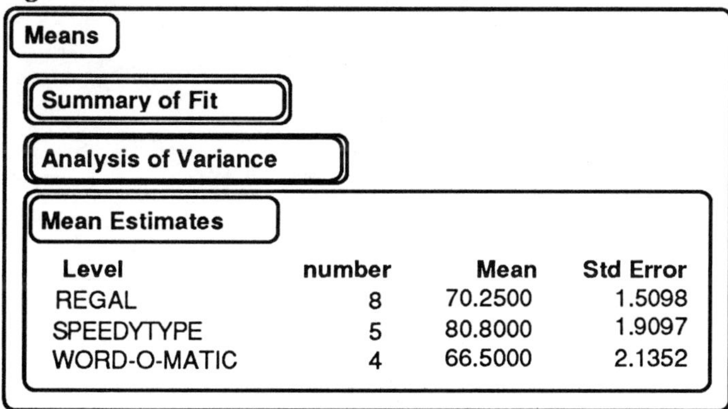

Means

Summary of Fit

Analysis of Variance

Mean Estimates

Level	number	Mean	Std Error
REGAL	8	70.2500	1.5098
SPEEDYTYPE	5	80.8000	1.9097
WORD-O-MATIC	4	66.5000	2.1352

Chapter Summary

In this chapter, a corporate statistician summarized the difference in mean typing scores for three brands of word processor. The **Fit Y by X** analysis platform was used to

- plot the typing scores for the three brands of word processor
- overlay a means diamond on each group of typing scores
- overlay a quantile box plot on each group of typing scores
- produce confidence circles to visualize the difference in mean typing scores
- compute and display a one-way analysis of variance table
- display a table of the group means and standard errors.

Using the scissors or cross cutting tools from the **Tools** menu, the statistician can copy the graphs or tables and prepare a report for the corporate decision makers. The report concludes that, in this typing trial, the SPEEDYTYPE word processor produced significantly higher scores than either of the other two brands.

Chapter 4.
Regression and Curve Fitting

Class Assignment

A statistics professor has given an exploratory regression assignment to a class. The assignment is to select a set of data suitable for regression analysis and to compare various regression model fits. One student knows about the interactive regression capabilities of JMP and is eager to start the class project.

The student chooses data from Eppright et al. (1972) as reported in Eubank (1988, p. 272). The study subjects are young males. The two variables in the data table are age group (in months) and the ratio of weight to height. The student's goal is to describe and model the growth pattern of the subjects for the range of ages in the data.

Look Before You Leap

The first step is to become familiar with the data. The student begins by reviewing the data to determine how to proceed with the regression assignment.

Open a JMP file

To follow along with the student, start JMP running and open the file called GROWTH in the SAMPLE DATA folder. Alternatively, you can double-click on the GROWTH file. JMP begins automatically and the GROWTH data table appears in spreadsheet form. The student sees there are 2 columns and 72 rows. The column called ratio contains the average weight/height ratio for each age group in the study. The age groups range from .5 months to 71.5 months as shown in **Figure 4.1**.

Figure 4.1 The **GROWTH** Data Table

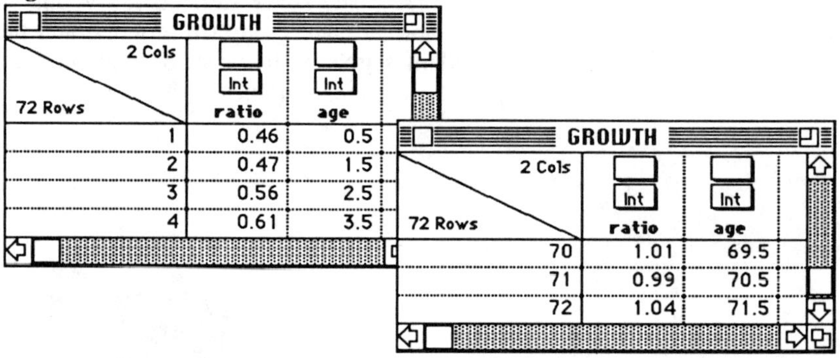

GROWTH

2 Cols / 72 Rows	Int ratio	Int age
1	0.46	0.5
2	0.47	1.5
3	0.56	2.5
4	0.61	3.5

GROWTH

2 Cols / 72 Rows	Int ratio	Int age
70	1.01	69.5
71	0.99	70.5
72	1.04	71.5

The level of measurement for each column is indicated in the *measurement level box* above the column name. Both columns have *interval* values, as needed for the regression assignment.

Choose Variable Roles

To continue with the regression and curve fitting assignment, the student chooses a role assignment for each variable as shown in **Figure 4.2**. The role assignment box at the top of each column accesses a pop-up menu which the student uses to assign the role of **X** for age and **Y** for ratio.

- **Y** identifies a response or dependent variable.
- **X** identifies a classification or independent variable.

Figure 4.2 Assign Variable Roles to ratio and age

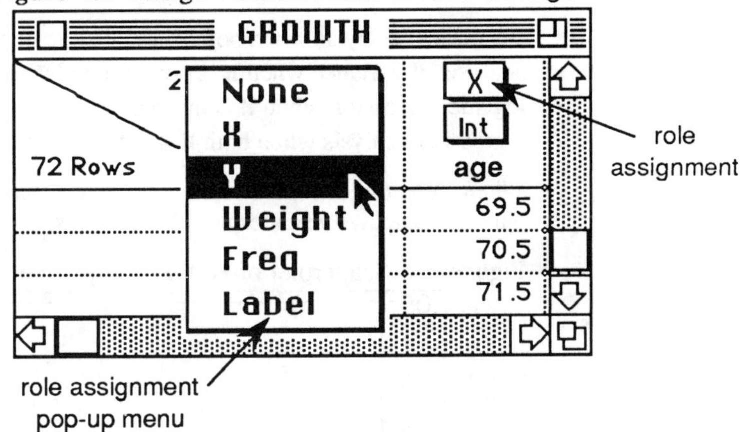

Note➜ If you do not assign variable roles in the spreadsheet, a dialog prompts you to assign roles when you select a command from the **Analyze** menu.

Select an Analysis Platform

JMP chooses the statistical analysis appropriate for the roles assigned to columns, the measurement level of each variable, and the analysis platform used. The student's next step is to choose a command from the **Analyze** menu that fits regression curves to interval data.

3 To fit regression curves, the student launches the **Fit Y by X** platform from the **Analyze** menu.

Figure 4.3 The **Fit Y by X** Command

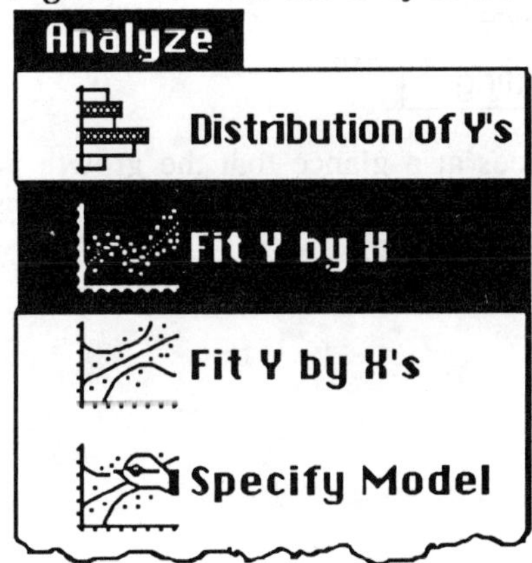

Fit Y by X can be used for four kinds of analyses.

- categorical analysis when both **X** and **Y** have nominal or ordinal values
- analysis of variance when **X** is nominal and **Y** has interval values
- logistic regression when **X** is interval and **Y** has nominal or ordinal values
- regression analysis when both **X** and **Y** have interval values.

When both the X and Y variables have interval values, the **Fit X by Y** platform displays a scatterplot of the values.

Figure 4.4 Scatterplot of Ratio by Age from the **growth** Data Table

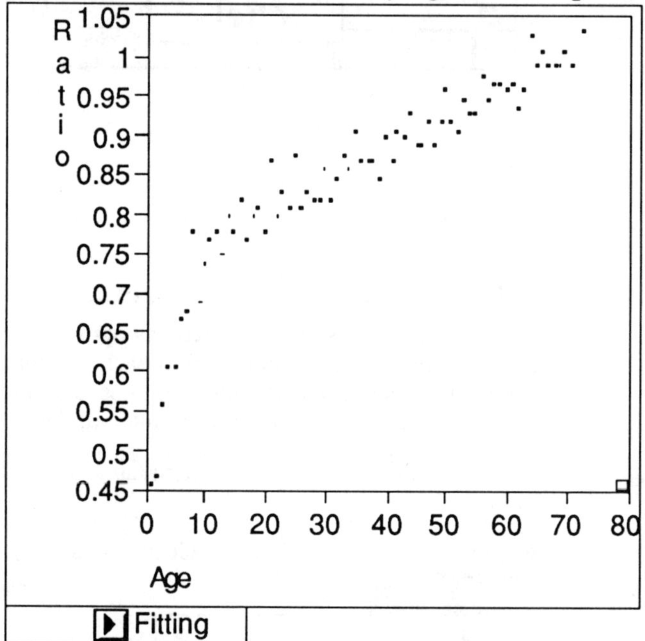

How do the data points look?

The student sees at a glance that the growth pattern is not random and decides that a straight line regression is a good baseline fit to compare with other regression curves.

Fitting Models to Interval Data

The Fitting pop-up menu, shown in **Figure 4.5**, lists a **Show Points** toggle command and five commands for regression fits.

Figure 4.5 Regression Fitting Pop-up Menu

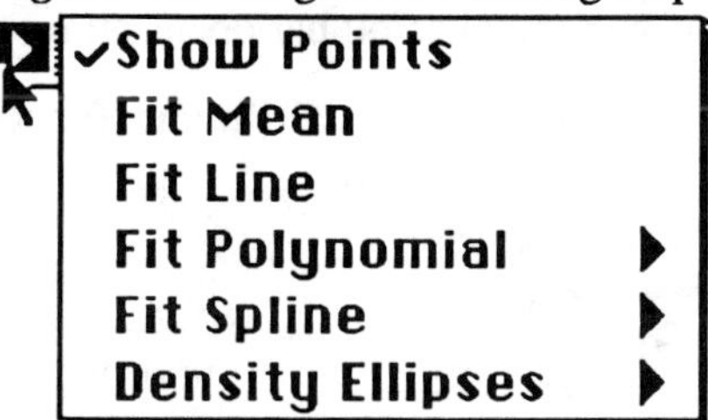

Note➜ The fit commands are not toggles. They can be selected repeatedly so that the student can try different fits on the same plot for comparison.

The Fit Line Command

4 The student first selects **Fit Line** from the Fitting pop-up menu to fit a simple regression line through the data points. The regression line minimizes the sum of squared distances from each point to the line of fit. In this example, the student also uses the corresponding Linear Fit pop-up menu (shown in **Figure 4.6**) to suppress the confidence curves which accompany the linear fit.

Figure 4.6 Linear fit of Ratio by Age

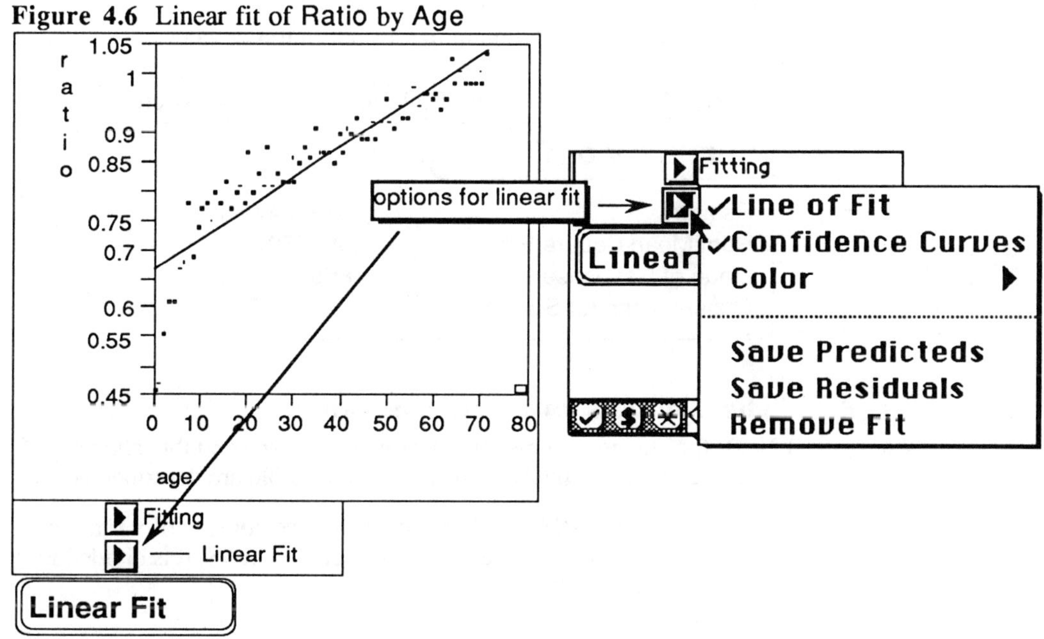

5 The regression line appears to fit the growth data fairly well. However, the student wants to evaluate the linear fit with statistical text reports. To follow along with the student, click on the linear fit reveal button beneath the scatterplot. Three more reveal buttons appear, each corresponding to a different statistical report. Note that the first two reports are already open. For expediency, the **Reveal All** command in the asterisk pop-up menu to the left of the horizontal scroll bar opens all reports at once.

Figure 4.7 Text Report Reveal Buttons for Regression

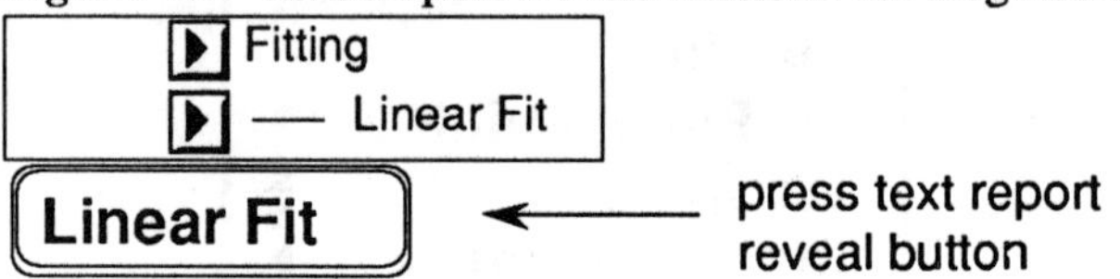

The Summary of Fit Table

6 The Summary of Fit table, shown in **Figure 4.8**, is a numeric summary of the linear fit.

- Rsquare quantifies the proportion of total variation in the growth ratios accounted for by fitting the regression line.
- Root Mean Square Error (RMSE) is a measure of the variation in the ratio values that is attributable to different people rather than different ages.
- Mean of Response is the mean (arithmetic average) of all the ratio values.
- Observations is the total number of non-missing values.

Figure 4.8 Summary of Fit Table for Linear Regression

Summary of Fit	
Rsquare	.8225347
Root Mean Square Error	.0516527
Mean of Response	.8555555
Observations (or Sum Wgts)	72

The Analysis of Variance Table

7 Next, the student reviews the Analysis of Variance table shown in **Figure 4.9**. The values in the Analysis of Variance table are described below.

- **Sum of Squares** (SS for short) identifies the sources of variation in the growth ratio values. C Total is the corrected total SS. It is divided into the SS for Model and SS for Error.

The model SS is the amount of the total variation in the ratio scores explained by fitting a straight line to the data.

The error SS is the remaining or unexplained variation.

- A **Mean Square** is a sum of squares divided by its associated degrees of freedom (**DF**).
- The **F ratio** is the regression mean square divided by the error mean square.
- **Prob > F** is the probability of a greater F value occurring if the ratio values differed only because of different subjects rather than because the subjects are different ages.

In this example the significance of the F value is .0000 which indicates that the linear fit to the weight/height growth pattern is significantly better than the horizontal line that fits the sample mean to the data.

Figure 4.9 Analysis of Variance Table for Linear Regression

Analysis of Variance

Source	DF	Sum of Squares	Mean Square	F Ratio
Model	1	0.8656172	0.865617	324.4433
Error	70	0.1867605	0.002668	**Prob > F**
C. Total	71	1.0523778		0.0000

The Parameter Estimates Table

8

The student now clicks the Parameter Estimates reveal button to open the remaining report shown in **Figure 4.10.**

- **Term** lists the terms in the regression model.
- **Estimate** lists the estimates of the coefficients in the regression line equation.
- **Std Error** lists estimates of the standard error of the parameters.
- **t ratio** is the parameter estimate divided by its standard error.
- **Prob > |t|** is the probability of a greater absolute t value occurring by chance alone if the parameter has no effect in the model.

Figure 4.10 Parameter Estimates Table for Linear Regression

Parameter Estimates

<table>
<tr><th>Term</th><th>Estimate</th><th>Std Error</th><th>t Ratio</th><th>Prob>|t|</th></tr>
<tr><td>Intercept</td><td>.66562308</td><td>.012175</td><td>54.67</td><td>0.0000</td></tr>
<tr><td>age</td><td>.00527590</td><td>.000292</td><td>18.01</td><td>0.0000</td></tr>
</table>

How well does the line fit the data?

The significant F ratio in the Analysis of Variance table tells the student that the regression line fits significantly better than the horizontal line at the mean (the simple mean model). However, while the regression line looks like a good fit for age groups above 7 months, it does not describe the data well for ages younger than 7 months.

The Exclude Command

Since the low-age points are the trouble spots for the linear fit, the student decides to remove them from the analysis and to try the model with the remaining values. To highlight these outliers and exclude them from the analysis, the student

9

- selects the brush tool from the **Tools** menu
- shift-drags the brush over the points to be excluded
- uses the **Exclude** command from the **Rows** menu to mark the points as excluded
- uses the **Markers** command from the **Rows** menu to assign the **x** marker to the excluded points.

12

You can follow these steps and then choose **Fit Line** again to see the results of excluding the low age points. **Figure 4.11** shows both regression lines. The points still show on the plot but are not included in the second regression line's computation.

Figure 4.11 Linear Fit with Outliers Excluded

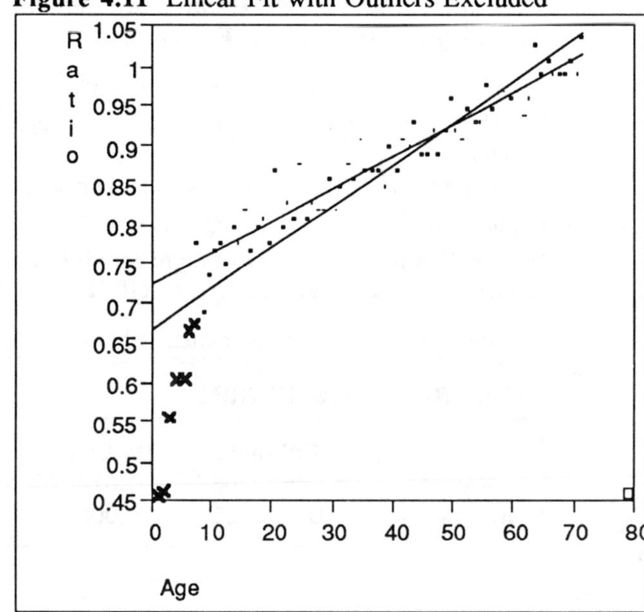

13

After completing this part of the exploratory regression analysis, the student reselects the arrow tool and then chooses the **Journal** command from the **Edit** menu. The first time the **Journal** command is selected during a JMP session, the dialog shown in **Figure 4.12** appears. The student names the journal and saves it in MacWrite® format. A journal window opens filled with the graphs and tables from the active report window.

Journaling JMP results to create a report

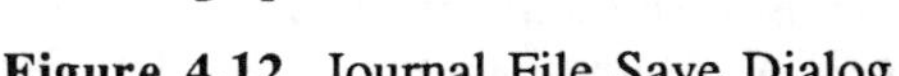

Figure 4.12 Journal File Save Dialog

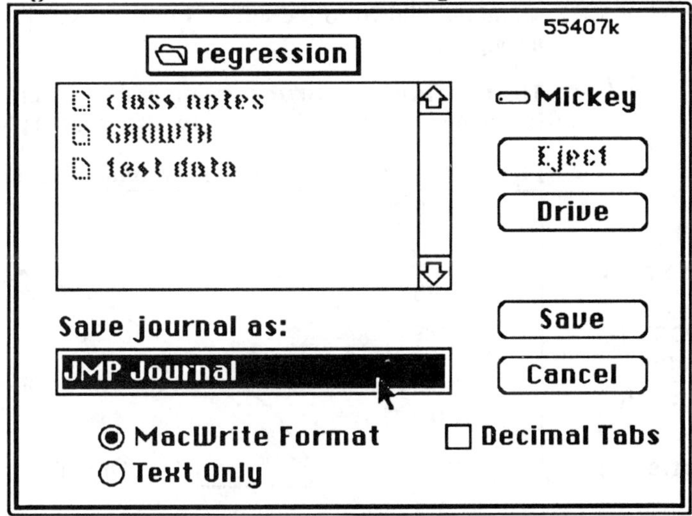

The Fit Polynomial Command

The student sees the full potential of using JMP for comparative exploratory analysis and proceeds by choosing a polynomial fit for comparison. A linear regression is actually a polynomial of degree 1. The **Fit Polynomial** command lets you choose degrees greater than 1 allowing the fit to have curvature.

14

Before proceeding, the student again includes all rows in the analysis. To follow along with the student,

- choose **Select Excluded** from the **Rows** menu. This highlights all currently excluded rows.
- choose the **Include/Exclude** toggle from the **Rows** menu.

15

The student also removes the line of fit which excluded the lower age groups. The **Remove Fit** option from the *second* (modified) regression line removes this line so that only the line fit to all the data points remains. See **Figure 4.6** for an illustration of the pop-up menu that accompanies each fit.

The student uses the Fitting menu twice more to overlay polynomial curves of degree 2 and degree 3 on the scatterplot. The **Polynomial** command submenu (shown to the left of **Figure 4.13**) lists the choice of degrees 2 through 6.

17

The **confidence curves** option in the pop-up menu for each polynomial fit suppresses the confidence curves so that the plot is not too cluttered.

Fit Polynomial

Fit Polynomial
Submenu

2-quadratic
3-cubic
4-quartic
5
6

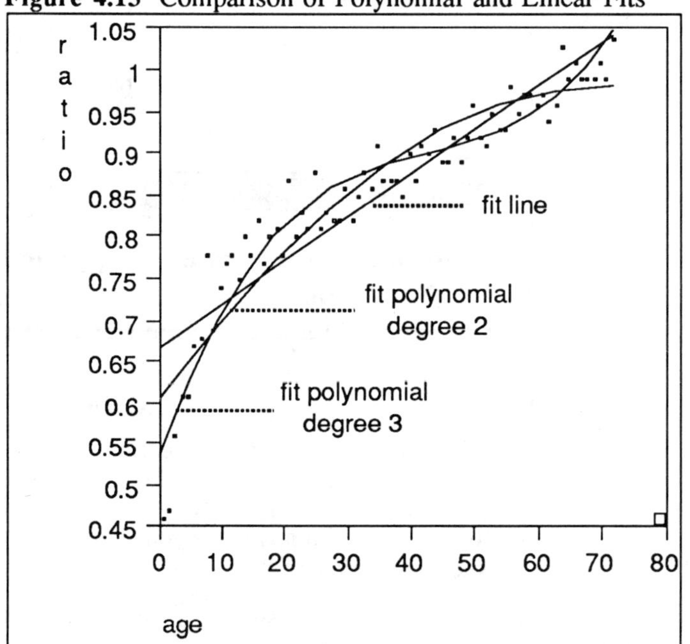

Figure 4.13 Comparison of Polynomial and Linear Fits

How much improvement is there?

Figure 4.14 shows a partial display of the Summary of Fit tables for the linear fit and each of the polynomial fits. As polynomial terms are added to the model, the regression curve appears to fit the data better and the Rsquare value increases.

Figure 4.14 Summary of Fit Tables for Linear and Polynomial Fits

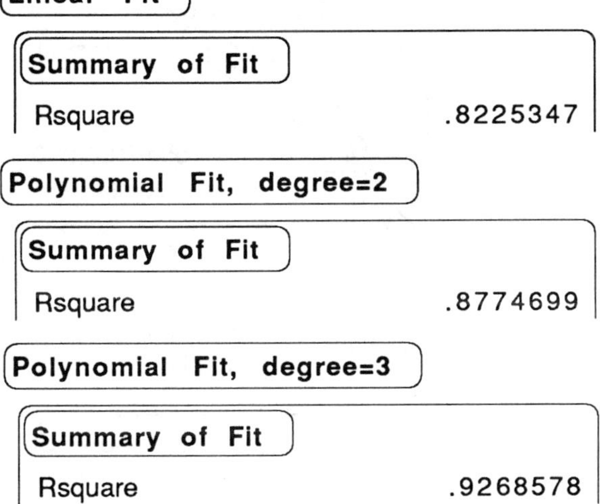

Linear Fit

Summary of Fit

Rsquare	.8225347

Polynomial Fit, degree=2

Summary of Fit

Rsquare	.8774699

Polynomial Fit, degree=3

Summary of Fit

Rsquare	.9268578

In preparation for writing a report to complete the class assignment, the student also journals these results. The **Journal** command appends these results to the open journal file.

The Fit Spline Command

Even the polynomial fit of degree 3 does not quite reach the outlying points for the very young subjects. The student needs a free-form function that acts like it smooths the data. The **Fit Spline** command fits a smoothing spline for this purpose.

18

Again, the student uses the **Remove Fit** options on both of the polynomial fits leaving only the linear regression line showing. See **Figure 4.6** for an illustration of the pop-up menu that includes this option. Next, the student uses the Fitting menu and chooses three splines with lambda values of 10, 1000, and 100,000. Lambda is a tuning factor that determines the flexibility of the spline. The **Fit Spline** command submenu (shown to the left of **Figure 4.15**) lists lambda values. The three new fits are overlaid on the scatterplot.

Fit Spline Submenu

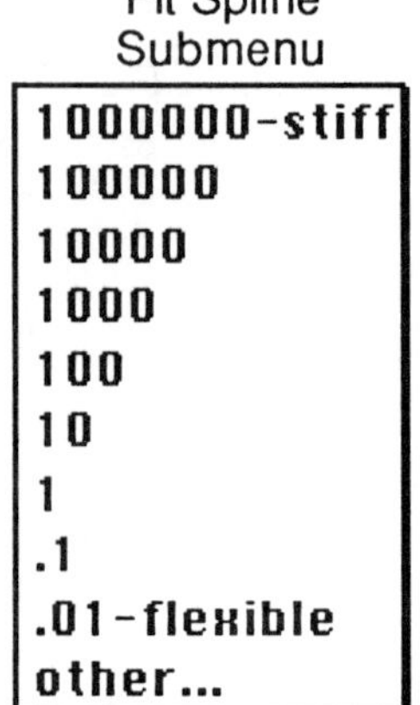

Figure 4.15 Comparison of Spline and Linear fits

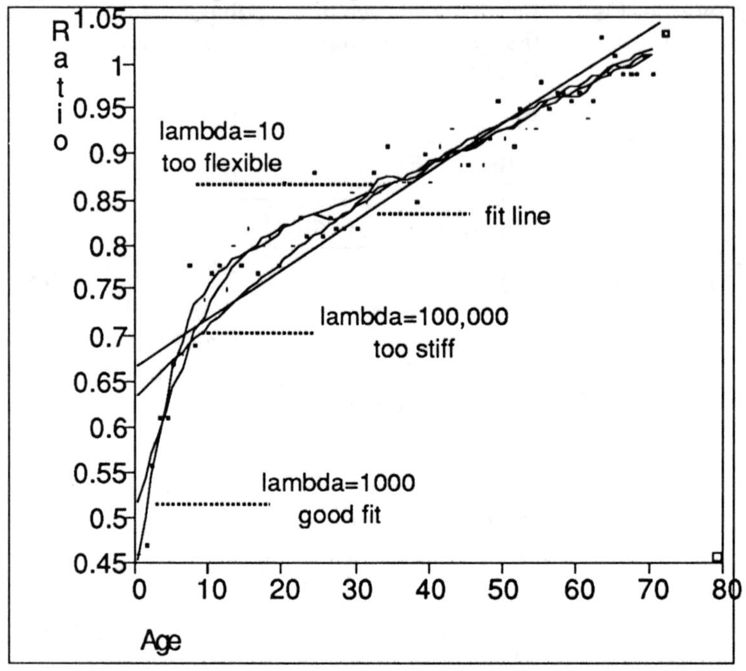

How well do the splines model the data?

By inspecting the plot, the student sees that the lambda = 10 curve is not smooth enough. It is too flexible and thus too affected by local error. The lambda = 100000 curve is too smooth. It is so straight that it does not reach down to model the lower ages closely. However, the lambda = 1000 curve fits well. Its shape is not influenced by local errors and it appears to be a smooth fit to the data.

Figure 4.16 Summary of Fit Tables for Spline Fits

Smoothing Spline Fit, lambda=10	
R-Square	.9700605
Sum of Squares Error	.0315076

Smoothing Spline Fit, lambda=1000	
R-Square	.9525186
Sum of Squares Error	.0499683

Smoothing Spline Fit, lambda=100000	
R-Square	.8710340
Sum of Squares Error	.1357208

The student again uses the **Journal** Command from the **Edit** menu to append this scatterplot and the smoothing spine text reports to the open journal file. After journaling the final analyses, the student types the following draft notes about the spline fitting technique at the end of the journal window.

```
"This fitting technique applies a cubic
polynomial to the interval between points; the
polynomial is joined such that the curve meets
at the same point with the same slope to form a
continuous and smooth curve. A small enough
lambda could make such a curve go through every
point, which would model the error, not the
mean. A moderate lambda value forces the curve
to be smoother, i.e. less curved. This is
accomplished by adding a curvature penalty to
the optimization that minimizes the sum of
squares error."
```

The regression exploration is done and the student has captured the graphs and text reports in a journal file. The journal file is ready for editing to complete the class assignment.

What can the student conclude about growth rate?

By comparing various regression fits, the student sees that both the polynomial fits and the spline fit with moderate flexibility best describe the data. These models show that infants grow most rapidly during the first months of life and that growth rate decreases significantly at approximately 7 months.

Chapter Summary

To complete an assignment for a statistics class, a student used the JMP **Fit Y by X** analysis platform to examine a variety of regression model fits. The assignment was to model and describe the growth pattern of subjects over a range of ages. Growth was measured by the ratio of weight to height. For this assignment, the student

- fit a straight line to use as a baseline comparison to other regression models and evaluated the fit using statistical text reports
- excluded outliers and again fit a straight line to compare the Rsquare values given by the Summary of Fit tables for both lines
- fit 2nd and 3rd degree polynomials to see if they modeled the growth pattern more realistically
- fit smoothing splines with lambda values of 10, 1000, and 100,000 and compared them with each other and with the linear fit.

The student used the journaling feature to append each of these regression reports and graphs to a journal file. This journal provided the information needed to write the paper on exploratory regression analysis for the statistics class.

Chapter 5. Analyzing Categorical Data

Market Watching

Survey data is frequently categorical data rather than measurement data. Analysis of categorical data begins by simply counting the number of responses in categories and subcategories. Counting is easy, but interpreting the relationship between categories based on counts is more complex. It requires computing probabilities and evaluating the likelihood of those probabilities compared to expectations.

For example, an American automobile manufacturer — feeling the pinch of competition from foreign auto sales — has decided to do a market analysis before proceeding with a multimillion dollar advertising campaign. With this in mind, the auto manufacturer contracts with a market researcher to survey a random sample of people and record each participant's age, sex, marital status and auto information. The auto information consists of the manufacturing country, the car's size, and whether it is a family, work or sporty car. Using this information, the market researcher hopes to provide the advertising experts with direction for the upcoming advertising campaign.

In particular, the market researcher wants to know more about relationships between car characteristics and the demographic information collected in the sample. In other words

Who buys what?

Look Before You Leap

The first step is to become familiar with the data. The car survey data are in a JMP file called CAR POLL in the SAMPLE DATA folder.

Open a Data Table

The market researcher begins a JMP session and opens the CAR POLL file. Alternatively, you can double-click on the CAR POLL icon to launch JMP automatically. The CAR POLL data table is displayed in spreadsheet form as shown in **Figure 5.1**.

Figure 5.1 The CAR POLL Data Table

CAR POLL

6 Cols / 303 Rows	sex (Nom)	marital status (Nom)	age (Int)	country (Nom)	size (Nom)	type (Nom)
1	Male	Married	34	American	Large	Family
2	Male	Single	36	Japanese	Small	Sporty
3	Male	Married	23	Japanese	Small	Family
4	Male	Single	29	American	Large	Family
5	Male	Married	39	American	Medium	Family

The car poll data were collected from a random sample of people in a specific geographic area. Age is a numeric variable assigned the interval measurement level. The other five columns are character variables with nominal measurement levels.

Address the Research Question

The basic research question is stated

> Is the response probability for manufacturing country, size of car, or type of car a function of the age, sex, or marital status of the owner?

The market researcher looks at the spreadsheet to see what specific relationships lend insight into this question. The relationships between the following automobile characteristics and demographics are of interest.

- manufacturing country by age
- manufacturing country by sex
- manufacturing country by marital status
- size of car by age
- size of car by sex
- size of car by marital status
- type of car by age
- type of car by sex
- type of car by marital status

Modify the Data Table

From past experience, the market researcher feels that age groups offer better summary information than specific ages. In fact, dividing people into two age groups often provides the basis for a valuable broad analysis.

Find the Median Age

The next step the market researcher takes is to find a reasonable way to divide the sample into two age groups. The distribution of a variable and its corresponding quantiles can reveal a good dividing point. To see the age distribution,

- assign age the role of Y using the role assignment pop-up menu at the top of the column
- select the **Distribution of Y's** platform from the **Analyze** menu as shown in **Figure 5.2.**

Figure 5.2 Look at the Distribution of age

2

3

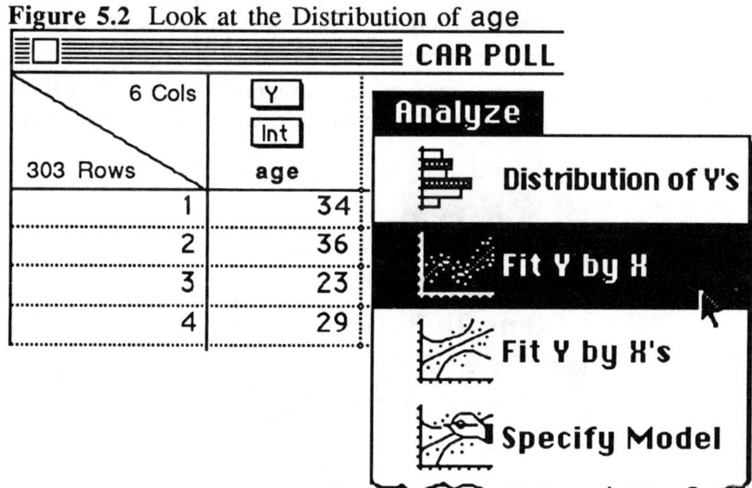

The **Distribution of Y's** platform shows a histogram with an accompanying quantile box plot and a quantiles table. The quantiles table is shown in **Figure 5.3**. This report shows 30 as the median age.

Figure 5.3 Quantile Report of age

Quantiles

100% maximum	60
99.5%	56.880
97.5%	44.400
90%	38
75% quartile	35
50% median	30
25% quartile	26
10%	24
2.5%	21.600
0.5%	9.360
0% minimum	0

age 30 divides the sample in half

4

The market researcher selects the **New Column** command from the **Columns** menu to create a new column in the data table.

create a new column

Figure 5.4 The **New Column** Command

5

The market researcher enters the new column's characteristics in the **New Column** dialog shown in **Figure 5.5.** The X in the **Use Formula** box indicates that the new column is to be computed. As shown, the computed column's values are automatically locked and cannot be edited in the spreadsheet. After completing the dialog, the researcher clicks on the **OK** button. The calculator shown in **Figure 5.6** appears.

Figure 5.5 The **New Column** Dialog

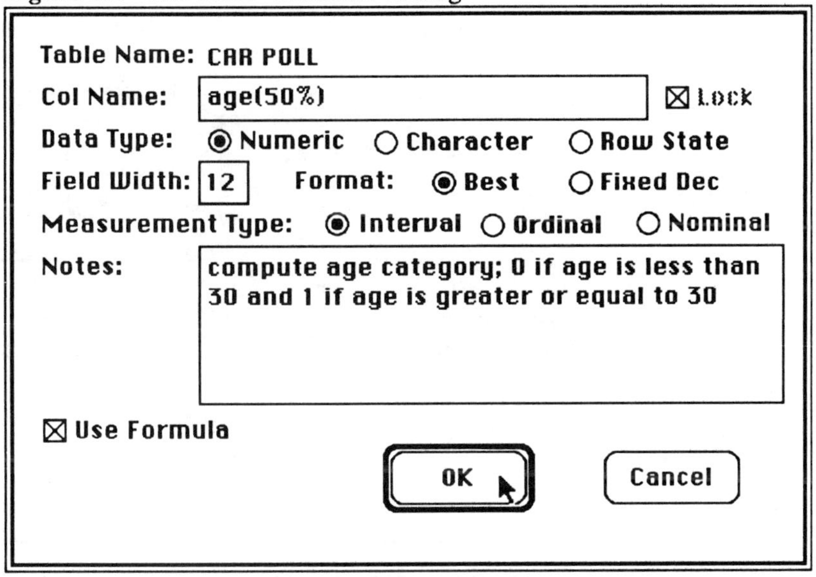

Table Name: CAR POLL
Col Name: age(50%) ☒ Lock
Data Type: ◉ Numeric ○ Character ○ Row State
Field Width: 12 Format: ◉ Best ○ Fixed Dec
Measurement Type: ◉ Interval ○ Ordinal ○ Nominal
Notes: compute age category; 0 if age is less than 30 and 1 if age is greater or equal to 30
☒ Use Formula
OK Cancel

To follow the steps of the market researcher, enter the formula as described below.

- Scroll through the function browser and click on Conditions.
- Click on New Condition in the list of Conditions.
- Click in the box below the **Constant** button, type 0, and click on the **Constant** button.
- Click on the missing term (the empty box) to the right of the word if.
- Click on age in the column selector list.
- Click on Comparisons in the function browser list and select the comparison x<y.
- Type 30 into the box below the **Constant** button and click on the **Constant** button. The 30 appears in the second comparison box.
- Click on the missing term (the empty box) to the left of the word otherwise.
- Type a 1 into the box below the **Constant** button and click on the **Constant** button. The 1 appears in the box to the left of the word otherwise.

Click on the calculator window's close box to fill the new column with calculated values.

For more information about the calculator, see Chapter 11, *Using the Calculator*.

Figure 5.6 The Calculator Window

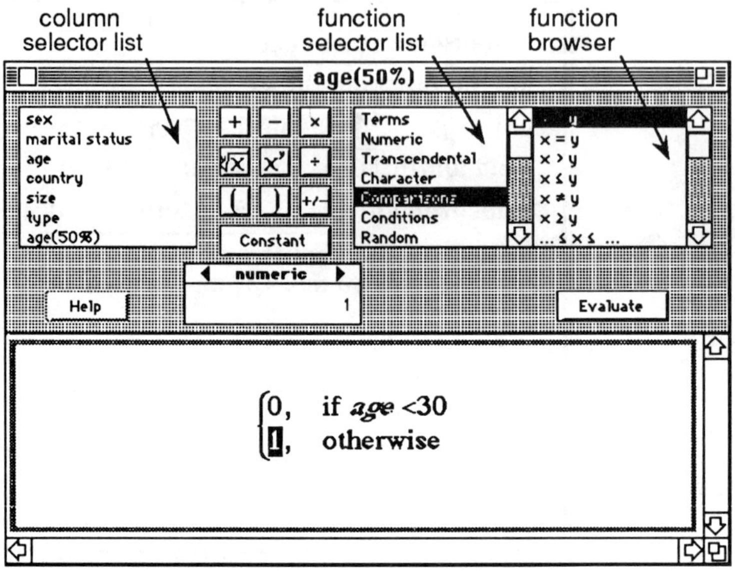

The market researcher is now ready to look at relationships between the columns.

Contingency Table Reports

Because the computed age classification is a grouping variable, the analyst changes the level of measurement from interval to ordinal as shown in **Figure 5.7**. JMP chooses the statistical analysis appropriate for a variable's measurement level and its role assignment.

Figure 5.7 Car Poll Data Table with New Column

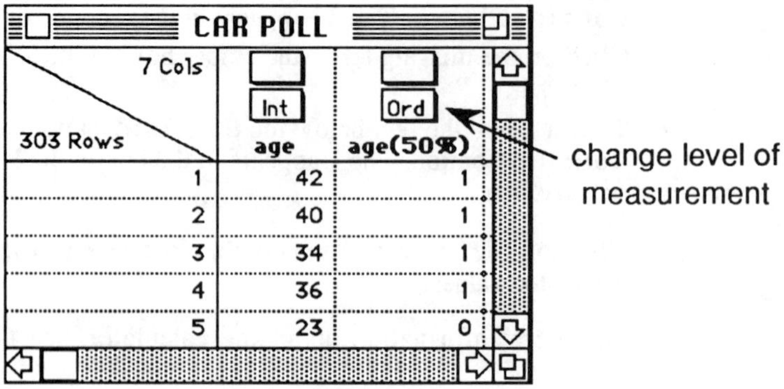

Assign Variable Roles

You can choose variable roles by using the role assignment pop-up menu at the top or each column or you can select an analysis platform from the **Analyze** menu and respond to a variable role dialog prompt.

The market researcher selects the **Fit Y by X** statistical platform from the **Analyze** menu and responds to the dialog prompt.

8

Figure 5.8 Select **Fit Y by X** analysis platform

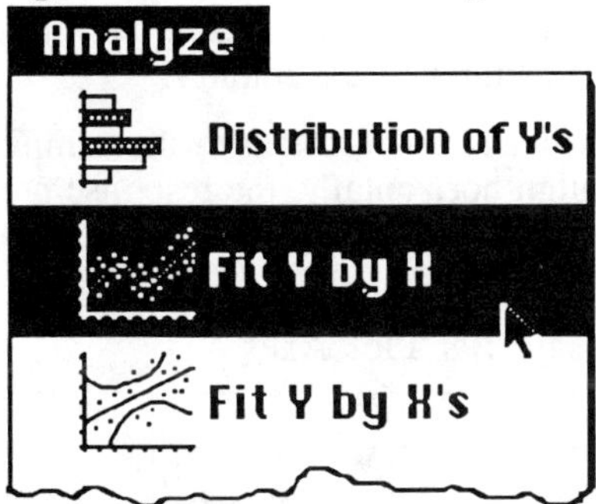

9

In this investigation, the country, size, and type columns are response variables and are assigned as Y variables. Sex, marital status and age(50%) are assigned as X variables.

The market researcher holds down the command key, clicks on each of the 3 Y variables, and then clicks on the **Y** button. The X variables are assigned in the same manner using the **X** button. The dialog in **Figure 5.9** shows the result. The market researcher clicks **Done** when the dialog is complete.

Figure 5.9 Role Assignment Dialog

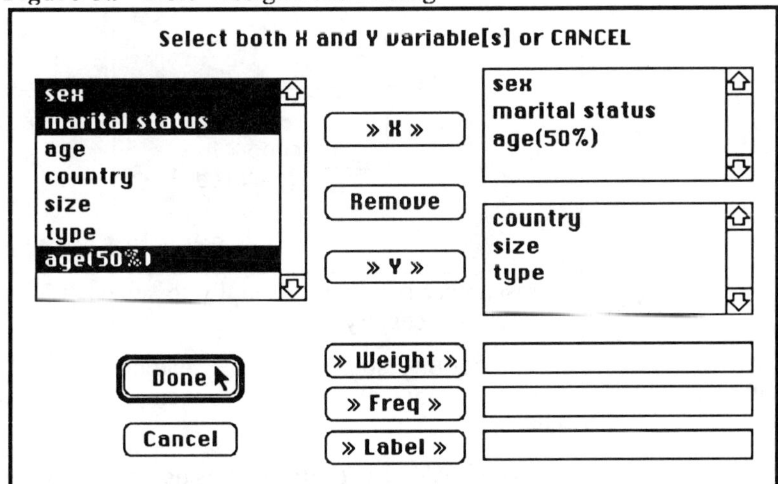

Contingency Table Mosaic Plots

If both X and Y have either nominal or ordinal values, the **Fit Y by X** platform displays a contingency table mosaic plot with accompanying text reports for each combination of X and Y columns. The market researcher can scroll across and down the report window to see each relationship at a glance.

A mosaic chart consists of side-by-side divided bars for each level of an X variable, where each bar is divided into proportional segments representing each discrete Y value. **Figure 5.10** shows the mosaic chart which relates marital status to the manufacturing country.

The width of each bar is proportional to the sample size. When the lines dividing the bars align horizontally, the response proportions are the same. When the lines are far apart, the response rates of the samples may be statistically different.

Figure 5.10 Mosaic Bar Plot Axes

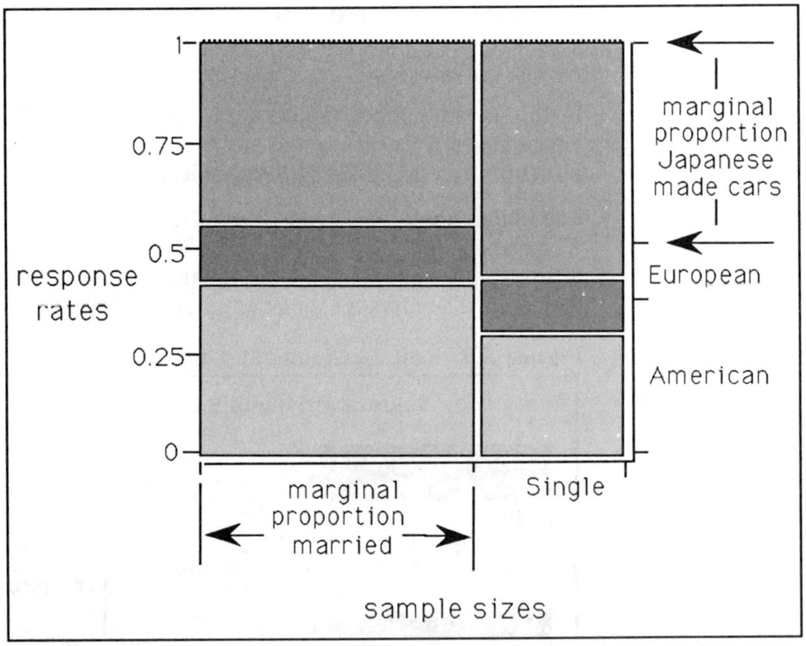

10

The researcher scrolls horizontally to see each X variable as it relates to manufacturing country.

What can be said about country?

Sex and country do not show any relationship at all. Each sex owns about the same number of automobiles from each of the three manufacturing countries.

The age(50%) by country mosaic indicates that the proportion of American car owners over 30 is only slightly greater than the proportion of American car owners under 30.

The most significant relationship occurs between marital status and country. The mosaic plot and supporting statistical reports shown in **Figures 5.11-12** indicate that married people are more likely to own American cars than single people.

Figure 5.11 Mosaic Plot of Country and Marital Status

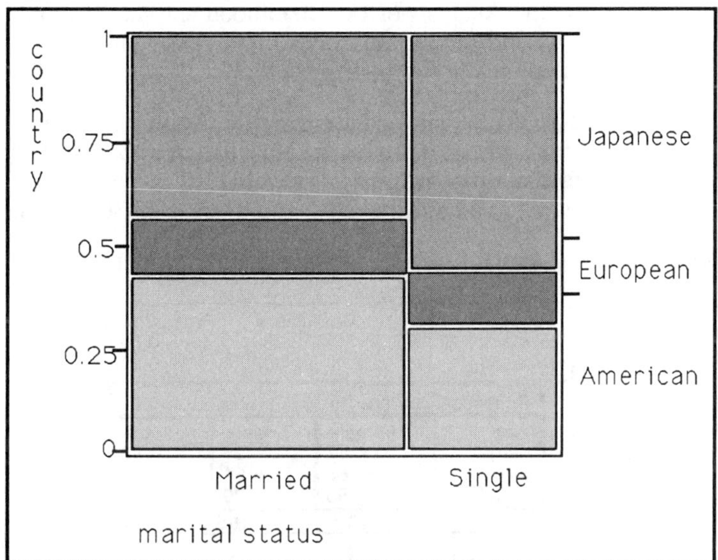

There are statistical text reports accompanying each mosaic plot. The market researcher opens and closes these reports as needed by clicking on the appropriate reveal/conceal buttons. The Analysis of LogLikelihood table is shown in **Figure 5.12.**

The Likelihood Ratio and Pearson Chi-Square tests evaluate the relationship between an automobile's manufacturing country and the marital status of the owner. If no relationship exists between country and marital status, a smaller Chi-square value would occur only 7 times in 100 similar surveys.

Figure 5.12 Analysis of LogLikelihood for Country and Marital Status

Analysis of LogLikelihood

Source	DF	-LogLikelihood
Model	2	2.57004
Error	299	295.88121
C. Total	301	298.45125

Test	ChiSquare	Prob>ChiSq
Likelihood Ratio	5.140	0.0765
Pearson	5.081	0.0788

The market researcher plans to advise the American auto manufacturer to emphasize married couples in advertising plans.

11

Next, the market researcher scrolls down to the first mosaic plot in the second row and then scrolls across to see the relationship between size of car and each of the X variables.

What can be said about size of car?

The second row of three mosaic plots shows no relationship between car size and gender, marital status, or age group. The market researcher clicks on the Analysis of LogLikelihood and the Response Counts reveal buttons for each of the mosaic plots in the second row. A partial view of these tables is shown in **Figure 5.13**.

The Chi-square values indicate that the purchase of large, medium, and small cars is not significantly different between the levels of each of the factors. The Chi-square probabilities range from .21 to .3, suggesting that a smaller Chi-square value would occur 20-30 times in 100 similar surveys.

Figure 5.13 Tables for Relationships with Size of Car

Gender and size of car

Test	ChiSquare	Prob>ChiSq
Likelihood Ratio	2.388	0.3030
Pearson	2.388	0.3030

Response Counts

size	Female	Male	Total
Large	17	25	42
Medium	63	61	124
Small	58	79	137
	138	165	303

Marital Status and size of car

Test	ChiSquare	Prob>ChiSq
Likelihood Ratio	2.753	0.2525
Pearson	2.743	0.2537

Response Counts

size	Married	Single	Total
Large	30	12	42
Medium	84	40	124
Small	82	55	137
	196	107	303

Age group and size of car

Test	ChiSquare	Prob>ChiSq
Likelihood Ratio	3.090	0.2133
Pearson	3.049	0.2177

Response Counts

size	0	1	Total
Large	15	27	42
Medium	59	65	124
Small	70	67	137
	144	159	303

The market researcher can tell the American automobile manufacturer that it makes no difference what size cars are shown in advertisements.

12

The market survey categorizes cars both on the basis of size and type, where a car's type is go-to-work, sporty, or family. The market researcher scrolls across the third row of plots to look at the relationship between type of car and each of these X variables.

What can be said about type of car?

The mosaic plots show that type of car relates significantly to both marital status and age class. As you might expect, a large proportion of the cars owned by married people are classified as family automobiles while the largest proportion of cars owned by single people are sporty cars.

Figure 5.14 Mosaic Plot with Chi-Square values for Type of Car and Marital Status

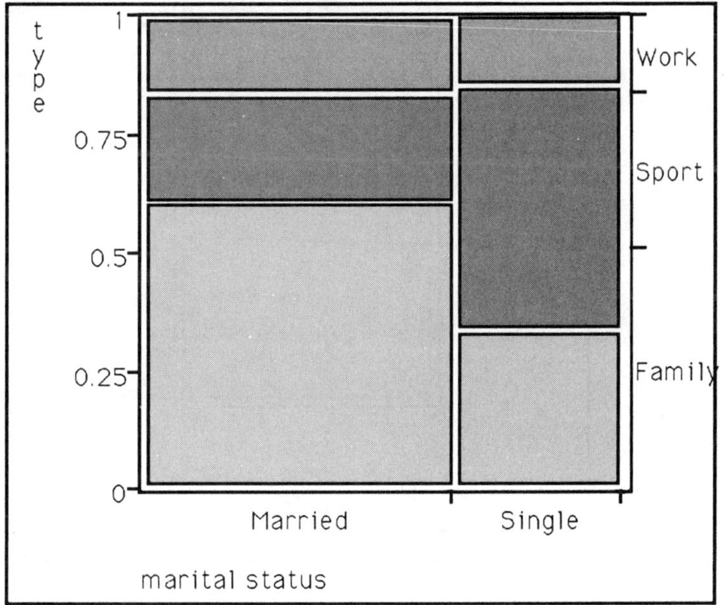

Test	ChiSquare	Prob>ChiSq
Likelihood Ratio	26.766	0.0000
Pearson	26.963	0.0000

The market researcher plans to advise the American automobile manufacturer to emphasize married couples buying family type automobiles.

It follows logically that a relationship between age group and type of car also exists because older people are more likely to be married. The proportion of people over 30 years old who own family cars is much greater than those under 30.

Figure 5.15 Mosaic Plot with Chi-Square Values for Type of Car and Age Group

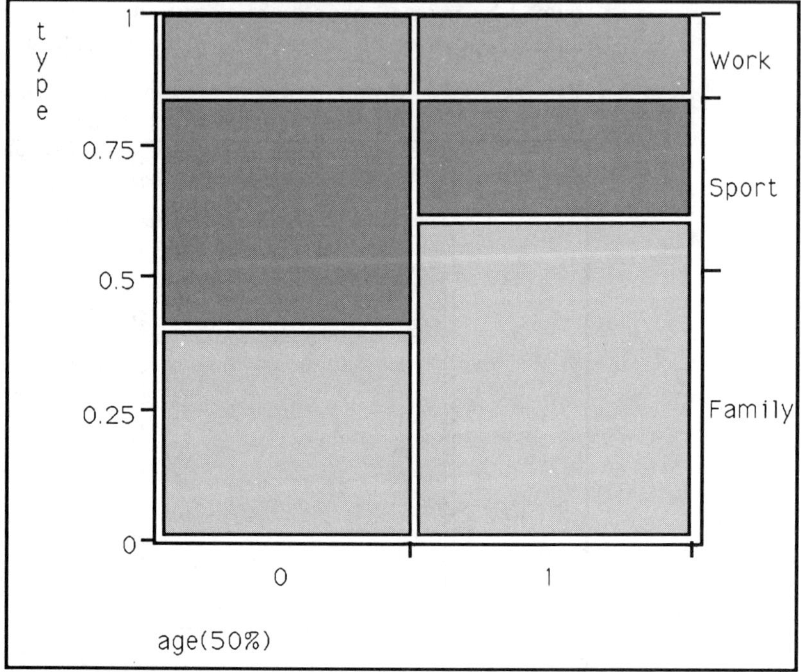

Test	ChiSquare	Prob>ChiSq
Likelihood Ratio	16.098	0.0003
Pearson	15.953	0.0003

Chapter Summary

In this chapter a market researcher looked at relationships between categorical variables obtained from a survey. The survey recorded age, sex, marital status and information about the type of automobile owned by a random sample of people in the same geographical area. The auto information included manufacturing country, size, and type of car. Car types were classified as go-to-work, sporty, and family. The market researcher investigated the question

> Is the size of car, type of car, or manufacturing country related to the age, gender, or marital status of the owner?

The **Fit Y by X** platform produced nine mosaic charts with supporting statistical summaries showing

- no relationship between either sex or age and manufacturing country
- a significant relationship between marital status and manufacturing country with married people more likely to own American cars than single people
- no relationship between sex, age, or marital status and size of car
- no relationship between sex and type of car
- significant relationships between marital status and type of car. As might be expected, married people over 30 years old were more likely to own family type cars than younger single people.

The market researcher prepared a final report to the American automobile manufacturer. The report recommended that the advertising campaign focus toward married couples who appear to be over 30 years old looking for a family car.

Chapter 6. A Factorial Analysis

Popcorn yield experiment

A food manufacturer wants to offer the public two qualities of popcorn. The plain everyday type has been serving the public for years, but now the company's research department claims to have discovered a special treatment of corn kernels. This new process supposedly increases the popcorn yield as measured by popcorn volume from a given measure of kernels.

Corporate decision makers want to know if this is really true and, if so, how much the yield increases. The special treatment raises the cost of the popcorn so the increase in yield must be significant enough to cover the higher processing costs.

Decisions must also be made about packaging, cost, and advertising for the new corn. The decision makers want to know if the increased yield (if indeed there really is one) is the same under all popping conditions or better under some than others.

(The popcorn data used in this chapter and for examples in subsequent chapters are artificial but the experiment was inspired by experimental data reported in Box, Hunter, and Hunter (1978).)

Look Before You Leap

The popcorn yield data is the result of a designed experiment. The same amounts of different kinds of corn were methodically popped under different conditions. A scientist from the research department first looks at the data to review the results of the popcorn experiment.

Open a Data Table

1 The popcorn research data is in a JMP file called POPCORN in the SAMPLE DATA folder. The research scientist double-clicks on the POPCORN file to begin a JMP session. The POPCORN data table displays in spreadsheet form. A partial listing of the POPCORN table is shown in **Figure 6.1**.

Figure 6.1 The POPCORN Data Table

POPCORN

5 Cols / 16 Rows	popcorn (Nom)	oil amt (Nom)	batch (Nom)	yield (Int)	trial (Int)
5	plain	little	small	9.9	1
6	gourmet	little	small	12.1	1
7	plain	lots	small	10.6	1
8	gourmet	lots	small	18.0	1
9	plain	little	large	8.8	2
10	gourmet	little	large	8.2	2
11	plain	lots	large	8.8	2
12	gourmet	lots	large	9.8	2

For the experiment, the scientist popped corn under controlled conditions. Plain popcorn and specially treated gourmet popcorn were each popped in large or small amounts of oil and in large or small batches. Two trials were done for both kinds of corn under all popping conditions.

This experimental design is called a factorial design and has three factors, usually called *main effects*, which are

- kind of popcorn (plain or gourmet)
- amount of cooking oil (little or lots)
- cooking batch size (large or small).

What Kind of Questions Can Be Asked?

The appropriate statistical analysis for a factorial design addresses the following questions about the main effects.

Is there an overall difference in yield between plain and gourmet popcorn?

Is there an overall difference in yield between cooking in lots of oil rather than a small amount of oil?

What is the difference in yield between cooking several small batches instead of one large batch?

Analysis of a factorial experiment also provides information about *interaction* between the main effects as addressed by the following questions.

Does the amount of oil used affect the yield of both types the same? In other words, is there an interaction effect between popcorn type and amount of cooking oil used.

Is there an interaction effect between cooking batch size and type of popcorn?

Is there an interaction effect between batch size and amount of oil used?

Is there an interaction effect between all three main effects?

The Specify Model Dialog

JMP has a special analysis command called **Specify Model** in the **Analyze** menu that lets you specify complex models such as the factorial described above.

Figure 6.2 Select the **Specify Model** Platform

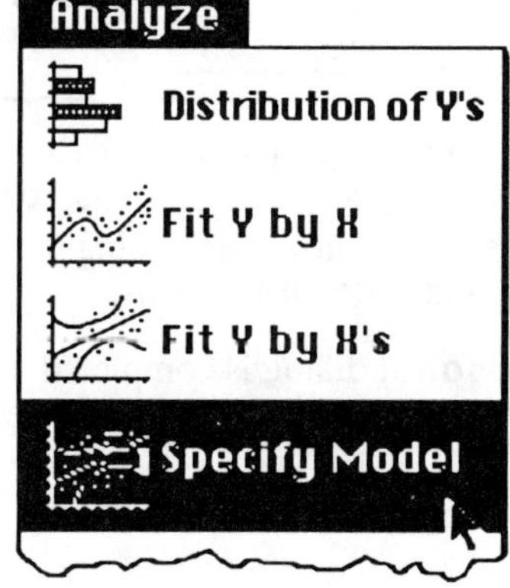

In this example, the columns are not preassigned X and Y roles in the spreadsheet. Instead, the **Specify Model** command displays the dialog shown in **Figure 6.3**. The scientist uses the dialog to designate the type of model, the model response variable and the effects in the model. The scientist

3

- selects (command-clicks) popcorn, oil amt, and batch from the column selector list
- clicks on the **Factorial** button
- selects yield from the column selector list as shown in **Figure 6.3**
- clicks on the **Y Response** button.

6

Figure 6.3 The **Specify Model** Dialog

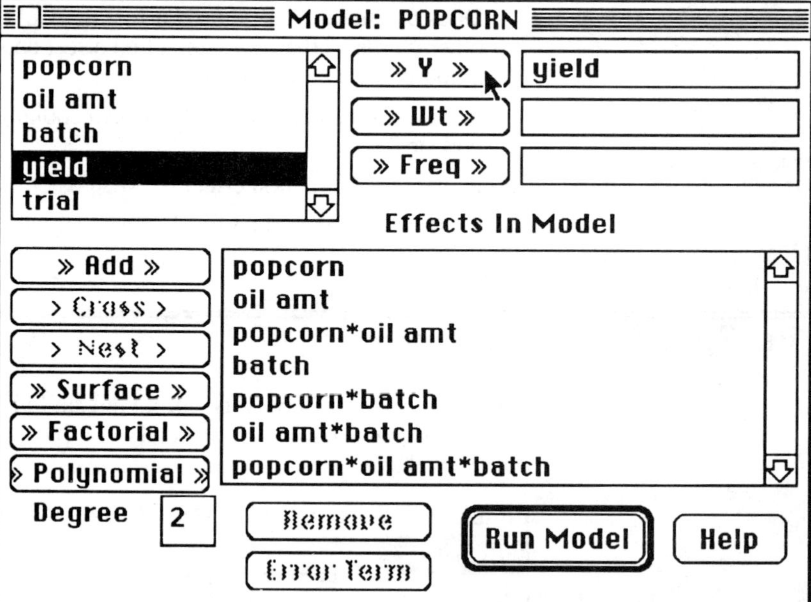

The factorial button adds all of the main effects and interactions (*crossed effects*) to the **Effects in Model** window. When the need arises, the scientist can tailor the model further by adding effects or removing unwanted effects with the **Add** and **Remove** buttons.

7

When the **Specify Model** dialog is completed the scientist clicks the **Run Model** button to estimate the model and view the results.

Graphical Display: Leverage Plots

The whole model and each effect in the model are displayed graphically as *leverage plots* as shown in **Figures 6.4-5**. The scientist can tell at a glance whether or not the factorial model explains the popcorn data, and which factors are most influential.

What can be said about the factorial model?

...the type of popcorn?

The whole-model plot shown on the upper left of **Figure 6.4** displays actual yield and predicted yield values with a regression line and 95% confidence curves. The plot to its right is a yield by popcorn leverage plot. In both of these displays the regression line and the 95% confidence curves cross the sample mean (shown as a horizontal line). This tells the scientist that the factorial model explains a significant proportion of the variation in popcorn yield and that there is a significant difference in yield between the two types of popcorn.

Figure 6.4 Leverage Plots for the Whole Model and popcorn Effects

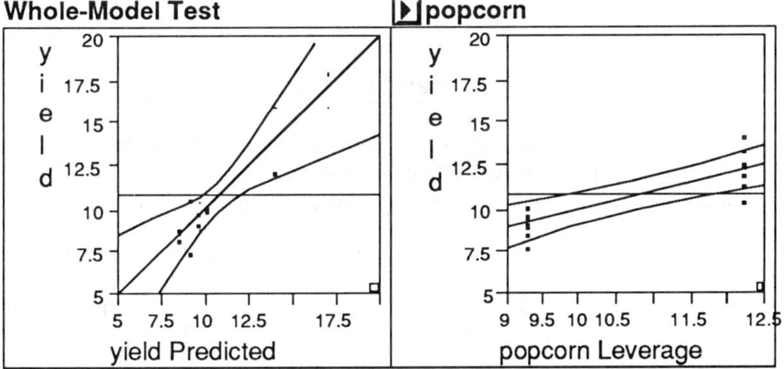

What can be said about oil amt?

...the popcorn by oil amt interaction?

The confidence curves for oil amt and the interaction, popcorn*oil amt, do not cross the horizontal mean line, This tells the scientist that neither of these factors are significant effects in the experiment.

Figure 6.5 Leverage Plots for the oil amt and Interaction Effects

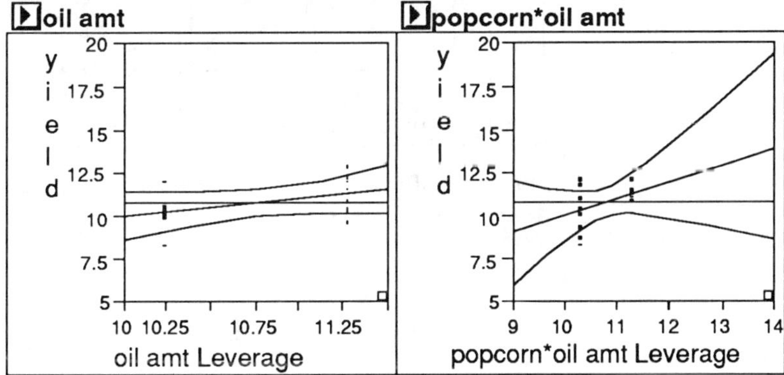

What can be said about batch size?

...about interaction between popcorn and batch size?

The scientist sees in the upper two leverage plots in **Figure 6.6** that both the batch effect and the interaction between popcorn type and batch size effect are significant effects in the factorial model. This means that the size of the batch makes a difference in the yield as well as the type of popcorn. Furthermore, the two types of popcorn are affected differently by different batch sizes.

Figure 6.6 Leverage Plots for the batch and Interaction Effects

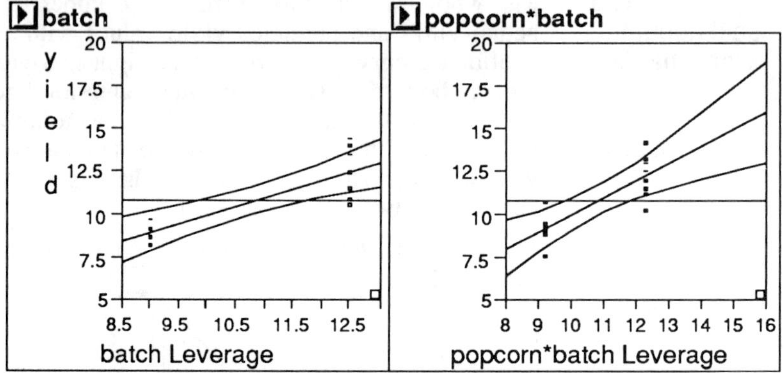

What can be said about the remaining interactions?

However, the two leverage plots shown in **Figure 6.7** indicate that there is no significant interaction between amount of oil and batch size and there is no *three-way* interaction effect between the model main effects.

Figure 6.7 Leverage Plots for Interaction Effects

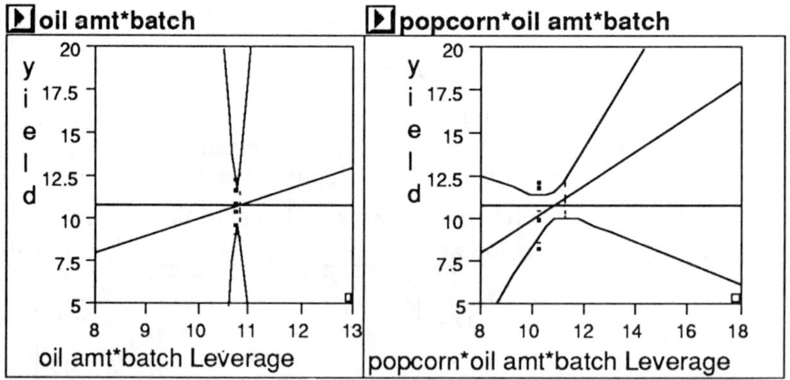

For more information about interpretation of leverage plots, see Chapter 16, *Analyze: Fit Y by X's*.

Quantify Results: Statistical Documents

Because oil amt and its interactions with other effects are insignificant, the scientist fits the popcorn data with the 2 significant factors, type of popcorn and batch size, and includes their interaction term. This approach condenses the statistical reports that show estimates of yield for the two types of popcorn under different conditions.

8 The scientist uses the same **Specify Model** dialog window as before. The new model can be specified in any of the following ways.

- Selecting unwanted terms from the previous model effects listed in the **Effects in Model** box and clicking on **Remove**.
- Removing all effects for the previous model, command-clicking to select only the **popcorn** and **batch** columns in the column selector list, and clicking on the **Factorial** button.
- Removing all effects for the previous model, selecting **popcorn** and **batch** from the column selector list and clicking **Add**. Then, deselecting them in the **Effects in Model** box, selecting them again in the column list, and clicking the **Cross** button as shown in **Figure 6.8**.

Note➜ If you closed the **Specify Model** dialog window used previously, select the **Specify Model** command again to open a new dialog. Use any of the methods listed above to specify the two-factor model.

Figure 6.8 Specify an Interaction Term

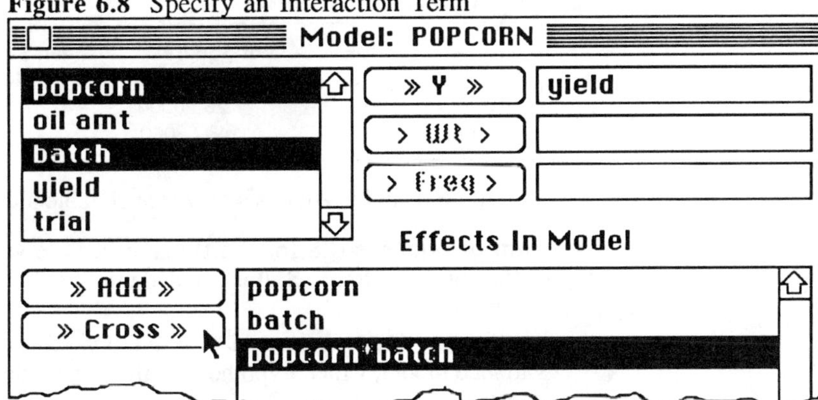

Analysis of Variance

9 When the scientist clicks **Run Model**, the leverage plots for the reduced model display. The whole model leverage plot shown in **Figure 6.9** shows that the two-factor model describes the popcorn experiment well. The scientist examines the tables that accompany the whole-model leverage plot.

The Analysis of Variance table, beneath the whole-Model leverage plot, quantifies the analysis results. It lists the *partitioning* of the total variation of the sample into components. The ratio of the Mean Square components forms an F statistic that evaluates the effectiveness of the model fit. If the probability associated with the F ratio is small, then the analysis of variance model is a statistically better fit than the overall response mean.

Figure 6.9 Analysis of Variance Tables for the Whole Model

Whole-Model Test

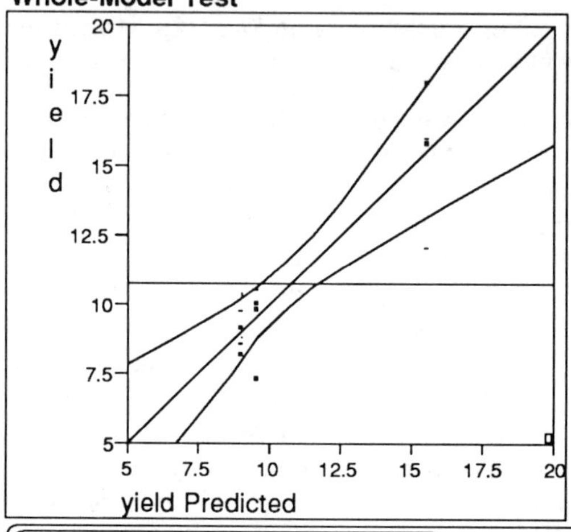

Analysis of Variance

Source	DF	Sum of Squares	Mean Square	F Ratio
Model	3	121.02000	40.3400	16.9853
Error	12	28.50000	2.3750	**Prob > F**
C Total	15	149.52000		0.0001

The columns in the Analysis of Variance table are described as follows.

- **Sum of Squares** (SS for short) quantifies the variation in yield. C Total is the corrected total SS. It is divided into the SS for Model and SS for Error.

  The SS for Model is the variation in the yield explained by the analysis of variance model, which hypothesizes that the model factors have a significant effect.

  The SS for Error is the remaining or unexplained variation.
- A **Mean Square** is a sum of squares divided by its associated degrees of freedom (**DF**).
- The **F ratio** is the model mean square divided by the error mean square.
- **Prob > F** is the probability of a greater F value occurring if the variation in popcorn yield resulted from chance alone rather than from the effect of the model factors.

In this example, the significance of the F value is .0001. This implies that the difference found in the popcorn yield produced by this experiment is expected only 1 time in 10,000 similar trials if the factors included in the model do not really affect the popcorn yield.

Model Summary Reports

10

The reveal buttons to the left of the leverage plots provide the scientist with a statistical summary of the response variable, yield, and each of the model effects. The Summary of Fit and Effect Test tables are automatically open as shown in **Figure 6.10**.

The Summary of Fit table shows the numeric summaries of the response for the factorial model.

- The Rsquare of .809 tells the scientist that the two-factor model explains nearly 81% of the variation in the data.
- Root Mean Square Error (RMSE) is a measure of the variation in the yield scores that can be attributed to random error rather than differences in the model's factors.
- Mean of Response is the mean (arithmetic average) of all the yield scores.
- Observations is the total number of recorded scores.

The F test probabilities shown in the Effect Test table tell the scientist that each factor in the model explains a significant proportion of the total variation.

Figure 6.10 Summary Tables for the Two-Factor Model Response

Response: yield

Summary of Fit

Rsquare	.8093900
Root Mean Square Error	1.541103
Mean of Response	10.75
Observations (or Sum Wgts)	16

Parameter Estimates

Effect Test

Source	Nparm	DF	Sum of Squares	F Ratio	Prob > F
popcorn	1	1	34.810000	14.6568	0.0024
batch	1	1	49.000000	20.6316	0.0007
popcorn*batch	1	1	37.210000	15.6674	0.0019

There are also additional tables showing beneath the leverage plots for each effect in the model.

Factor Summary Reports

11

The scientist now looks below the leverage plots to see the summary tables for each effect in the model. The tables for the main effects are shown in **Figure 6.11-12**. The Least Squares Means table lists the least square means and their standard errors for each level of the model factor, without consideration for any interaction between them. In this balanced example, the least squares means are simply the sample means within each level of the factor.

Figure 6.11 Tables for the Popcorn Type Effect

Effect Test

Sum of Squares	F Ratio	DF	Prob > F
34.810000	14.6568	1	0.0024

Least Squares Means

Level	Least Sq Mean	Std Error	Mean
gourmet	12.22500000	0.5448623679	12.2250
plain	9.27500000	0.5448623679	9.2750

Figure 6.12 Tables for the Batch Size Effect

Effect Test

Sum of Squares	F Ratio	DF	Prob > F
49.000000	20.6316	1	0.0007

Least Squares Means

Level	Least Sq Mean	Std Error	Mean
large	9.00000000	0.5448623679	9.0000
small	12.50000000	0.5448623679	12.5000

The character of the interaction has an important affect on the interpretation of the popcorn experiment. The tables beneath the leverage plot for the popcorn*batch effect (see **Figure 6.13**) tell the whole story.

12

The scientist sees from these least squares means that the batch size does not make much difference in yield for plain popcorn. However, gourmet popcorn yields significantly more volume than the plain type when popped in small batches.

Figure 6.13 Tables for the Interaction Effect

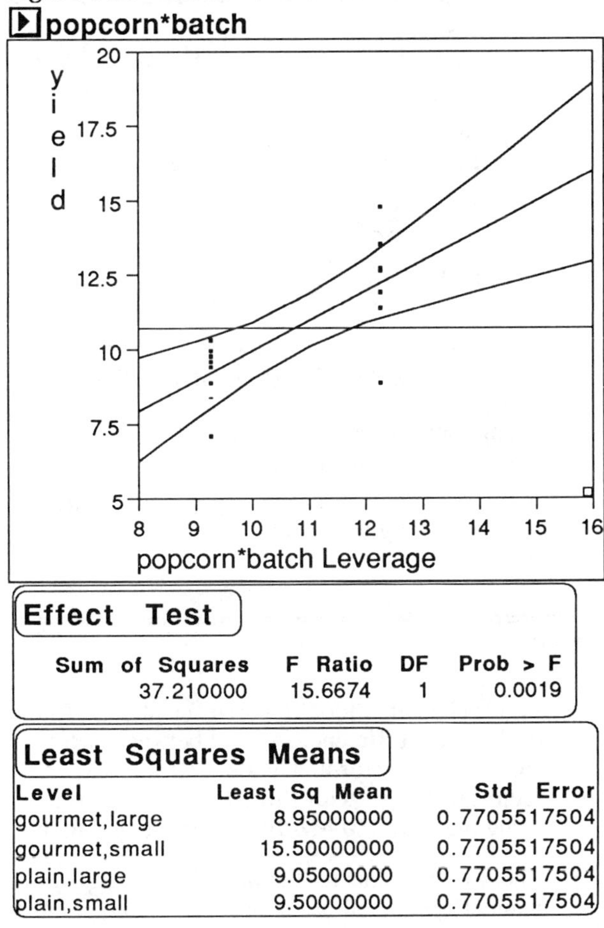

Effect Test

Sum of Squares	F Ratio	DF	Prob > F
37.210000	15.6674	1	0.0019

Least Squares Means

Level	Least Sq Mean	Std Error
gourmet,large	8.95000000	0.7705517504
gourmet,small	15.50000000	0.7705517504
plain,large	9.05000000	0.7705517504
plain,small	9.50000000	0.7705517504

13

Because the factorial model with two-factors is a good prediction model for the popcorn experiment, the scientist selects the **Save Prediction Formula** command from the save ($) pop-up menu at the lower left of the horizontal scroll bar. This command creates a new column in the POPCORN data table called **Pred Formula yield** that contains the predicted values for each experimental condition. The prediction formula, shown in **Figure 6.14**, is saved as part of the column information. The scientist chooses the **Column Info** command for the new column and clicks on the formula box in the dialog to see the formula in a calculator window. The scientist then copies the formula into a report to the corporate decision makers with the description of the analysis.

Figure 6.14 Prediction Equation for the Factorial Popcorn Model

$$10.75+\left(\begin{cases}-1.475, & \text{if } \textit{popcorn} = \text{“plain”} \\ 1.475, & \text{if } \textit{popcorn} = \text{“gourmet”} \\ \bullet, & \text{otherwise}\end{cases}\right)+\left(\begin{cases}1.75, & \text{if } \textit{batch} = \text{“small”} \\ -1.75, & \text{if } \textit{batch} = \text{“large”} \\ \bullet, & \text{otherwise}\end{cases}\right)+$$

$$+\left(\begin{cases}\begin{cases}-1.525, & \text{if } \textit{batch} = \text{“small”} \\ 1.525, & \text{if } \textit{batch} = \text{“large”} \\ \bullet, & \text{otherwise}\end{cases}, & \text{if } \textit{popcorn} = \text{“plain”} \\ \begin{cases}1.525, & \text{if } \textit{batch} = \text{“small”} \\ -1.525, & \text{if } \textit{batch} = \text{“large”} \\ \bullet, & \text{otherwise}\end{cases}, & \text{if } \textit{popcorn} = \text{“gourmet”} \\ \bullet, & \text{otherwise}\end{cases}\right)$$

What can the scientist conclude?

In the report to the corporate decision makers, the scientist intends to advise that the new gourmet popcorn be placed on the market

- in small packages so the yield will be good
- in family size packages with smaller packets inside
- in family size packages with popping instructions that clearly state the best batch size for good results.

Chapter Summary

In this chapter, a food research scientist performed a designed experiment to evaluate the difference in yield between two types of popcorn. Two popping trials were performed under various experimental conditions, giving a factorial design with three-factors. The scientist analyzed the results by using the **Specify Model** command and found the following results.

- The leverage plots for the factorial analysis of three factors showed one main effect and all of its associated interactions to be insignificant.
- A more compact two-factor analysis with interaction adequately described the variation in yield for the popcorn trials.
- The interaction between the two main effects was significant. The Least Squares Means table for the interaction showed how the two types of popcorn behaved under different popping conditions.

The scientist concluded that the new, more expensive gourmet popcorn had better yield than the plain everyday type only if popped in small batches. The scientist advised the corporate decision makers to market the new, more expensive type of gourmet popcorn in small packets or in larger quantities with carefully worded popping instructions.

Chapter 7.
Multiple Regression
Advanced Example

Multiple Regression is the technique of fitting or predicting a response by a linear combination of several regressor variables. The fitting principle is like simple linear regression, but the space of the fit is in 3 or more dimensions, making it more difficult to visualize. With multiple regressors, there are more opportunities to model the data well, but the process is more complicated.

This chapter begins with an example of a 2-regressor fit that includes 3 dimensional graphics for visualization. The example is then extended to include 6 regressors (but unfortunately no 7-dimensional graphics to go with it).

Aerobic Fitness Data

Aerobic fitness can be evaluated using a special test that measures the oxygen uptake of a person while running on a treadmill for a prescribed distance. However, it would be nice to evaluate fitness more economically using a formula which could predict oxygen uptake using simpler measurements.

In an effort to identify such an equation, run time and fitness measurements were taken for 31 participants who ran 1.5 miles. The participants' ages were also recorded. The data are in the JMP file called FITNESS (SAS Institute 1987) in the SAMPLE DATA folder. For purposes of illustration, certain values of maxpulse and runpulse have been changed from data reported by Rawlings (1988, p.105).

Figure 7.1 The Oxygen Uptake Data Table

FITNESS

7 Cols / 31 Rows	Int age	Int weight	Int oxy	Int runtime	Int rstpulse	Int runpulse	Int maxpulse
1	45	87.66	37.39	14.03	56	186	192
2	54	91.63	39.20	12.88	44	168	172
3	57	73.37	39.41	12.63	58	174	176
4	44	81.42	39.44	13.08	63	174	176
5	51	69.63	40.84	10.95	57	168	172

The first step in this analysis is to select the **Fit Y by X's** command from the **Analyze** menu and to respond to the assign roles dialog.

The oxy column is then assigned as a response variable (Y), and both age and runtime are assigned as factors (X). Clicking **Done** launches the platform with graphics and reports which are appropriate for a response and factors which both have interval values.

Figure 7.2 Statistical Text Reports for **Fit Y by X's** Platform

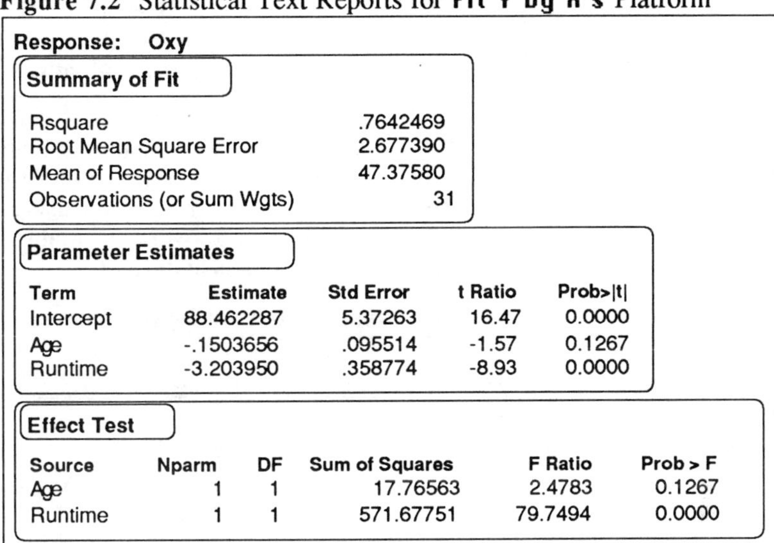

Response: Oxy

Summary of Fit

Rsquare	.7642469
Root Mean Square Error	2.677390
Mean of Response	47.37580
Observations (or Sum Wgts)	31

Parameter Estimates

Term	Estimate	Std Error	t Ratio	Prob>|t|
Intercept	88.462287	5.37263	16.47	0.0000
Age	-.1503656	.095514	-1.57	0.1267
Runtime	-3.203950	.358774	-8.93	0.0000

Effect Test

Source	Nparm	DF	Sum of Squares	F Ratio	Prob > F
Age	1	1	17.76563	2.4783	0.1267
Runtime	1	1	571.67751	79.7494	0.0000

Before examining the results in detail, you can save the prediction equation using the **Save Prediction Formula** command in the save ($) pop-up menu to the left of the horizontal scroll bar. This command creates a new calculated column in the FITNESS data table called Pred Formula oxy. To see the new column's formula, select the new column, choose **Column Info** from the **Columns** menu, and click in the formula display box in the **Column Info** dialog. The calculator window opens and displays the formula,

$$88.4622874+-.15036566 \cdot Age+-3.2039505 \cdot Runtime \quad .$$

This formula defines a plane of fit for oxy as a function of age and runtime.

The Fitting Plane

You can use JMP to see the relationship between oxy, runtime, and age in three dimensions and also to see the fitting plane represented by a grid of values which satisfy the above formula. To do this, you need to

- add rows to the data table to contain the grid values for the fitting plane
- define new variables to calculate the grid values for plotting.

Compute a Grid for Plotting

This fitting plane is a 7 by 7 grid, with age values increasing from 30 to 60 in increments of 5, and runtime values ranging from 8 to 14 in increments of 1.

Table 7.1 shows formulas for computing additional new variables. The variable called oxy|grid is assigned either the value of oxy or a value computed from the prediction equation using the new grid values. The computed column, mark, is a row state variable used to distinguish between observed and predicted points in the plot.

Table 7.1 Computations for New Columns

New Variable	Formula
age|grid	$\begin{cases} Age, & \text{if } Age \neq \bullet \\ \left\lfloor \left(\frac{(i-32)}{7} \right) \right\rfloor \bullet 5+30, & \text{otherwise} \end{cases}$
runtime|grid	$\begin{cases} Runtime, & \text{if } Runtime \neq \bullet \\ (i-32) \bmod 7 \bullet 1+8, & \text{otherwise} \end{cases}$
oxy|grid	$\begin{cases} Oxy, & \text{if } Oxy \neq \bullet \\ Pred\ Formula\ Oxy, & \text{otherwise} \end{cases}$
mark	$\begin{cases} [\times], & \text{if } Oxy \neq \bullet \\ [\cdot], & \text{otherwise} \end{cases}$

Note➡ In order to reference the grid versions of oxy and runtime, the formula for Pred Formula oxy must be changed to

88.4622874+-.15036566•*Age/Grid*+-3.2039505•*Runtime/Grid* .

See Chapter 11, *Using the Calculator,* for information about modifying formulas in the calculator window.

To activate the new row state column, select **Copy to RowState** from the [«] pop-up menu at the top of the mark column. The **Spin** command from the **Analyze** menu can then launch a spin plot of the three grid columns called age|grid, runtime|grid, and oxy|grid. The result is a graphic dramatization of the actual points (marked as x's) and the plane of fit (marked as dots). You can spin the plot to see how the observed points scatter about the plane.

Figure 7.3 Observed Points with the Predicted Plane of Fit

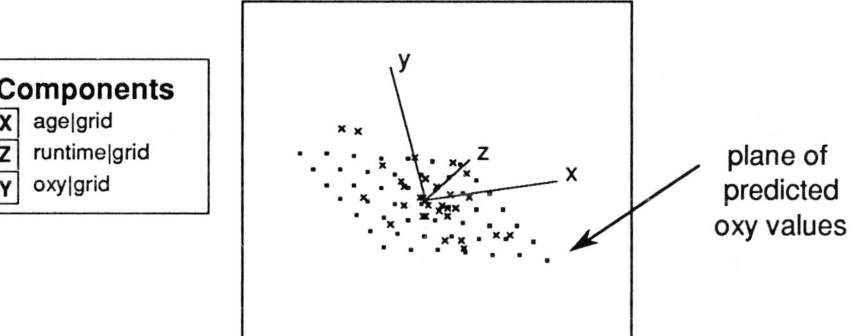

When the axes are in the home position, as shown to the left in **Figure 7.4**, you see a plot of oxy by age. By rotating the plot on the Y-axis to the position shown on the right, you see a plot of oxy by runtime, which illustrates a stronger relationship.

Figure 7.4 Rotating to See Relationships

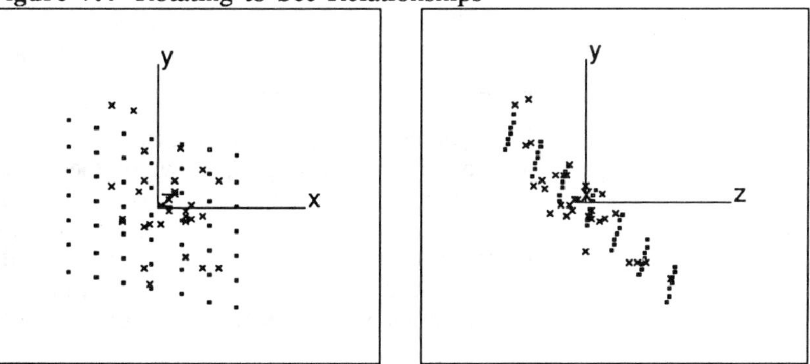

Fit Planes to Test Effects

You can use JMP to look at hypothesis tests for each regressor and to test whether the regressor's parameter is significantly different from zero. One way to view this test is to evaluate the difference between the current fit and the fit that occurs if you remove the regressor variable from the model. For example, consider removing the runtime variable from the model. Fit just the subset model, oxy to age, and save the prediction formula as before. The new predicted column is calculated using the formula,

62.2206385+-.31135980•*Age/Grid* .

To compare this fitted line with the plane, add a second grid of points using this new prediction formula. Rows 32 to 80 are the original bivariate regression grid. Observations 81 to 129 are the new grid, which slopes along age but is flat along runtime. The formulas for the variables are shown in **Table 7.2**.

Table 7.2 Computations for New Columns

New Variable	Formula
mark	$\begin{cases} \text{[x marker]}, & \text{if } Oxy \neq \bullet \\ \text{[dot marker]}, & \text{if } i \leq 80 \\ \text{[small square marker]}, & \text{otherwise} \end{cases}$
oxy|grid	$\begin{cases} Oxy, & \text{if } Oxy \neq \bullet \\ Pred\ Formula\ Oxy, & \text{if } i \leq 80 \\ Pred\ Formula\ Oxy2, & \text{otherwise} \end{cases}$
age|grid	$\begin{cases} Age, & \text{if } Age \neq \bullet \\ \left\lfloor \left(\frac{((i-32) \bmod 49)}{7} \right) \right\rfloor \cdot 5+30, & \text{otherwise} \end{cases}$
runtime|grid	$\begin{cases} Runtime, & \text{if } Runtime \neq \bullet \\ (i-32) \bmod 7 \cdot 1+8, & \text{otherwise} \end{cases}$

Now you can spin the two regressor grid variables, shown on the X and Z axis, and the response grid, shown on the y axis. As before, the actual data are marked as x's, and the diamonds make up the bivariate regression plane. The dots mark the regression plane for the age variable alone. Both grids represent least squares regression planes, but the dots plane has a slope of zero in the orientation of runtime. **Figure 7.5** shows the spinning plot from an angle with the **Box** display option in effect.

Components

- X age|grid
- Z runtime|grid
- Y oxy|grid

Figure 7.5 Three-dimensional Plot with Regression Planes

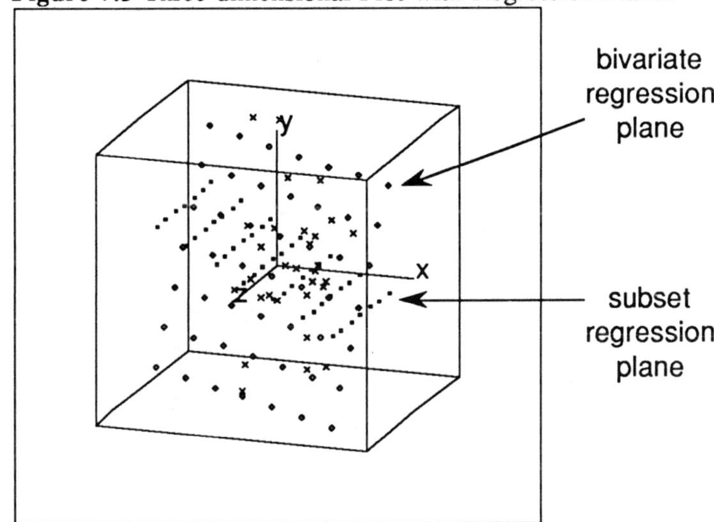

If you look at the plot in home position for age by oxy as shown on the left in **Figure 7.6**, you see the subset regression *edge-on* as a line rather than a plane. If you rotate the plot toward the edge-on view of runtime, you see the diamond grid reduced as shown on the right. If you fit a regression using runtime only, this view shows a straight-line fit of the x-marked points.

Figure 7.6 Comparison of Two-Dimensional Views

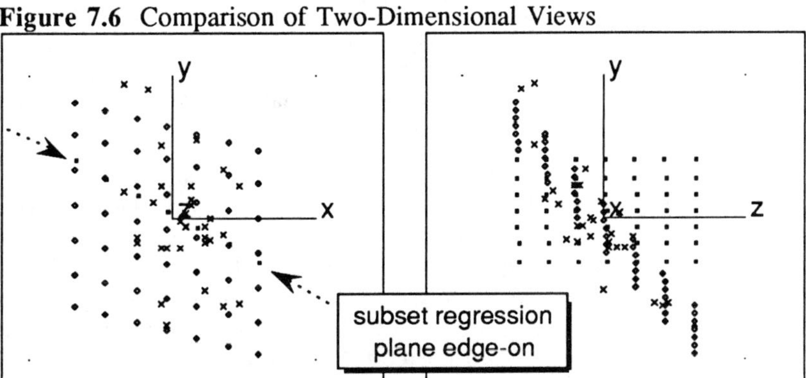

The view shown in **Figure 7.7** brings the bivariate regression plane edge-on and represents the linear combination of the effects fit by the plane.

Figure 7.7 Rotating to See Relationships

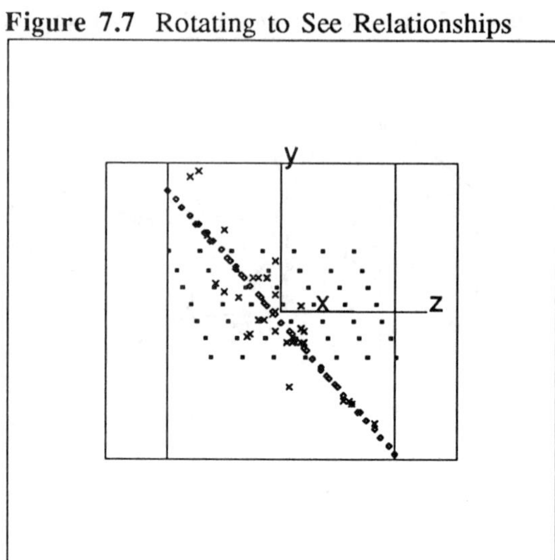

Leverage Plots: The Hypothesis-Eyed View

Suppose that you tilt the view from the previous example so that both the bivariate and subset regression planes are viewed edge-on. **Figure 7.8** shows the subset regression plane oriented in the X direction. This view best dramatizes the difference between the fit of the bivariate regression and the subset regression. You can see how much better the fit is for the diamond-marked bivariate regression plane than the dot-marked plane which represents the age variable only.

Now suppose that this view is altered slightly so that the Y-axis is vertical but the dot-marked plane is still the horizontal edge. This view of the regression and data is so revealing that it has a special name and a place in the Fit Y by X's text report for each effect. It is called a *leverage plot* and is an hypothesis-eyed view of the data. The three-dimensional plot, adjusted for the direction of the response, is the leverage plot for runtime which is shown on the right in **Figure 7.8.**

Figure 7.8 Compare Regression Planes and Leverage Plots

Components

- X age|grid
- Z runtime|grid
- Y oxy|grid

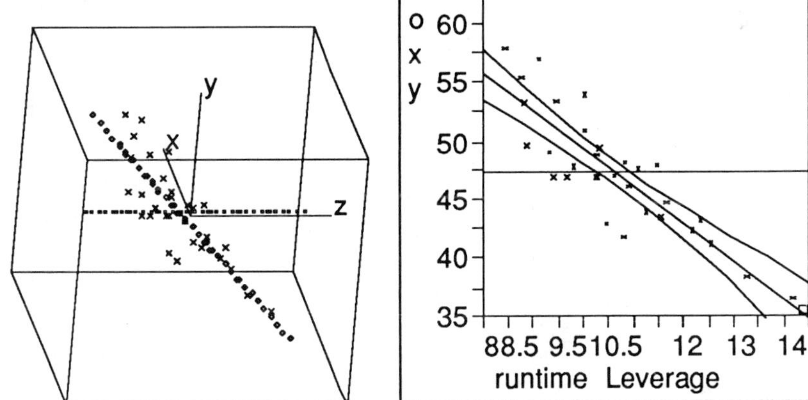

A leverage plot is important because

- the distance from each point to the line of fit represents the residual for the full model
- the distance from each point to the horizontal line represents the residual for a model constrained by the hypothesis.

In the example shown in **Figure 7.8**, the parameter for runtime is hypothesized to be zero.

You can form a leverage plot for any linear hypothesis. **Figure 7.9** shows the leverage plot for testing whether age is significant. Note that age does not relate as strongly to the response as runtime. The 95% confidence curves fully contain the horizontal line in the leverage plot for age, indicating that the line of fit is not significantly different from the horizontal line representing the simple mean model.

Figure 7.9 Leverage Plot for the age Variable

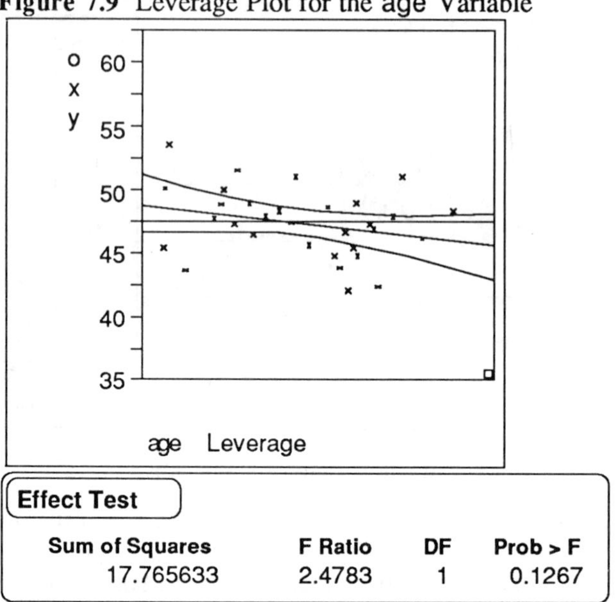

Effect Test

Sum of Squares	F Ratio	DF	Prob > F
17.765633	2.4783	1	0.1267

Whole Model Tests

The leverage plot in **Figure 7.10** shows the joint test of both the age and runtime effects in the model. This compares the full model with the model containing the intercept which fits the overall response mean only. This leverage plot is formed by plotting the actual observed values on the Y axis and the values predicted by the whole model on the X axis. The residual for the subset model is the distance from a point to the horizontal line drawn at the sample mean.

Figure 7.10 Leverage Plot for the Whole Model (age and runtime)

Whole-Model Test

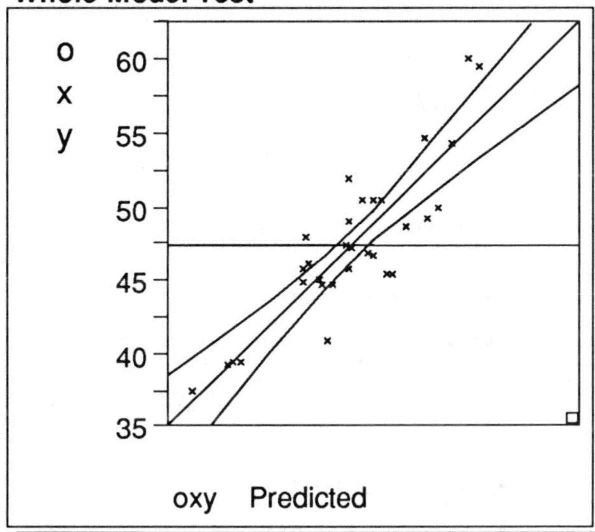

Analysis of Variance

Source	DF	Sum of Squares	Mean Square	F Ratio
Model	2	650.66573	325.333	45.3842
Error	28	200.71581	7.168	**Prob > F**
C. Total	30	851.38154		0.0000

More and More Regressors

It's easy to visualize two regressors predicting a response using fitting planes, but how can this be done with more regressors when the analysis requires more than 3 dimensions? In actuality, the fitting, testing, and leverage plot analyses still work for more regressors.

Continuing with the previous example, you can add weight, runpulse, and maxpulse to the regression to obtain the prediction formula,

$$\text{oxy} = 102.2 - .2196\ \text{age} - .0723\ \text{weight} - 2.68\ \text{runtime} - .3734\ \text{runpulse} + .3049\ \text{maxpulse}\quad .$$

You can look at the t ratios in the Parameter Estimates report or the F ratios in the Effects Tests report (see **Figure 7.11**) to examine the significance of each regressor. Because each effect has only one parameter, the F ratios are the squares of the t ratios, and the significance probabilities are the same.

The age variable seems significant, but weight does not. The runtime variable seems highly significant. Both runpulse and maxpulse also seem significant, but maxpulse is less significant than runpulse.

Figure 7.11 Statistical Text Reports for Multiple Regression

Summary of Fit

Rsquare	.8480018
Root Mean Square Error	2.275156
Mean of Response	47.37580
Observations (or Sum Wgts)	31

Analysis of Variance

Source	DF	Sum of Squares	Mean Square	F Ratio
Model	5	721.97309	144.395	27.8951
Error	25	129.40845	5.176	Prob > F
C. Total	30	851.38154		0.0000

Parameter Estimates

Term	Estimate	Std Error	t Ratio	Prob>|t|
Intercept	102.20427	11.9792	8.53	0.0000
Age	-.2196213	.095502	-2.30	0.0301
Weight	-.0723023	.053310	-1.36	0.1871
Runtime	-2.682522	.340985	-7.87	0.0000
RunPulse	-.3734008	.117141	-3.19	0.0038
MaxPulse	.30490783	.133936	2.28	0.0316

Effect Tests

Source	Nparm	DF	Sum of Squares	F Ratio	Prob > F
Age	1	1	27.37429	5.2884	0.0301
Weight	1	1	9.52157	1.8394	0.1871
Runtime	1	1	320.35968	61.8892	0.0000
RunPulse	1	1	52.59624	10.1609	0.0038
MaxPulse	1	1	26.82640	5.1825	0.0316

Interpreting Leverage Plots

The leverage plots for this multiple regression model let you visualize the contribution of each effect. First, look at the whole-model leverage plot of observed versus predicted values which illustrates the test for the whole set of regressors. The Analysis of Variance table shown in **Figure 7.12** reveals a highly significant F of 27.89 corresponding to this plot. The confidence curves in this plot show the strong relationship because they cross the horizontal line.

Figure 7.12 Whole-Model Leverage Plot

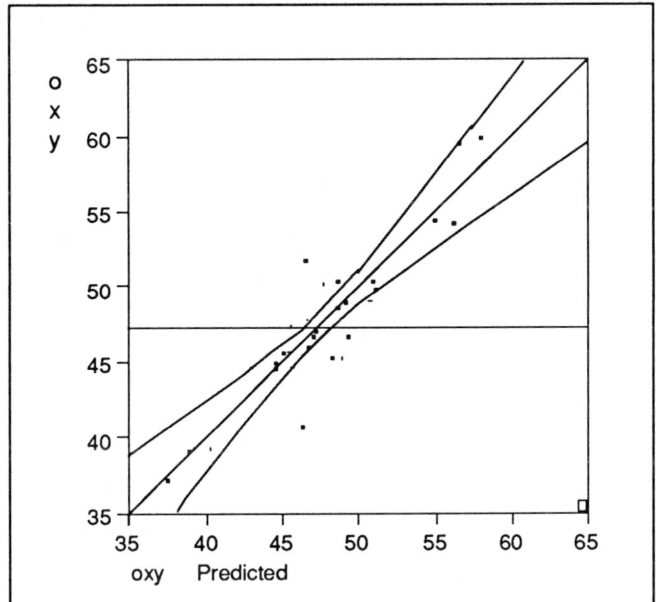

Now you can examine the leverage plots for the regressors. Each plot illustrates the residuals as they are and as they would be if that regressor were removed from the model.

The confidence curves in the leverage plot for age, shown on the left in **Figure 7.13**, indicate that age is borderline significant because the curves barely cross the horizontal line of the mean. Note that the significance of the age effect is .03 in the text reports (shown in **Figure 7.11**), which is only slightly different from the .05 confidence curves drawn by JMP.

The leverage plot for weight shows that the effect is not significant. The confidence curves do not cross the horizontal line of the mean.

Figure 7.13 Leverage Plots for the age and weight Effects

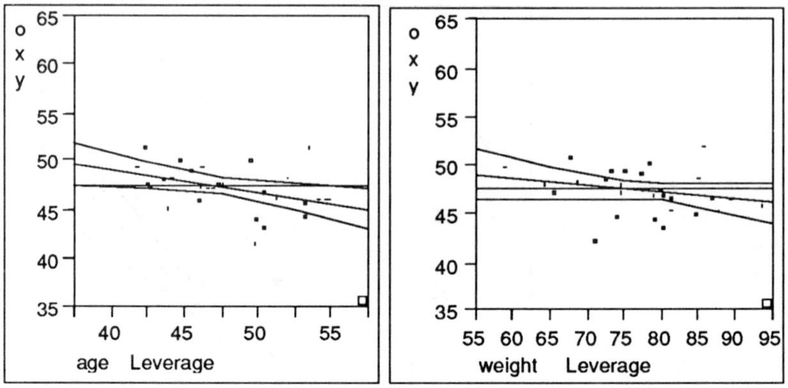

The leverage plot for runtime (**Figure 7.14** below) shows that runtime is the most significant of all the regressors. The runtime leverage line and its confidence curves cross the horizontal mean at a steep angle.

Figure 7.14 Leverage Plots for the runtime Effect

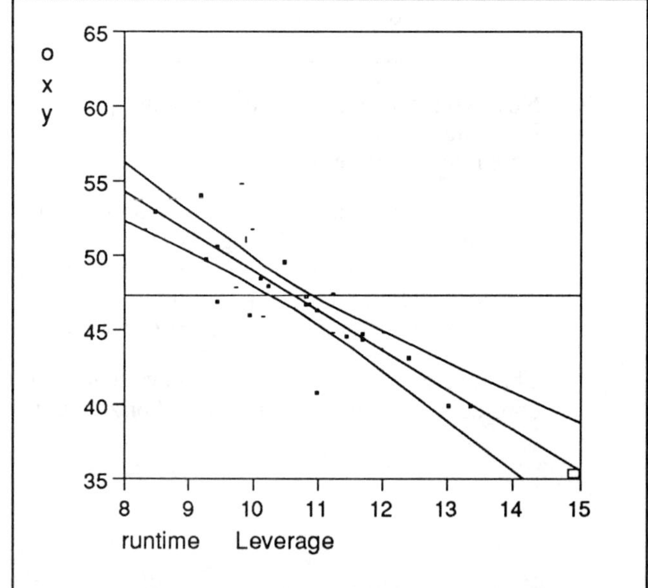

The leverage plots for runpulse and maxpulse are similar. Each is somewhat shrunken on the X axis. This indicates that other variables are related in a strong, linear fashion to these two regressors. This means that the two effects are strongly correlated with each other.

Figure 7.15 Leverage Plots for the runpulse and maxpulse Effects

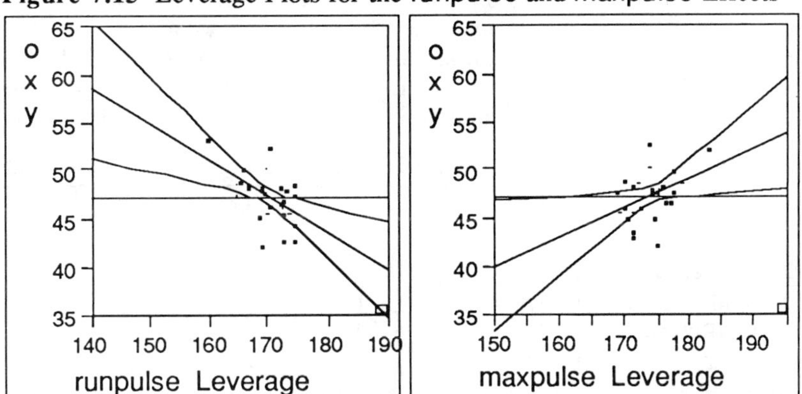

Colinearity

When two or more regressors have a strong correlation, they are said to be colinear. These regression points occupy a narrow band showing their linear relationship.

When you fit a plane representing colinear regressors to a response, the plane fits the points well in the direction where they are scattered widely. However, in the direction where the scatter is very narrow, the fit is weak and the plane is unstable.

In text reports, this phenomenon translates into high standard errors for the parameter estimates and potentially high values for the parameter estimates themselves. This occurs because a small random error in the narrow direction can have a huge effect on the slope of the corresponding fitting plane. You can see an indication of colinearity in leverage plots when the points tend to collapse toward the center of the plot in the X direction.

The fitness example shows colinearity geometrically in the strongly correlated regressors, runtime and maxpulse. You can repeat the steps which produced the 2 grids, modifying them so that one grid represents the fit of oxy to both runpulse and maxpulse. The other grid is for the subset model that excludes runpulse in order to show its leverage. **Figure 7.16** shows these two views. If you spin the plot, you notice that most of the points are near the intersection of the two planes. If you turn the plot so that both planes are edge-on, as shown on the right of **Figure 7.16**, then you are looking at a view where most of the scatter is hidden. The result is that the fitting plane representing both variables holds no better than the subset plane from that angle. The angle of the fitting plane is steep, but the hold is unstable.

Geometrically, colinearity between two regressors means that the points they represent do not spread out in X space enough to provide stable support for a plane. Instead, the points cluster around the center causing the plane to be unstable. The regressors act as substitutes for each other to define one direction redundantly. This can be cured by dropping one of the colinear regressors from the model. Since maxpulse and runpulse measure essentially the same thing, dropping either is a reasonable solution.

Figure 7.16 Comparison of runpulse and maxpulse Effects

Components

X RunPulse|Grid
Z MaxPulse|Grid
Y Oxy|Grid

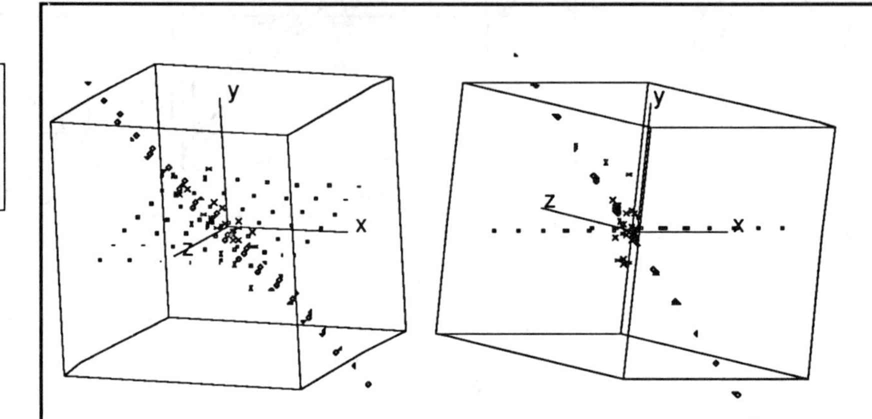

Chapter Summary

Multiple regression uses the same fitting principle as simple regression, but accounting for significance is more subtle. Each regressor opens a new dimension for fitting a hyperplane, and its significance is tested by how much the fit suffers in its absence. Graphically, leverage plots provide the best view of each effect's partial contribution. When regressors correlate to each other, they are said to be colinear, and they define directions where the fitting hyperplane is not well supported.

Chapter 8.
Exploring Data
Advanced Example

Exploration is the search to find something new...the endeavor to make some discovery. For data analysis, exploratory study is often the most fruitful part of the analytical process because it the most open to serendipity. Something you notice about your data can be the seed of some important advance.

There are two important aspects of exploration.

- What is the pattern or shape of the data?
- Are there points unusually far away from the bulk of the data (*outliers*)?

When you explore data composed of many variables, the great challenge is dealing with this high dimensionality. You can have many variables that you hope have some interesting relationships, but it's hard to visualize more than a few at a time.

Solubility Data

A chemist interested in studying solubility is looking for compounds with unusual solubility patterns in various solvents. Data from an experiment (Koehler, Grigoras and Dunn (1988)) are in the SOLUBILITY DATA file in the SAMPLE DATA folder. There are 72 compounds which have been tested with six solvents, labeled **eth**, **oct**, **cc14**, **c6c6**, **hex**, and **chc13**. To follow the exploratory path the chemist takes, double-click on the SOLUBILITY DATA file to begin a JMP session. Next, assign the role of Label to the column called Labels as shown in **Figure 8.1** so that the names instead of row numbers identify points in plots.

Figure 8.1 The Solubility Data Table

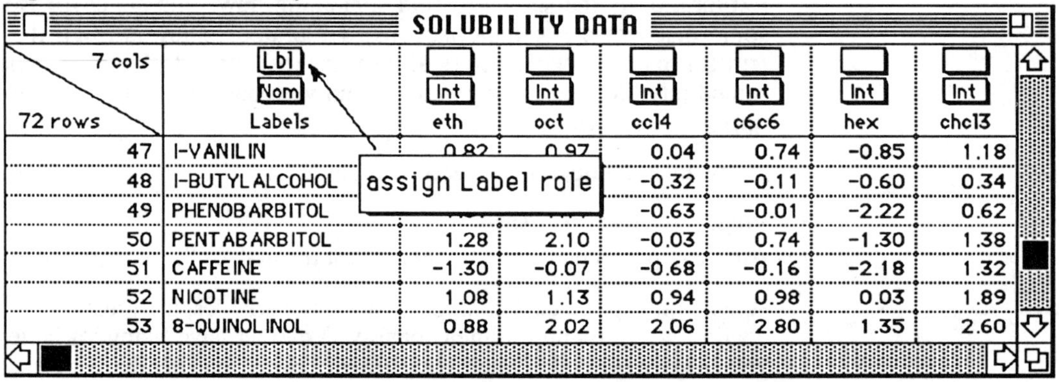

SOLUBILITY DATA

7 cols / 72 rows	Labels (Lbl, Nom)	eth (Int)	oct (Int)	ccl4 (Int)	c6c6 (Int)	hex (Int)	chcl3 (Int)
47	I-VANILIN	0.82	0.97	0.04	0.74	-0.85	1.18
48	I-BUTYLALCOHOL	[illegible]	[illegible]	-0.32	-0.11	-0.60	0.34
49	PHENOBARBITOL	[illegible]	[illegible]	-0.63	-0.01	-2.22	0.62
50	PENTABARBITOL	1.28	2.10	-0.03	0.74	-1.30	1.38
51	CAFFEINE	-1.30	-0.07	-0.68	-0.16	-2.18	1.32
52	NICOTINE	1.08	1.13	0.94	0.98	0.03	1.89
53	8-QUINOLINOL	0.88	2.02	2.06	2.80	1.35	2.60

assign Label role

There are 6 variables but no 6-dimensional graphics. However, you can look at 6 one-dimensional graphs, 15 two-dimensional graphs, and 20 three-dimensional spinning graphs. Using principal components, you can even display a representation of higher dimensions.

One-dimensional Views

The **Distribution of Y's** platform shows you data one column at a time. You don't see any relationships between variables, but the shape of the individual distributions helps you identify the one-dimensional outliers.

You can begin exploring the solubility data by launching the **Distribution of Y's** platform and selecting the six solubility columns. The histograms, resized and trimmed of other output, are shown in **Figure 8.2**. When you click on any histogram bar, the bar and all other representations of that data are highlighted in all related windows. In this example, click on the outlying bars in each histogram to see how those values are distributed in the other histograms. Shift-click the outlying bars of each histogram to extend the selection so that it includes the outlying rows in each single dimension. You can use the **Marker** submenu in the **Rows** menu to assign x's (or any marker) to these selected rows. These markers are displayed in the data table and in subsequent plots.

Figure 8.2 One-Dimensional Views

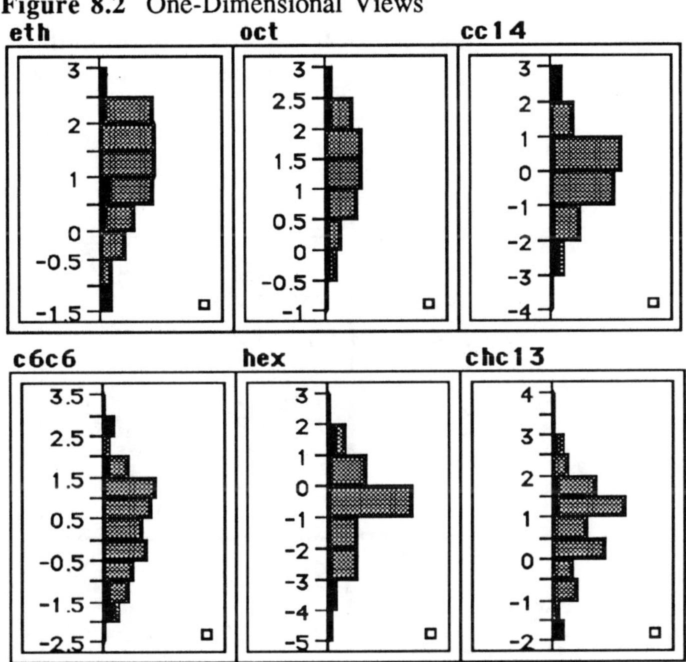

Once you have highlighted and marked the one-dimensional outliers, a new data table can be created containing only those rows using the **Transform, Subset** command from the **Edit** menu. Scroll through the new table as illustrated in **Figure 8.3** to see the 15 compound names selected as one-dimensional outliers.

Figure 8.3 Subset of One-Dimensional Outliers

	Lbl Nom Labels
1	METHANOL
2	ACETIC_ACID
3	CHLOROACETICACID
4	ETHYLAMINE
5	2 NAPHTHOL
6	HYDROQUINONE
7	P-AMINOBENZOICACID
8	CAFFEINE
9	8-QUINOLINOL
10	ANISOLE
11	2-6-DIMETHYLPHENOL
12	M-CHLOROPHENOL
13	SULFATHIAZOLE
14	BENZOYLACETONE
15	O-METHOXYBENZOICACID

Two-Dimensional Views

You can use the **Y's by Y's** platform to display all 30 two-dimensional scatterplots arranged in a matrix. The one-dimensional outliers appear as x's in each of the scatterplots.

Figure 8.4 Two-Dimensional Views

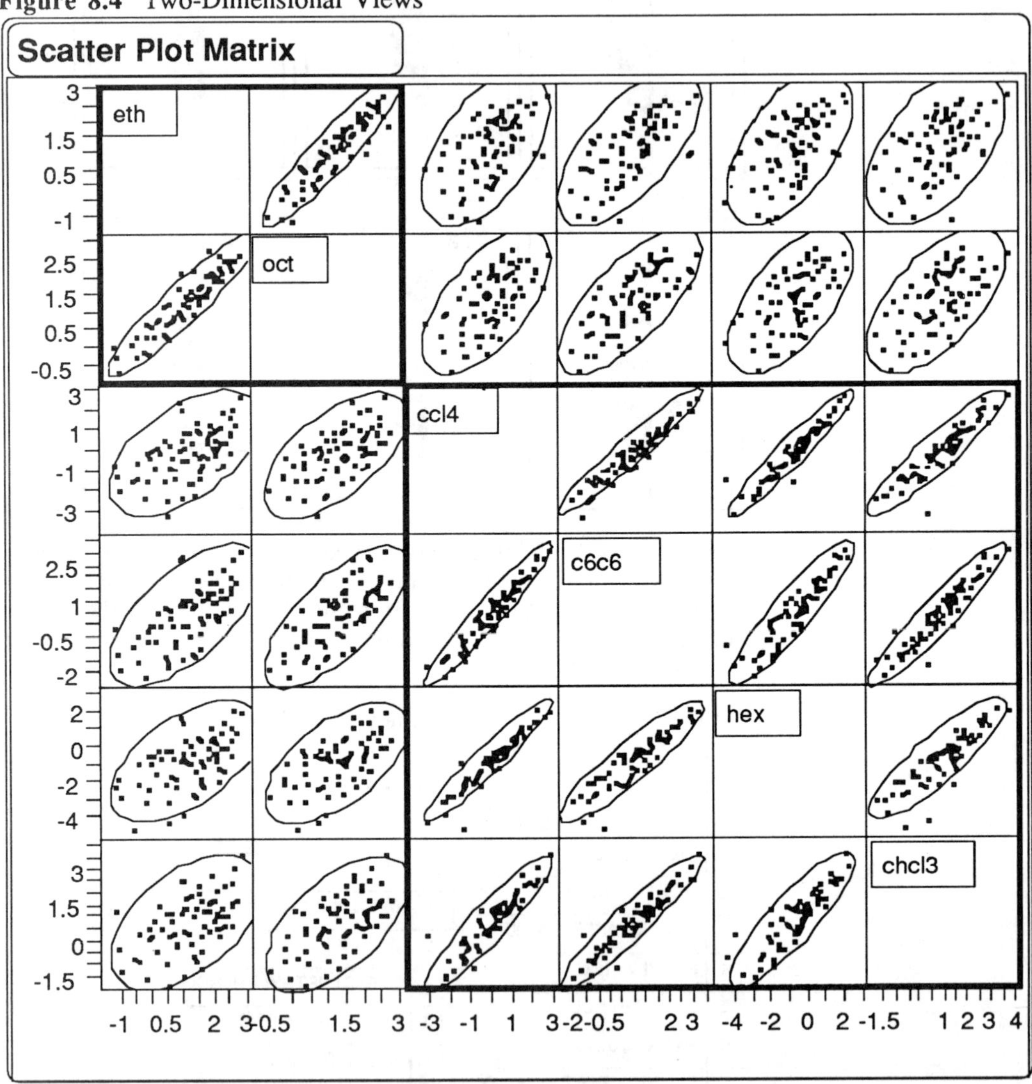

Note that many of the variables appear to be correlated in the above matrix, as evidenced by the diagonal flattening of the normal bivariate density ellipses. There appear to be two groups of variables which correlate among themselves but are not very correlated with variables in the other group.

One group is made of **eth** and **oct**, and the other group consists of the remaining four variables as illustrated in **Figure 8.4.**

You can scan these plots looking for outliers of a two-dimensional nature and identify them with squares markers using the **Marker** submenu from the **Rows** menu. Now, you have identified both one- and two-dimensional outliers.

Three-Dimensional Views

To see points in three dimensions, use the **Spin** command from the **Analyze** menu which launches a three-dimensional spinning platform. You can add all six variables to the spin list and then drag the X, Y, and Z axis tags to various combinations of 3 variables. For each combination, use the spin controls or select the grabber (hand) tool from the **Tools** menu to rotate the plot and look for points away from the point cloud.

In the view of **cc14** (X) by **hex** (Y) by **chc13** (Z) shown in **Figure 8.5**, there appear to be two outlying points. Shift-clicking on these points causes their labels, Sulfathiazole and Hydroquinone, to show on the plot.

Figure 8.5 Spotting Outliers in One Three-dimensional View

Components

- [ ] eth
- [ ] oct
- [X] ccl4
- [ ] c6c6
- [Y] hex
- [Z] chcl3

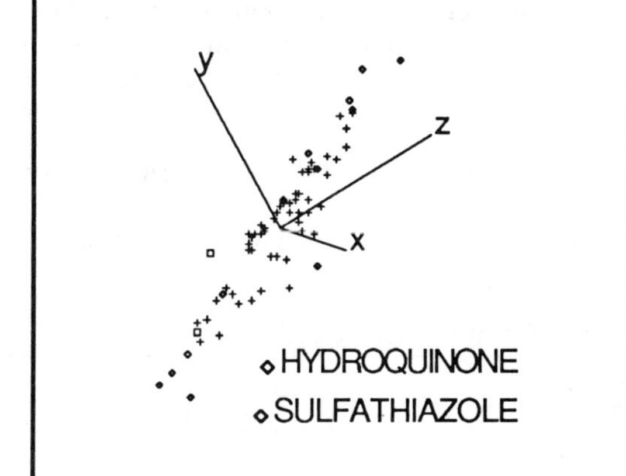

Principal Components and Biplots

Because many of the variables in the SOLUBILITY DATA table are highly correlated, there is not as much scatter in six dimensions as you might expect. The scatter is oriented in some directions but is flattened in other directions.

To illustrate this, launch the spin platform again and add only the two highly correlated variables, **eth** and **oct,** as spin components. Select the principal components command from the check (√) pop-up menu at the lower left of the platform window. The results are shown in **Figure 8.6.** Note that because the data are highly correlated, the scatter in the points runs in a narrow ellipse whose principal axis is oriented in the direction marked P1.

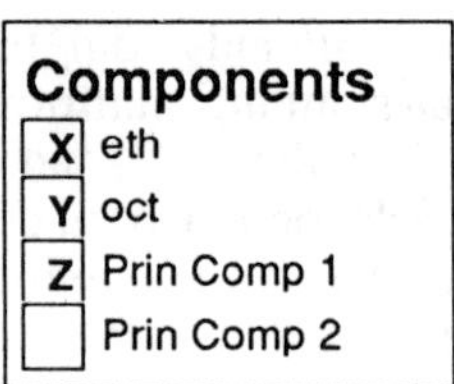

Figure 8.6 Two Correlated Variables with Principal Components

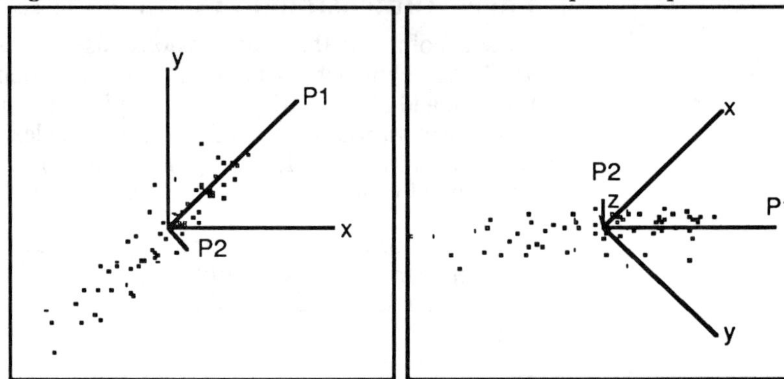

To see of the greatest variation of the data in one dimension, rotate the axis so that first principal component, P1, is horizontal. The technique of extracting these orientations that capture the highest variance is called principal components analysis. The technique of principal components captures the most variation possible in the smallest number of dimensions.

If you use the **Principal Components** command in the 6-variable spin platform, the Principal Components table in **Figure 8.7** shows that 97.8% of the 6-dimensional variation is accounted for by the first three principal components.

Figure 8.7 Principal Components Text Report

Principal Components						
EigenValue:	4.7850	0.9408	0.1427	0.0630	0.0464	0.0221
Percent:	79.7503	15.6804	2.3779	1.0495	0.7738	0.3680
CumPercent:	79.7503	95.4307	97.8087	98.8582	99.6320	100.0000
Eigenvectors:						
eth	0.34887	0.64095	0.17653	0.63780	-0.12841	0.11419
oct	0.37314	0.56277	-0.17684	-0.67519	0.23816	0.01361
ccl4	0.43187	-0.29274	0.17821	-0.20373	-0.46785	0.66003
c6c6	0.44552	-0.14827	-0.21629	-0.01927	-0.52424	-0.67642
hex	0.42231	-0.27033	0.68072	-0.00820	0.47344	-0.24696
chcl3	0.41919	-0.30285	-0.62900	0.30884	0.45709	0.18056

On the spinning plot, principal components appear as additional *rays* labeled P1, P2, and so on as shown in **Figure 8.8**. These rays are the projections of the six-dimensional direction of the principal component in the 3 dimensions shown. P1 tends to be the longest ray because it shows the direction of greatest variance. P2 appears as the second longest ray and describes the next most prominent direction. The other rays become shorter and shorter.

Figure 8.8 Principal Component Rays in Three of the Variables

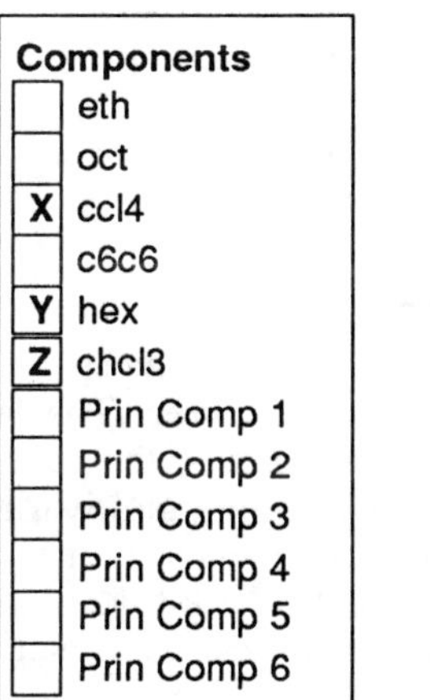

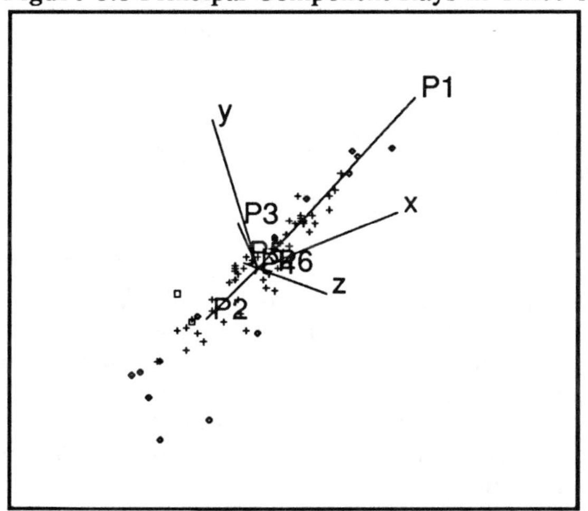

If the spinning plot is based on the first three principal components, it produces the best three-dimensional representation of the six-dimensional space. In this space you can see Caffeine appears as an outlier. In the principal component space, the variables are shown as rays. A plot showing both variable rays and points in an approximation of a high-dimensional space is called a *biplot* (Gabriel 1971). The configuration of the variable rays in **Figure 8.9** shows how the variables are related. Note that **eth** and **oct** seem to dominate one direction, while the other four variables cluster to define the other directions. As more dimensions are squeezed into principal components, the angles between variables become indicators of their correlation. If you do a factor analysis, these directions are further refined, mapping the variables into clusters called factors.

Figure 8.9 A Biplot of Both Variables and Observations

Components
- eth
- oct
- ccl4
- c6c6
- hex
- chcl3
- X Prin Comp 1
- Y Prin Comp 2
- Z Prin Comp 3
- Prin Comp 4
- Prin Comp 5
- Prin Comp 6

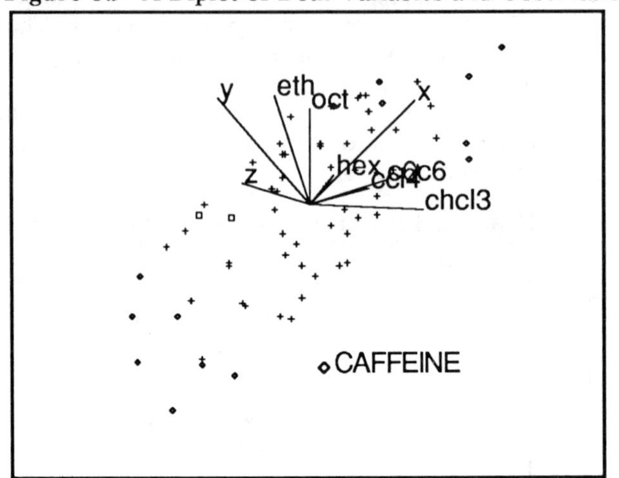

If you are looking for outliers, it is often revealing to examine the space of the last three principal components rather than the first three. The last principal components define the least popular directions of the scatter (directions with the least variation). If a point is unusual in a multivariate sense, then its prominence in the least popular direction suggests it is an outlier. Using this strategy, **Figure 8.10** shows Sulfathiazole as the most unusual value. The other points are in a tight cluster near the center.

Figure 8.10 Outlier in the Last Principal Components

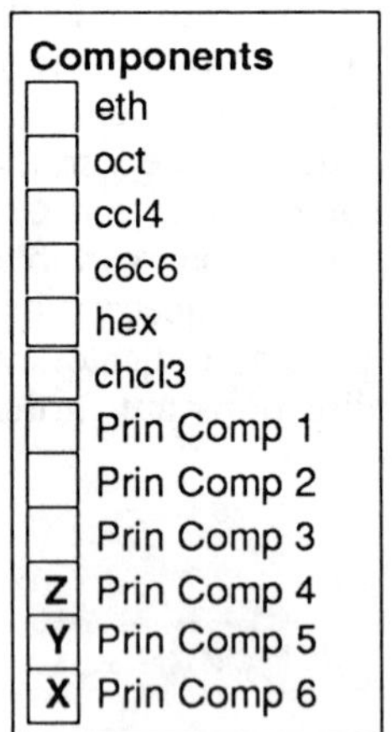

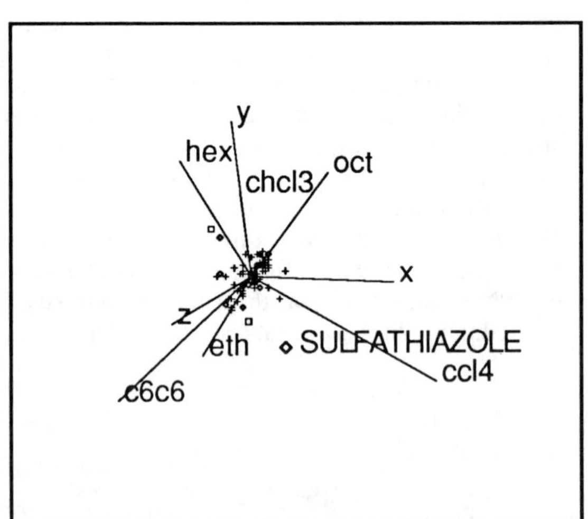

Another strategy is to plot each variable with the first two principal components and then with the last two. In the plot shown in **Figure**

8.11, you see **hex** on the Z axis and the first principal components plotted on the X and Y axes.

Figure 8.11 Plotting a Variable and First Two Principal Components

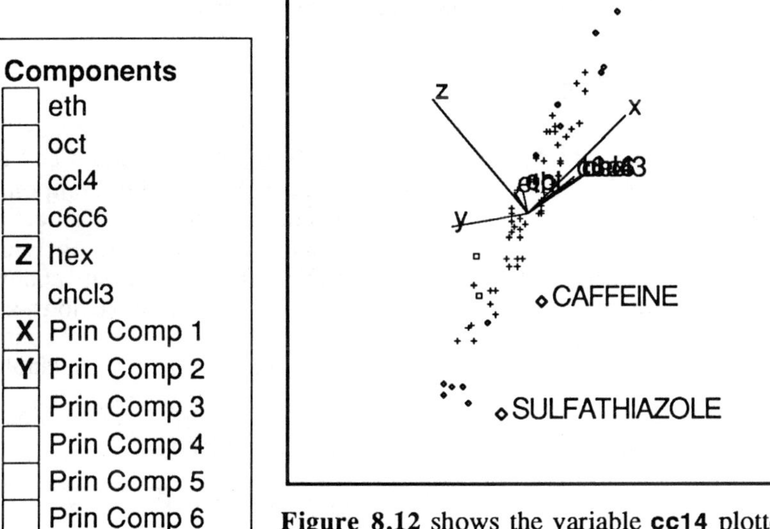

Figure 8.12 shows the variable **ccl4** plotted with the last two principal components.

Figure 8.12 Plotting a Variable and Last Two Principal Components

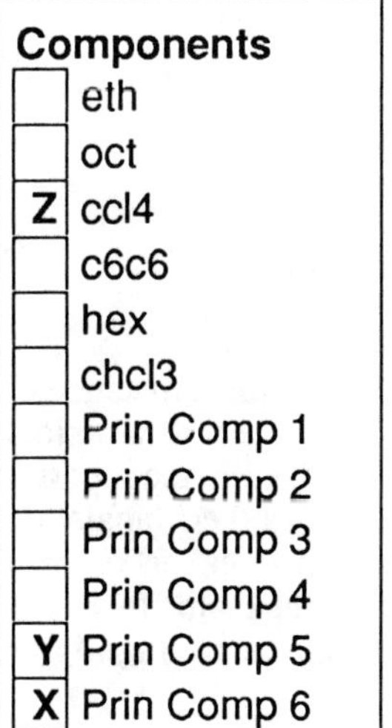

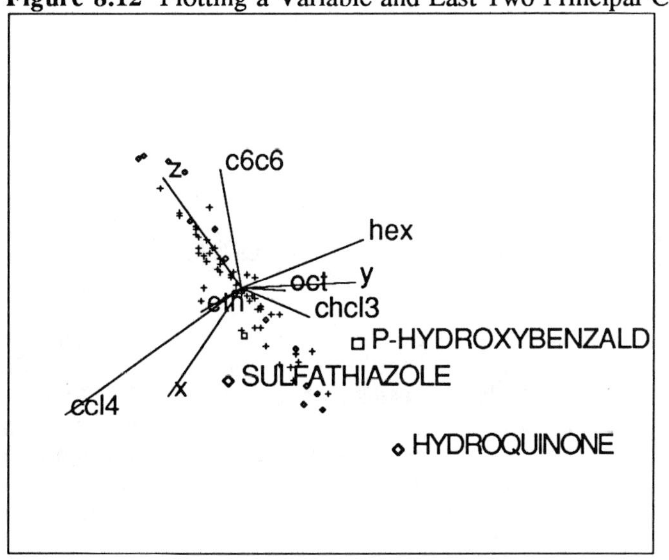

Using Color, Markers, and the Brush Tool

Though a given plot shows two or three dimensions geometrically, you can add aspects of other dimensions by plotting the points with different markers and colors. In addition, you can use the brush tool from the **Tools** menu to see where points in areas of one plot show up on other plots.

Multivariate Distance

The basic concept of distance in several dimensions is related to the correlation of the data. For example, in the **Y's by Y's** scatterplot cell for **c6c6** by **chc13** shown in **Figure 8.13**, you see the compound Hydroquinone. It is not particularly unusual in either the X or Y direction alone, but is an outlier because of its distance from the strong linear relationship between the two variables. The ellipse is a 95% density contour for a bivariate normal distribution with the means, standard deviations, and correlation estimated from the data. The concept of distance that takes into account the multivariate normal density contours is called Mahalanobis distance.

Figure 8.13 Scatterplot of Variables **c6c6** and **chc13**

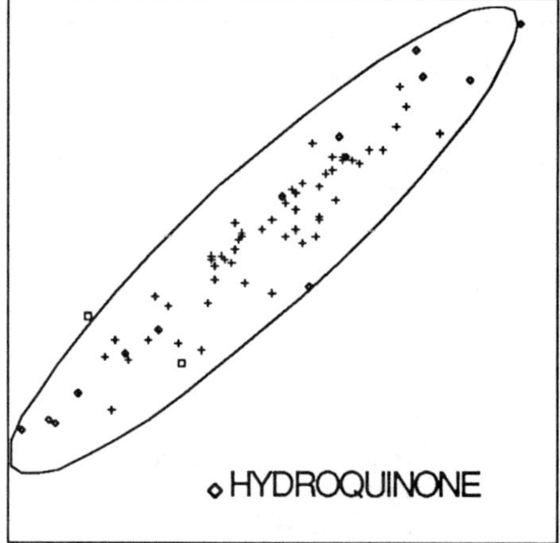

Though you can visualize only three dimensions, Mahalanobis distance can be calculated for any number of dimensions. The six-dimensional Mahalanobis distance for the SOLUBILITY data is included in the report for the **Y's by Y's** platform. The outlier distance plot shows the Mahalanobis distance by the row number for each point. You can click on or brush points which are outliers in six-dimensions to highlight them. In **Figure 8.14**, the 5 points with the greatest Mahalanobis distances are highlighted and labelled on the plot.

Figure 8.14 Mahalanobis Distance Plot to Identify Multivariate Outliers

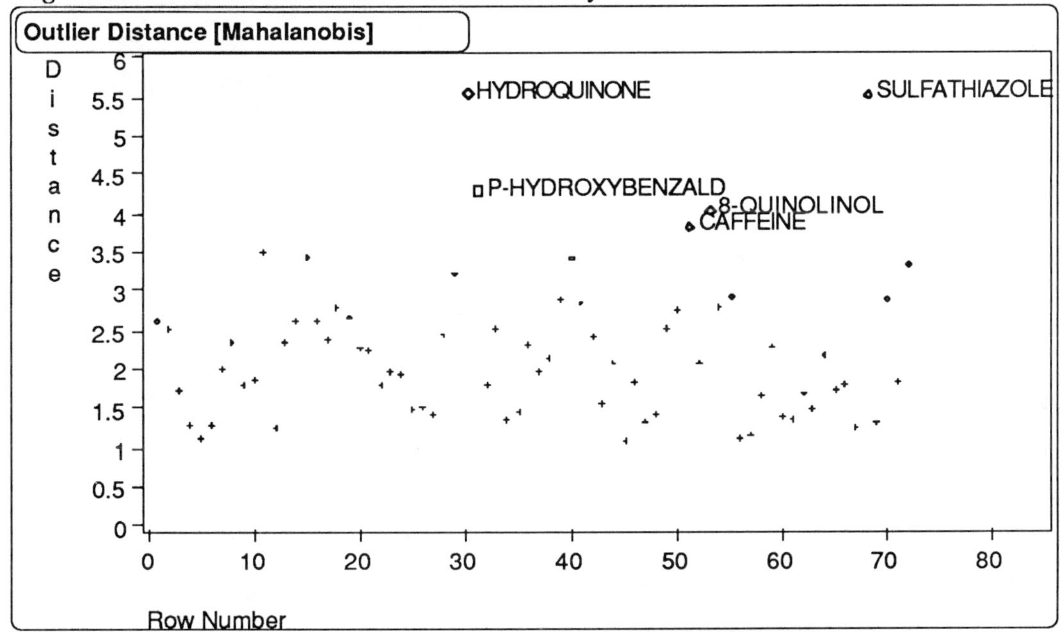

Chapter Summary

In this example, commands from the **Analyze** menu were used for data exploration to locate and identify unusual points. The data were first examined in one dimension using the **Distribution of Y's** command and then in two dimensions using the **Y's by Y's** command to look for unusual points in histograms and scatterplots.

Next, the **Spin** command in the **Analyze** menu was used to plot three columns at a time. Using the technique of principal components, six dimensions were summarized and principal component rays were plotted. The Principal Components table showed that more than 97% of the total variation in the sample was accounted for by the first three principal components. To locate multivariate outliers, each column was plotted with the first two principal components and then with the last two principle components.

Finally, the Mahalanobis outlier distance plot, produced by the **Y's by Y's** platform, summarized the points in 6-dimensions, and the multivariate outliers were highlighted and labelled.

Chapter 9.
The Menu Bar

This chapter is a reference to the JMP menu bar shown in **Figure 9.1**.

Figure 9.1 The Menu Bar

File	Edit	Rows	Columns	Analyze	Tools	Window

The chapter begins with a brief section describing the organization of JMP data both in memory and on disk. The terms described in this section are used throughout the chapter to explain the scope of JMP commands.

The remainder of the chapter contains documentation for all the commands in the JMP menu bar organized to facilitate easy reference. The menus and commands are presented in order, and each menu title is included in the appropriate page headers to help you locate specific commands.

The chapters following this chapter provide more information about data table management and details about all of the JMP statistical platforms.

Chapter Contents

JMP Processing

JMP processing takes place through the use of standard Macintosh windows. These windows display data values, results of analyses, help information, and more. Each window is open or closed, displayed or hidden, and active or inactive. You work with only one window at a time, called the active window, which is open and displayed. It is the window affected by selections from the menu bar.

The definitions below describe the four basic elements of the JMP system. They are an important basis for understanding how menu commands affect data in the active window. These terms are used throughout JMP reference documentation.

Data Tables

JMP data are organized in memory as rows and columns of a table which is referred to as the *data table*. Each column has a name and type, and each row is numbered. You can have as many data tables in memory during a JMP session as available memory allows.

Spreadsheets

Reference to an *active data table* implies a view of the table, called the *spreadsheet*, which appears in the active window. Like the data table, the spreadsheet is also organized in rows and columns of data. You communicate with the data table through the active spreadsheet window.

Reports

Analyzing the the active data table produces a *report*. The report consists of a *graphical display* and a statistical *text report*. Reference to a *current data table* refers to the data table which corresponds to the active report window.

Files

JMP data tables can be stored in *files* on disk. These files may also be referred to as JMP documents. When you save a data table, you write its information to a JMP file on disk.

You can hide or close a JMP spreadsheet or report window without losing access to the corresponding data table in memory. When you close a data table, you are prompted to save your changes to a JMP file.

The relationship of the four components of a JMP session are illustrated in **Figure 9.2.**

Figure 9.2 JMP Data Processing

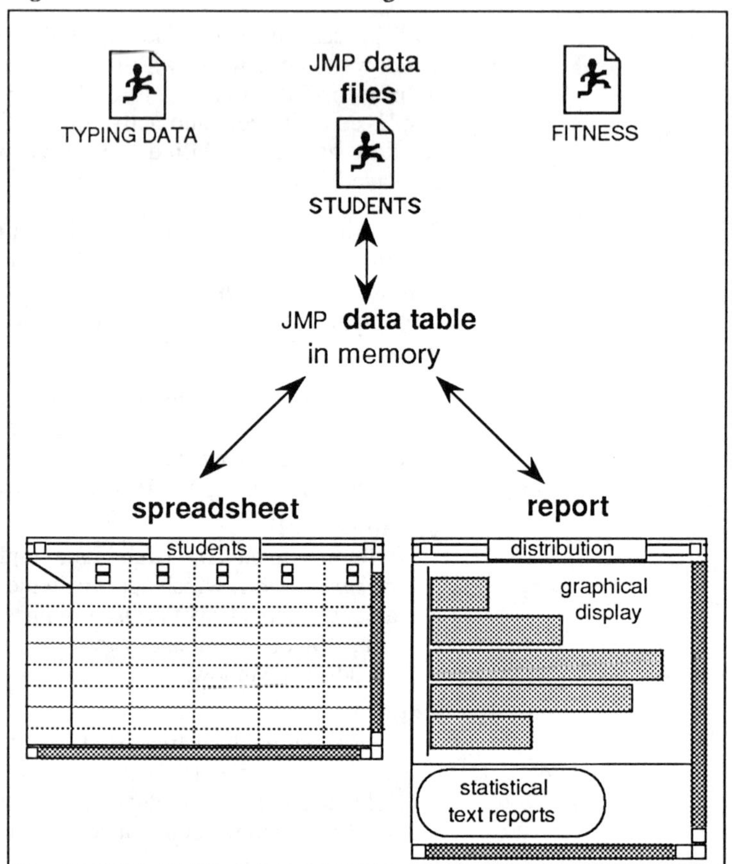

The File Menu

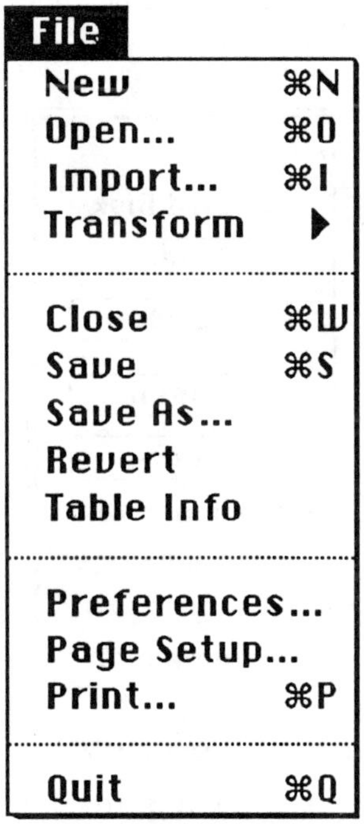

File
New ⌘N
Open... ⌘O
Import... ⌘I
Transform ▶
Close ⌘W
Save ⌘S
Save As...
Revert
Table Info
Preferences...
Page Setup...
Print... ⌘P
Quit ⌘Q

The **File** menu contains standard Macintosh commands and other selections that affect general data table management and the JMP environment. Related commands are grouped according to their functions. The four groups contain commands that

- open data files, create new data tables, and rearrange data tables
- close windows, close and save data tables, retrieve a saved data table, and view and record basic information about the current data table
- set global options and print spreadsheets or reports
- end a JMP session.

You modify a data table by working with its spreadsheet view as the active window. In the description of **File** commands, *data table*, *table* and *spreadsheet* are used synonymously unless specifically noted otherwise.

New ⌘N

The **New** command creates a new data table and displays an empty untitled spreadsheet with one column labeled Column 1. You can create more columns using the **Columns** menu and type or paste data into the spreadsheet. Use **Save as** from the **File** menu to name a new data table and to save it as a JMP file. To reopen the file after it has been saved, use the **Open** command.

Open ⌘O

The **Open** command displays the standard Macintosh open file dialog. You can open an existing JMP file using this command, or you can open a file by double clicking on it in the Finder. When you open a file, you have access to the spreadsheet view of the data table.

Import... ⌘I

The **Import** command displays the dialog shown in **Figure 9.3**. You can create a new JMP data table by reading data from a text-format file, a SAS transport file (described below), or a text file created by JMP. When you import a file in text format, you can use options in the **Import** dialog to identify the field and line delimiters contained in the incoming file.

Figure 9.3 The **Import** Dialog

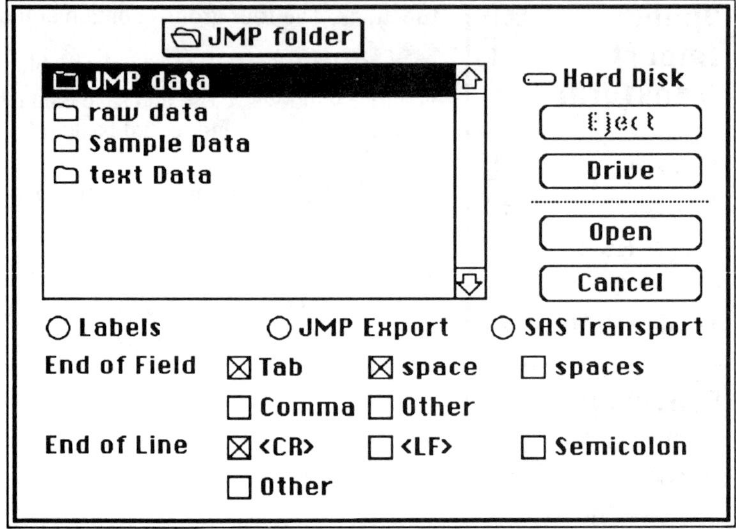

By default, the **Import** command reads in a text file of rows and columns without header information and names the JMP data table columns Column 1, Column 2, and so on. If the file contains labels, JMP header information, or SAS transport formatting, use the radio buttons in the **Import** dialog described below.

◉ **Labels**
indicates that the incoming data are from a text-format file that contains column labels as the first line of the data. These labels are used as column names in the new JMP data table.

◉ **JMP Export**
indicates that the incoming data are from a text-format file that contains column header information as exported from a JMP file.

◉ **SAS Transport**
indicates that the incoming data are in SAS transport format.

Note➜ A SAS transport file is a portable file created by the SAS System. You can also create a SAS transport file by saving a JMP data table using the **SAS Transport** option in the **Save As** dialog.

The check boxes in the **Import** dialog identify the end of field and end of line delimiters.

End of Field
designates the character that identifies the end of each field in the incoming file. This delimiter can be a tab, one or more blank spaces, any character, or any combination of the above. When you choose the **Other** option, an editable text box appears where you can enter a character or its hexadecimal representation.

End of Line
selects the character or key that identifies the end of each line. You can use the return key, the line feed character, a semicolon or another character of your choice. When you choose the **Other** option, an editable text box appears. You can enter a character or its hexadecimal representation if you prefer.

Note➜ To enter a hexadecimal value, precede the value with '0x'.

Click on **Open** when you have completed this dialog.

Transform ▶

- Sort...
- Subset
- Join...
- Stack...
- Concat...

The **Transform** command creates a new data table by modifying the active data table or combining tables. The **Transform** command has a submenu that lists the five commands described below.

Sort...

sorts a JMP data table by one or more columns. The **Sort** command displays the dialog shown in **Figure 9.4**, which you use to specify columns as sort fields. Press the **Sort** button when you have completed the dialog.

Figure 9.4 Sort Field Selection Dialog

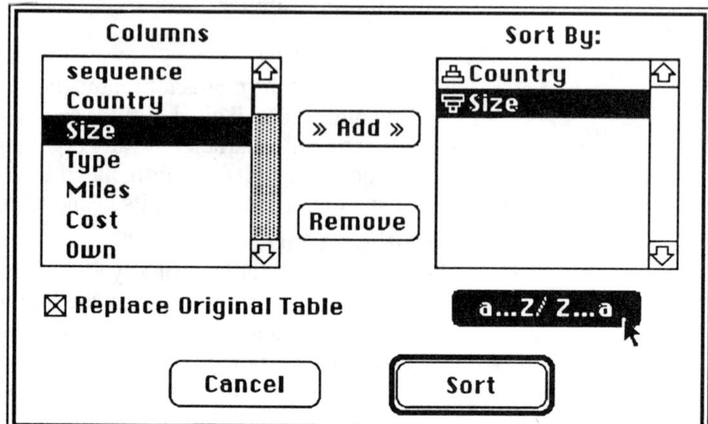

Use components and options in the **Sort** dialog as described below.

Columns

Select sort fields from the **Columns** list and add them to **Sort By** list with the **Add** button. The order in which you build this list establishes the sorting order.

Sort by

The **Sort By** list shows the columns to be sorted in order of precedence. You can remove sort fields by selecting them in the sort list and clicking the **Remove** button.

a..Z/Z..a

You can sort any column in either ascending or descending order by selecting the sort column and clicking on the **a..Z/Z..a** toggle. By default, columns are sorted in ascending order.

Replace Original Table

By default, **Sort** creates a new data table. You can replace the original data table instead by checking **Replace Original Table** as shown in **Figure 9.4.**

Subset produces a subset consisting of the highlighted rows and columns from the active spreadsheet. Highlighting rows is done by

- selecting them on the spreadsheet
- highlighting histogram bars or points on a plot
- using one of the **Select** commands from the **Rows** menu
- highlighting histogram bars or points on a plot.

All graphical displays are linked to their corresponding spreadsheet so that highlighting bars or points selects the corresponding spreadsheet rows.

Join... creates a new data table by merging (joining) two tables side-by-side. Tables can be joined

- by row number
- by matching the values in one or more columns that exist in both data tables
- in a Cartesian fashion where all values in a column of one data table are merged with all values in a column of another table.

When you select the **Join** command, the dialog shown in **Figure 9.5** displays the name of the active table next to the word **Join**. You select an open data table to join **With** from the selector list in the **Join** dialog. Optionally, specify the name of the new data table. By default, the new table name is the concatenation of the original data table names, separated by a plus sign.

Figure 9.5 The Join Dialog

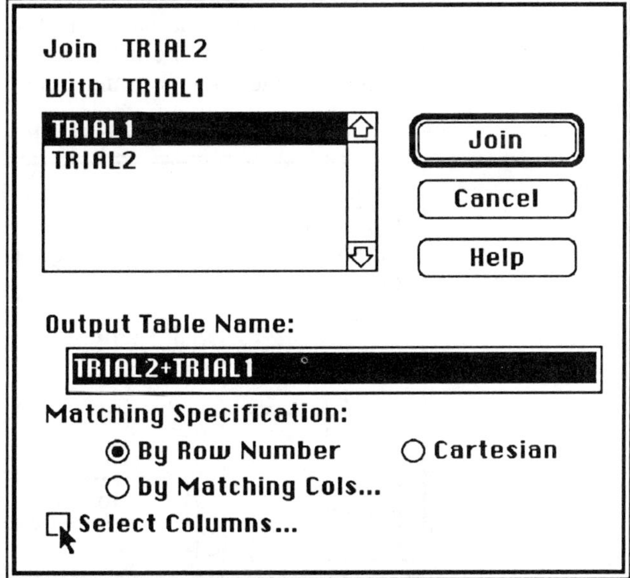

The **Matching Specification** options are described below.

◉ By Row Number

Joining by row number joins tables side-by-side matching by row number. If the tables don't have the same number of rows, the columns from the shorter table have missing values in the extra cells. By default, tables are joined by row numbers.

Cartesian

If you choose the Cartesian join, each row in the **Join** table is joined with each row of the **With** table.

Note➜ The number of rows in the resulting table is the product of the rows in the two original tables. Be sure not to create a data table too large for available memory.

By Matching Cols...

To join rows with matching column values, click on **By Matching Cols** and complete the dialog that appears. The dialog is described later.

If you don't use the **Select Cols** option, all columns from both tables are included in the joined table. The dialog in **Figure 9.6** asks you to select columns from the original tables to include in the new table. If a name is unique, it is written directly to the new table. Columns with the same names in both tables are not overwritten. Instead, both columns are included in the new table. Each is named by appending the original column name to its data table name.

Figure 9.6 The **Select Cols** Dialog

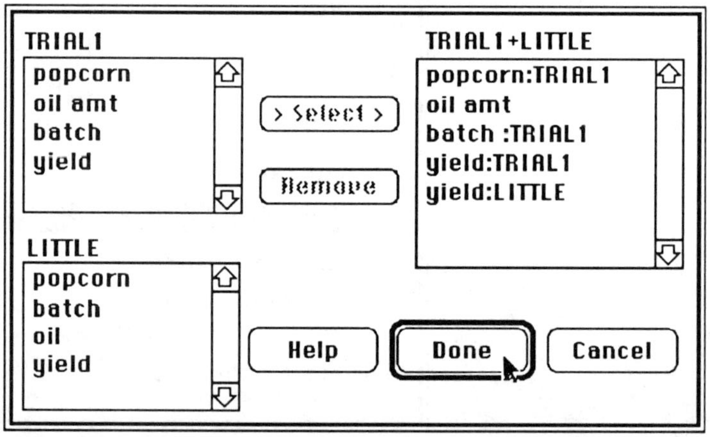

If you join **By Matching Cols** (see **Figure 9.5**), the dialog shown in **Figure 9.7** then prompts you to select those columns whose values must match to complete a join from both tables. After selecting a column from each list, click on the **Match** button. This displays the columns in the lower boxes. Select additional pairs of columns as needed. If you want to remove columns, select the same number of columns from each of the lower lists and click on the **Remove** button.

Matching columns do not have to have the same names and do not have to be in the same relative column position in both tables. When you click on the **Done** button, the first column in the lefthand list is paired with the column in the righthand list. Likewise the second columns are paired, and so on. Rows are joined only if values match for all the column pairs.

Figure 9.7 The Specify Matching Columns Dialog

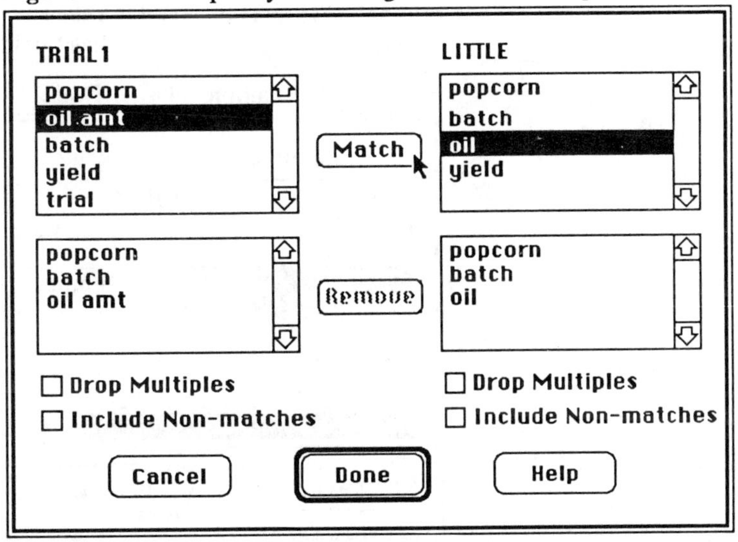

TRIAL1
popcorn
oil amt
batch
yield
trial
LITTLE
popcorn
batch
oil
yield
Match
popcorn
batch
oil amt
Remove
popcorn
batch
oil
Drop Multiples
Include Non-matches
Drop Multiples
Include Non-matches
Cancel
Done
Help

After you choose columns whose values must match, the additional matching options shown beneath the column lists display for both tables. If you do not select any of these options, a Cartesian join is performed within each group of matching column values. The matching columns options are described below.

Drop Multiples

You can specify **Drop multiples** in either or both of the data tables being joined. If you drop multiples in both tables, only the first match is written to the new table. If you specify this option on only one table, the first match value is joined with all matches in the other table.

Include Non-Matches

When you **Include non-matches** from a data table, each row from that data table is included in the new data table even if there is no matching value. Because there is no matching row in the other data table, cells for columns from the other data table have missing values for that row. You can specify this option for either or both data tables being joined.

Stack... creates a new data table from the active table by stacking specified columns into a single new column. The example dialog shown in **Figure 9.8** specifies that two columns called **yield1** and **yield2** be stacked into a single column called **yield**. The values in the other columns are preserved in the new data table.

Stack creates a *type* column to identify each row in the new table. The value in the type column is the name of the column in the original table that contained the stacked value. In the example below, the type column, named **trial**, has 'yield1' and 'yield2' as its values.

Figure 9.8 Example **Stack** Dialog

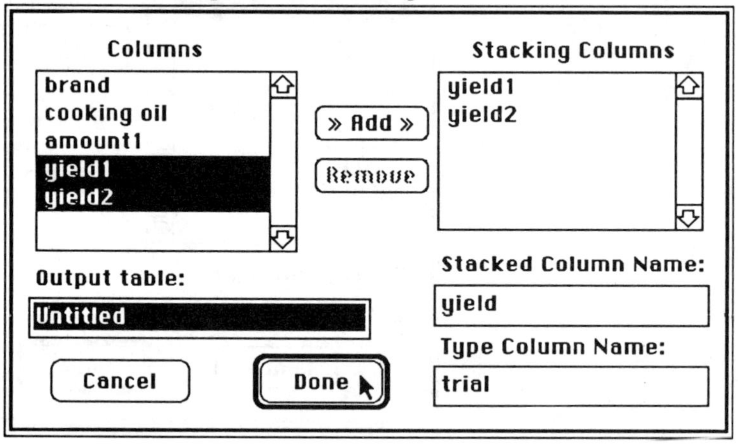

Columns
brand
cooking oil
amount1
yield1
yield2
» Add »
Remove
Stacking Columns
yield1
yield2
Stacked Column Name:
yield
Output table:
Untitled
Type Column Name:
trial
Cancel
Done

The elements of the **Stack** dialog are described below.

Columns

You select each column to be stacked from the **Columns** list and add it to the **Stacking Columns** list by clicking on the **Add** button.

Note➜ The number of rows in the new table is the total number of stacked columns multiplied by the number of rows. Remember, a data table must fit in available memory.

Stacking Columns

All columns in the **Stacking Columns** list are stacked or piled one after another in the order they appear in the list. You can remove a column from the list by selecting its name and clicking on the **Remove** button

Stacked Column Name

Optionally, enter a name in the **Stacked Column Name** box to specify a name for the new column of stacked values. By default, this column is named _Stacked_. If you delete the default name and do not specify a new name, the new stacked column is not written to the data table.

Type Column Name
Optionally, enter a name for the type column that identifies the original column for each stacked cell in the new table. By default, this column is named _Type_. If you delete the default name and do not specify a new name, the new *type* column is not written to the data table.

Concat... appends two or more tables to each other end-to-end. The arrangement of the new data table depends on the columns in the appended tables. The **Concat** command creates one column in the new table for each unique column name in all of the appended tables.

The **Concat** dialog shown in **Figure 9.9** lists all the open JMP data tables in the selector list on the left. Select the tables you want to concatenate and click on the **Add** button to build a list of tables in the list on the right. When you finish adding data table names, click on the **Concat** button. This creates a new untitled data table consisting of all the rows in the first table in the new list, followed by all of the rows from the second table, and so on. The new table contains a column for each column found in all tables listed.

Figure 9.9 Example **Concat** Dialog

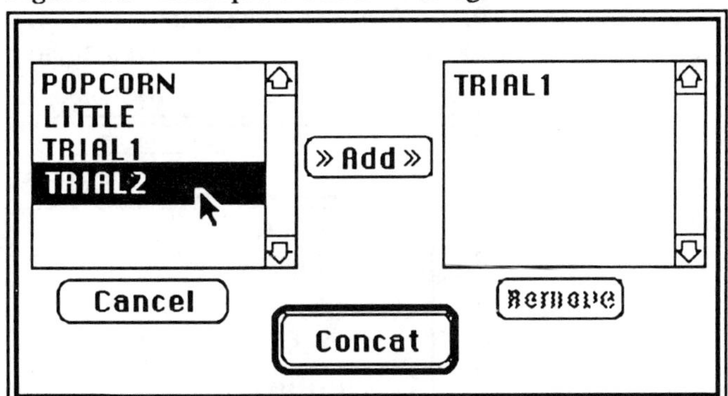

See Chapter 10, *JMP Data Tables,* for examples of the **Transform** submenu table operations.

Close ⌘W

The **Close** command closes the active window. If you close the spreadsheet window while any of its analysis windows are open, a close dialog asks if you are closing the data table only or all related analysis windows as well. If you close the data table, you are also prompted to save any changes. Using the **Close** command is the same as clicking in the close box of an active window.

Save ⌘S

The **Save** command writes the current form of the active data table to a file. If the data table has been saved before, it is rewritten to the same file name and location replacing the old information. If the data table is new, the **Save** command has the same effect as the **Save As** command. Saving a data table does not automatically close it.

Note ➡ JMP report windows are not saved with the data table. However, you can use JMP tools to copy reports to other applications, or you can use the journaling feature to save reports in text or word processing format.

Save As...

The **Save as** command saves the current data table to a file after prompting you for a name and disk location as shown in **Figure 9.10**. It saves the data table as a JMP file or converts it to another format if you click on a different radio button in the dialog.

Figure 9.10 The **Save as** Dialog

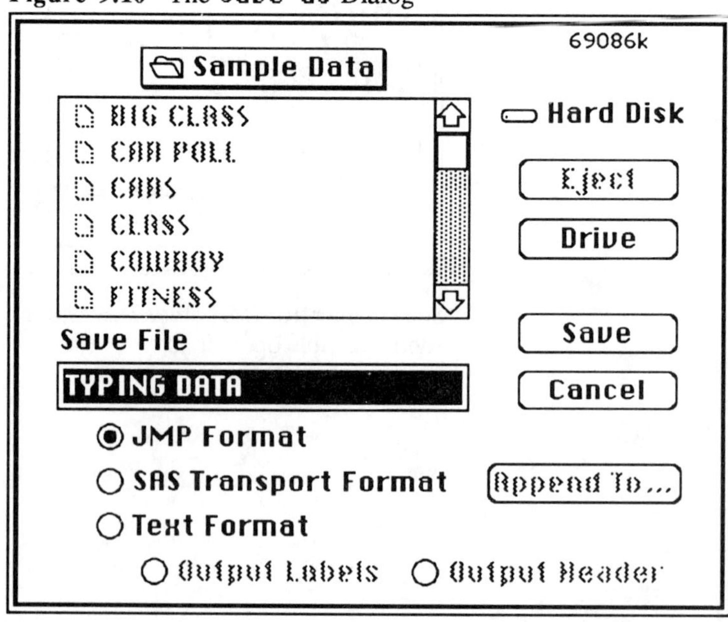

Each of the options for saving JMP files is described below.

◉ JMP Format
saves the JMP data table in JMP format.

◉ SAS Transport
converts a JMP data table to SAS transport file format and saves it in a SAS transport library.

Note➜ If you don't use the **Append To** option, a new SAS transport library is created using the name in the **Save File** field. If you do not specify a new file name, the SAS transport library replaces the existing JMP data table.

Append To...
lets you append the current data table in SAS transport file format to an existing SAS transport library.

◉ Text Format
converts the data from a JMP file to standard text format maintaining the rows and columns. Tab characters delimit fields and a carriage return delimits rows.

Output Labels
specifies that the names of the columns in the JMP file be written as the first line of the text format file.

Output Headers
exports JMP header information and data to a text file. The header information includes the number of rows and columns in the file and information about each column. If you need the data in JMP format again, JMP can import this kind of text file efficiently using the **JMP Format** radio button in the **Import** dialog.

Revert

The **Revert** command restores the data table to the form it had when last saved. If it was last saved as a SAS Transport file or a text format file, the **Revert** command is not available.

Table Info

The **Table Info** command displays the dialog shown in **Figure 9.11.** You can use this dialog to make notes about the current data table or to examine its attributes.

You can also **Lock** the the data table if you have not modified it since you last saved it. If you have made changes, the **Lock** box is dimmed and the **Dirty** box is checked.

You can also use the **Table Info** dialog to create a new data table by entering a new table name. The new data table is a duplicate of the original table. The original data table remains as last saved and the new table is not saved until you select the **Save** or **Save As** command.

Figure 9.11 The **Table Info** Dialog

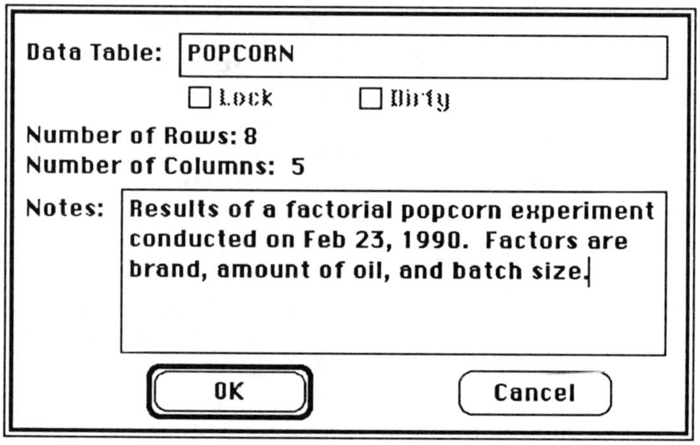

Data Table: POPCORN
☐ Lock ☐ Dirty
Number of Rows: 8
Number of Columns: 5
Notes: Results of a factorial popcorn experiment conducted on Feb 23, 1990. Factors are brand, amount of oil, and batch size.
OK Cancel

Preferences...

The **Preferences** command provides the dialog shown in **Figure 9.12** for setting default text and background conditions. The background for all JMP graphical displays except the **Spin** platform can be either white or black (the **Spin** plot has an independent background setting). Text options include any font, size, and format available in your system's resources.

Figure 9.12 The **Preferences** Dialog

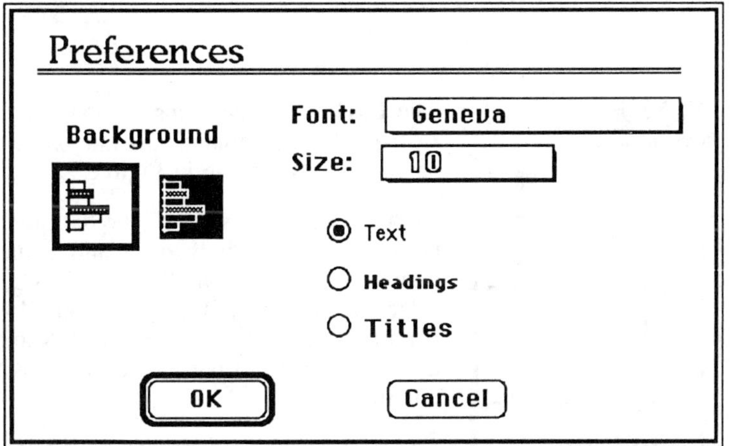

Page Setup...

The **Page Setup** command displays the standard Macintosh dialog for setting the characteristics of printed pages. The form of the dialog depends on the current printer driver.

Print... ⌘P

The **Print** command displays the standard Macintosh dialog for printing the active spreadsheet or report window. The appearance of this dialog depends on the printer driver.

Quit ⌘Q

The **Quit** command closes all JMP windows (prompting you to save changes) and ends the JMP session.

The Edit Menu

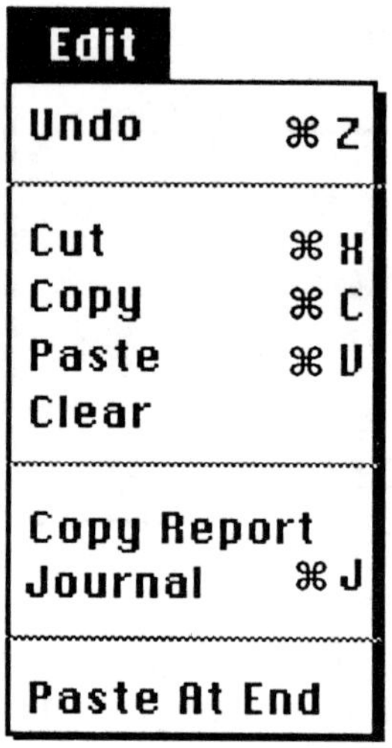

The **Edit** menu contains standard commands including **Cut** and **Copy.** These commands operate on *selected data values* in the JMP spreadsheet as shown in **Figure 9.13** or selected areas of reports.

In the spreadsheet view of the data table, you can select rows only, columns only, or both rows and columns at the same time. **Edit** commands operate on all values in selected rows if no columns are selected. Likewise, they operate on all values from selected columns if no rows are selected. If both rows and columns are selected, **Edit** affects the subset of values defined by the intersection of rows and columns.

To select a row in a spreadsheet, click on the space that contains the row number. To select a column, click on the background area above the column name. The row number or column selection area highlights.

To select multiple rows or columns, drag the mouse across them. You can also shift-click the first and last rows or columns of a range to select it. To make a discontinuous selection, command-click on the rows or columns you want. To select a block of cells, drag the cross cursor diagonally across the subset of cells.

Figure 9.13 Selecting Rows and Columns

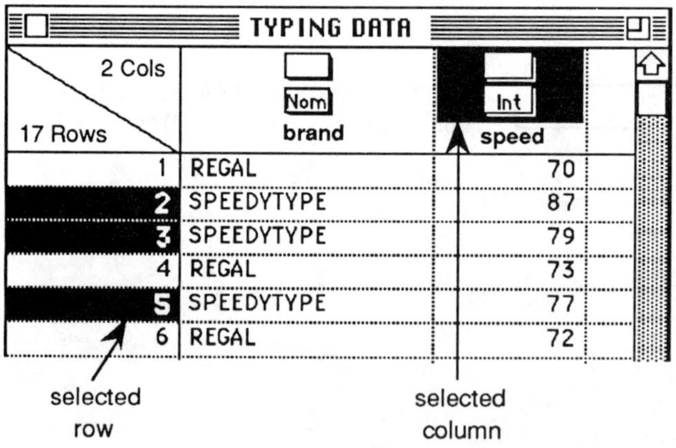

The **Edit** menu can also be used in conjunction with JMP tools to copy all or part of active report windows. See the **Tools** menu section for more information.

Undo ⌘Z

The **Undo** command cancels the effect of the most recent reversible command selected from the **Edit**, **Rows**, or **Columns** menu. If **Undo** is available, its display in the **Edit** menu is accompanied by the name of the command it can cancel. Most destructive spreadsheet operations (such as cut, paste, or delete rows) are reversible. **Undo** is dimmed when the most recent command is irreversible. Once you select **Undo**, the menu item changes to **Redo *command***, where ***command*** is the action that was undone.

Cut ⌘X

The **Cut** command copies selected fields from the active spreadsheet to the clipboard and replaces them with missing values. It is equivalent to **Copy**, then **Clear**. You can also use the **Cut** command to copy all or part of a report window. However, **Cut** works like **Copy** in graphical displays (it does not clear the copied image).

Copy ⌘C

The **Copy** command copies the values of selected data cells from the active data table to the clipboard. You can copy all values from selected rows if no columns are selected. Likewise, you can copy all values from selected columns if no rows are selected. If both rows and columns are selected, **Copy** copies the subset of cells defined by the intersection of those rows and columns.

The **Copy** command can also be used to capture graphical displays or text reports using the scissors tool in the **Tools** menu. The copied image is stored in PICT file format.

Note➜ Data you cut or copy to the clipboard can be pasted into JMP tables or into other applications. Pictures can be pasted into any application that accepts PICT format graphics.

Paste ⌘V

The **Paste** command copies data from the clipboard into a JMP data table. **Paste** can be used with the **Copy** command to duplicate rows, columns or any subset of cells defined by selected rows and columns.

To duplicate an entire row or column

- Select and **Copy** the row or column to be duplicated.
- Select an existing row or column to receive the values.
- Use the **Paste** command.

To duplicate a subset of values defined by selecting both rows and columns, follow the steps above, making sure to select the same arrangement of rows and columns to receive the copied values as originally contained them.

Clear

The **Clear** command clears all selected cell values from the active data table and replaces them with missing values. The values are not copied to the clipboard.

Copy Report

The **Copy Report** command copies all text from the active report window (no graphical displays) and stores the text in the clipboard.

Journal ⌘J

The **Journal** command displays the dialog shown in **Figure 9.14** if no journal file is open. This dialog asks you to name a journal file for saving JMP reports. When you click the **Save** button, the journal file displays as an open window, and contains the contents of the active JMP window. Each subsequent selection of the **Journal** command writes information from the active JMP window to the journal file. Also, you can append your own notes to the current end of the open journal file.

You can save in text only format or use the **MacWrite**™ option and save both text reports and graphical displays. After the journal file is opened, the journal window persists until you close it.

Figure 9.14 Open Journal File Dialog

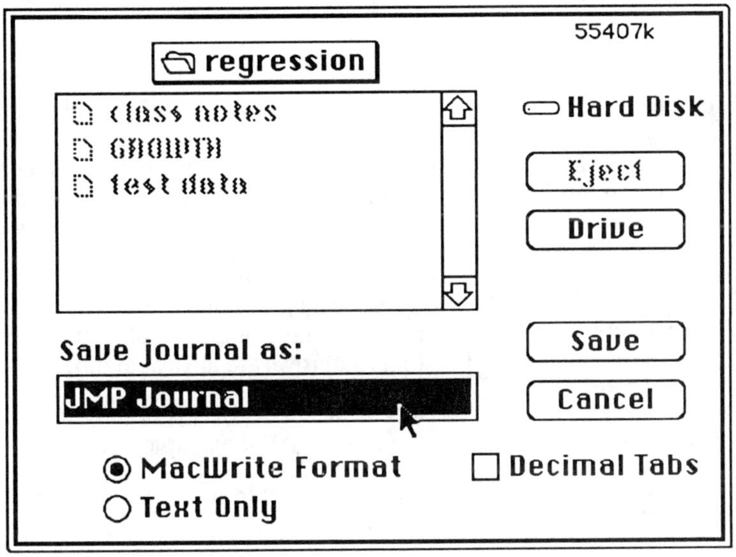

55407k
regression
class notes
GROWTH
test data
Hard Disk
Eject
Drive
Save journal as:
JMP Journal
Save
Cancel
MacWrite Format
Text Only
Decimal Tabs

Paste at End

The **Paste at End** command is used to extend a JMP data table by

- adding rows to a data table
- adding columns to a data table.

If you select rows before choosing this command, the effect is similar to joining data tables by row number. If you highlight columns, the **Paste at End** command adds cells to the bottom of the data table filling them with values from the clipboard.

To transfer data from another application into a JMP data table, first copy the data to the clipboard from within the application and then use the **Paste at End** command to copy the values to a JMP data table. Rows and columns are automatically created as needed when you **Paste at End**.

The **Rows** Menu

Rows	
Exclude/Include	⌘E
Hide/ Unhide	⌘H
Label/Unlabel	
Colors	▶
Markers	▶
Select	▶
Invert Selection	
Locate Next	⌘A
Locate Previous	⌘B
Add Rows...	
Delete Rows	

To select a row in JMP, highlight the space that contains the row number by clicking on it. You can highlight multiple contiguous rows by dragging down the row numbers or by shift-clicking the first and last rows of a desired range. To make a discontinuous selection use command-click on the rows want to select.

Rows are also highlighted (and selected) when points or bars in a corresponding graphical display are highlighted. Because the JMP data table and any graphical display of that table are linked, selected points and rows are always highlighted simultaneously.

Once a subset of rows is highlighted, you can choose from a variety of commands. For example you can

- exclude rows from further analysis
- hide points in current graphical displays
- search for highlighted rows
- assign special colors or markers for graphical display of points from highlighted rows
- delete rows from the data table
- create a subset of highlighted rows.

To deselect all rows, click in the upper left corner of the data table where the number of rows in shown; to deselect a single row, command-click on the space that contains the row number.

Exclude/Include ⌘E

Exclude/Include is a toggle command used to exclude selected rows from statistical analyses. To do this, select the rows that contain extreme values and choose **Exclude/Include** from the **Rows** menu. Data remain excluded until **Exclude/Include** is chosen again for selected rows. Excluded rows are marked by the symbol shown in **Figure 9.15**.

The **Exclude/Include** status is a *row state* characteristic. It can be saved permanently with the data table in a special row state column. If you do not save this row state, it is lost when you close the data table.

Figure 9.15 Excluded Rows

CLASS

5 Cols / 45 Rows	name (Nom)
1	JOYCE
⊘ 2	GAIL
⊘ 3	DORIS
⊘ 4	THOMAS

excluded rows

Hide/Unhide ⌘H

Hide/Unhide is a toggle that suppresses the display of points in all scatter plots. To hide data, select appropriate rows in the spreadsheet and choose **Hide/Unhide** from the **Rows** menu. Data remain hidden until **Hide/Unhide** is chosen again for selected rows. Hidden rows are marked by the symbol shown in **Figure 9.16**.

The **Hide/Unhide** status is a *row state* characteristic. It can be saved permanently with the data table in a special row state column. If you do not save this row state, it is lost when you close the data table.

Figure 9.16 Hidden Rows

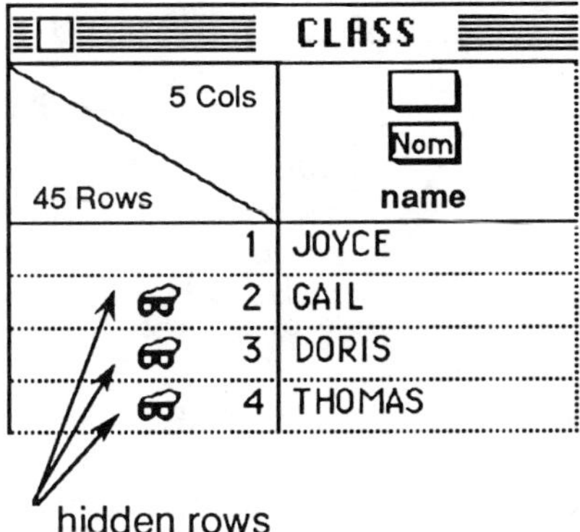

Note ➜ Hidden points are included in statistical analyses even though they do not display in plots.

Label/Unlabel

Label/Unlabel is a toggle that labels points on all scatter plots. To label points, select the rows containing the points and use the **Label/Unlabel** command. The label row state characteristic remains in effect until **Label/Unlabel** is chosen again for the selected rows.

By default, the row number is used as the label value on scatter plots. However, you can use the role assignment box or the **Assign Roles** command in the **Columns** menu to designate any column from the spreadsheet as a label column. The values of a label column are used to identify points in plots.

Data remain labeled until **Label/Unlabel** is chosen again for selected rows. Labelled rows are marked by the tag shown in **Figure 9.17**.

The **Label/Unlabel** status is a *row state* characteristic. It can be saved permanently with the data table in a special row state column. If you do not save this row state, it is lost when you close the data table.

Figure 9.17 Labelled Rows

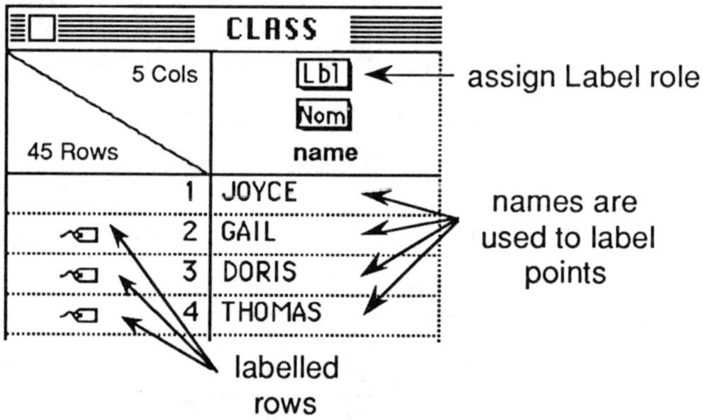

Colors

Colors changes points in all scatter plots to the colors you choose if you have a color monitor. To color a group of related points, select the appropriate rows in the spreadsheet or the points in a plot and the **Colors** submenu in the **Rows** menu.

There are 15 colors available for on screen viewing in JMP. The default color is black. If you choose other colors, you can save them permanently with the data table in a special row state column.

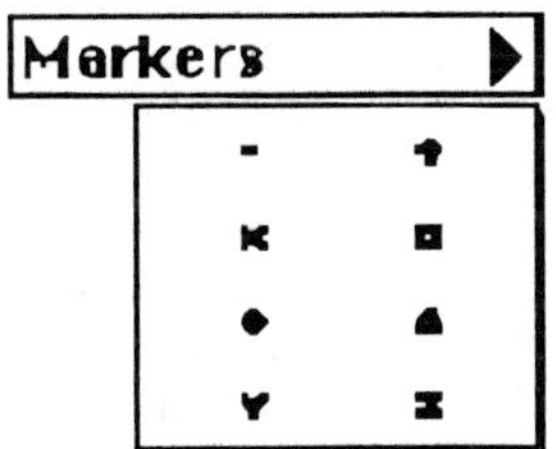

Markers assigns a plot character to replace the standard points in scatter plots and spinning plots. To assign one of the markers shown at left, select appropriate rows in the spreadsheet or the corresponding points in a plot and choose a marker from the **Markers** submenu in the **Rows** menu.

There are 7 markers available for on screen viewing in JMP. Marker is a *row state* characteristic. It can be saved permanently with the data table in a special row state column. The default marker is a dot.

Select ▶

All Rows
Excluded
Hidden
Labelled

The **Select** command has a submenu with options for selecting all rows in a data table, or a subset of rows based on row states.

All Rows
selects the entire JMP data table. This command is a shortcut for shift-clicking the first and last rows in the spreadsheet.

Select Excluded
selects all excluded rows regardless of their current selection status. All previously selected rows are deselected.

Select Hidden
selects all hidden rows regardless of their current selection status. All previously selected rows are deselected.

Select Labelled
selects all labelled rows regardless of their current selection status. All previously selected rows are deselected.

Invert Selection

The **Invert Selection** command deselects all currently selected rows and selects all unselected rows.

Locate Next ⌘A

The **Locate Next** command locates the next selected row after the *current row* and makes it blink briefly. The current row is the first row when you open a data table. The current row changes to the last row you edited or identified by the **Locate Next** or **Locate Previous** command. You can also set the current row by option-clicking anywhere in a row. Each time you choose the **Locate Next** command, the next selected row is found and blinks. A beep tells you when the last selected row has been located.

Locate Previous ⌘B

The **Locate Previous** command locates the next selected row after the *current row* and makes it blink briefly. The current row is the first row when you open a data table. The current row changes to the last row you edited or identified by the **Locate Previous** or **Locate Next** command. You can also set the current row by option-clicking anywhere in a row. Each time you choose the **Locate Previous** command, the next selected row is found and blinks. A beep tells you when the last selected row has been located.

Add Rows...

The **Add rows** command increases the number of rows in the current data table using the dialog shown in **Figure 9.18**. The new rows contain missing values that can be replaced by typing or pasting in data.

Figure 9.18 The **Add Rows** Dialog

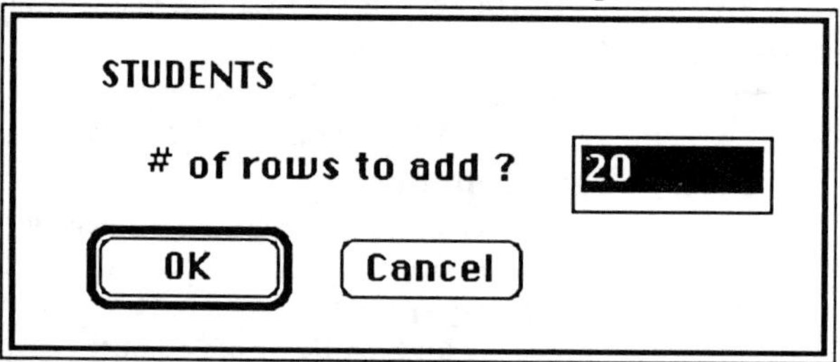

Delete Rows...

The **Delete Rows** command deletes all selected rows from a JMP data table. Use the **Undo** command on the **File** menu to undo a mistaken deletion.

The Columns Menu

Columns
Assign Roles...
Clear All Roles ⌘K
New Column...
Move to First ⌘F
Move To Last ⌘L
Hide Columns
Unhide...
Column Info...
Delete Columns

To select a column in JMP, click in the background area above the column name to highlight it. This area is called the column selection area. You can highlight multiple contiguous columns by dragging across their column selection areas or by shift-clicking the first and last columns of a desired range. To make a discontinuous selection, use command-click on the columns you want to select.

Once a subset of columns is highlighted, you can choose from a variety of commands. For example you can

- add or delete columns
- hide columns or show hidden columns
- compute values for new columns
- rearrange the order of columns.

You can also assign roles to columns for display and analysis using the **Columns** menu instead of using the role assignment box at the top of each column.

Assign Roles...

The **Assign Roles** command displays the dialog shown in **Figure 9.19**, which is used for assigning analysis roles to columns. You can also specify variable roles either by using the role assignment box at the top of each column in the spreadsheet or by selecting an analysis platform and responding to an assign roles prompt.

Figure 9.19 The Assign Roles Dialog

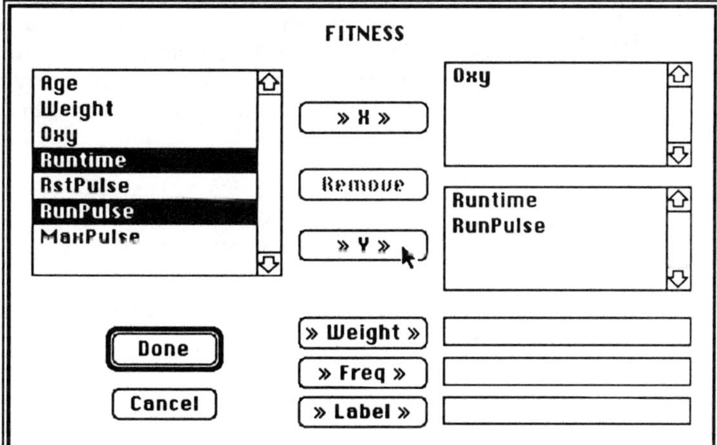

If you use the **Assign Roles** command or the role assignment boxes in the spreadsheet, your choices remain in effect until you specifically change them or close the data table. If you assign a sufficient number of roles, the **Analyze** platforms use these assignments to determine the contents of the resulting displays.

If you choose an analysis that requires more role assignments than you provided, a dialog displays requiring more role assignment information. Your role selections in response to the specific analysis dialog affect only that analysis.

When you use **Assign Roles,** you can assign one or more columns as either X or Y variables. You can also assign a column the role of **Weight**, **Freq**, or **Label**.

Each role assignment is described below.

- **X** identifies columns as classification or independent variables whose values divide the rows into sample groups.
- **Y** identifies columns as response or dependent variables whose distributions are to be studied.
- **Weight** identifies a column whose values are used to supply weights for each response variable.
- **Freq** identifies a column whose values assign a frequency to each row for an analysis.
- **Label** identifies a column whose values are used to specify a labelling convention for points in plots.

Clear All Roles ⌘K

The **Clear All Roles** command removes role assignments from all variables in the active data table.

New Column...

The **New Column** command adds a new column to the current data table using the dialog shown in **Figure 9.20**. The dialog asks you to name the new column and provide column characteristics. Optionally, you can specify a formula to use for computing column values.

Figure 9.20 The New Column Dialog

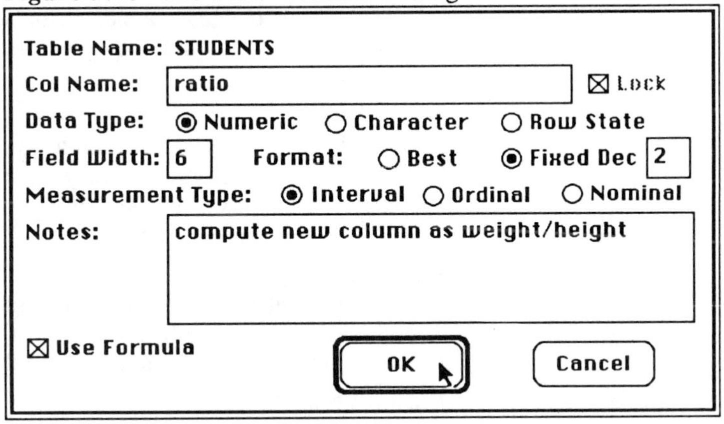

Table Name: STUDENTS
Col Name: ratio ☒ Lock
Data Type: ◉ Numeric ○ Character ○ Row State
Field Width: 6 Format: ○ Best ◉ Fixed Dec 2
Measurement Type: ◉ Interval ○ Ordinal ○ Nominal
Notes: compute new column as weight/height
☒ Use Formula OK Cancel

Below is a description of each element in the **New Column** dialog.

Table Name
displays the name of the active data table that cannot be changed.

Col Name
lets you type in the new column's name using up to 32 keyboard characters. If you choose a long name, you can expand the width of the column in the spreadsheet by dragging the column boundary with the mouse.

Lock
renders a column *uneditable*. If you click on **Use Formula**, the column is automatically locked. You can click in the **Lock** box to secure any column's values that should not be modified.

Data Type
assigns a data type to the the new column according to your selection.

◉ **Numeric** columns must contain numbers.

◉ **Character** columns can contain any characters including numbers. In character columns, numbers are treated as discrete values instead of continuous values.

◉ **Row State** columns contain special information that can affect the appearance of graphical displays. Row states include

- color
- marker
- selection status (highlighted or not highlighted)
- include/exclude status
- hide/unhide status
- label/unlabel status.

Active row state information displays in the row selection area in the spreadsheet. Row state data can be copied to or from a row state column using the pop-up menus located at the top of the column.

Field Width
specifies the field width needed accommodate to the largest number of characters or digits you plan to enter in the new column. The maximum field width for numeric values is 40 and for character values is 255.

Format
specifies a format for displaying numeric column values. The default format is **Best.** This means the JMP system considers the precision of the values and chooses the best display for them.

Measurement Type
specifies the measurement type for the values of the new column according to your choice.

- **Interval** columns must contain numeric values.
- **Ordinal** columns can contain characters or numbers and are treated by analysis platforms as though the values are ordered. If the values are numbers, they are ordered by magnitude. If the values are characters, they are ordered is in collating sequence.
- **Nominal** columns can contain either numeric or character values. All values are treated by analysis platforms as though they are discrete values.

Notes
is a text entry area you can use to document information about a new column. The **Column Info** command includes this text along with other column information.

Use Formula
specifies that a column is to be calculated. When you click on **OK**, the calculator window shown in **Figure 9.21** displays. You can use this window to construct a formula that computes values for the new column. The formula can include existing columns, constants, conditional logic, and a variety of functions. When the **Use Formula** box is checked, the new column is automatically locked.

Figure 9.21 The Calculator Window

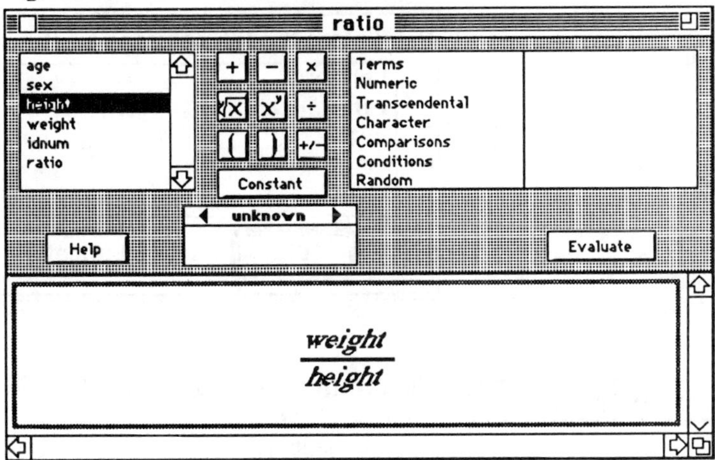

Move to First ⌘F

The **Move to First** command moves the selected column(s) to the left-most position on the spreadsheet. If you mistakenly move one or more columns, use the **Undo** command in the **Edit** menu to restore the previous order.

Move to Last ⌘L

The **Move to Last** command moves the selected column(s) to the right-most position in the spreadsheet. If you mistakenly move one or more columns, use the **Undo** command in the **Edit** menu to restore the previous order.

Hide Columns

The **Hide** command suppresses selected column(s) in the spreadsheet but does not remove them from the data table. Use the **Unhide** command to reshow columns that have been hidden.

Unhide...

The **Unhide** command displays the dialog shown in **Figure 9.22** containing a selector list of all hidden columns. You can select any subset of the hidden columns to reshow on the spreadsheet.

Figure 9.22 The **Unhide** Columns Dialog

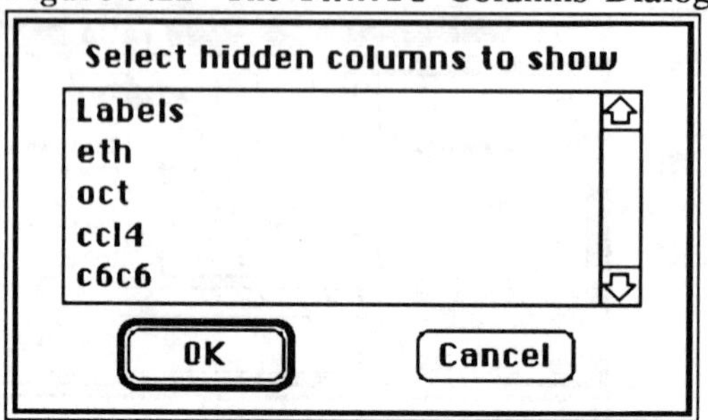

Column Info...

The **Column Info** command displays the dialog shown in **Figure 9.23**. It lists the column's name, type, field width, format, measurement type, and general text notes about the column. If the column is computed, its formula is displayed in the **Use Formula** box in the lower left corner of the dialog.

Note➜ If you click on **Use Formula** in the **Column Info** dialog of a column that is not calculated, the calculator window opens when you click on **OK**. If you click on an existing formula in the **Column Info** dialog, the calculator window opens so you can modify the formula.

Figure 9.23 The **Column Info** Dialog

Table Name: STUDENTS
Col Name: ratio ☒ Lock
Data Type: ◉ Numeric ○ Character ○ Row State
Field Width: 6 Format: ○ Best ◉ Fixed Dec 2
Measurement Type: ◉ Interval ○ Ordinal ○ Nominal
Notes: ratio of weight to height of students participating in a local health study.
☒ Use Formula

$$\frac{weight}{height}$$

OK Cancel

Delete Columns

The **Delete Columns** command removes the selected columns from the data table. If you mistakenly remove columns, you can use the **Undo** command in the **Edit** menu to restore them.

The Analyze Menu

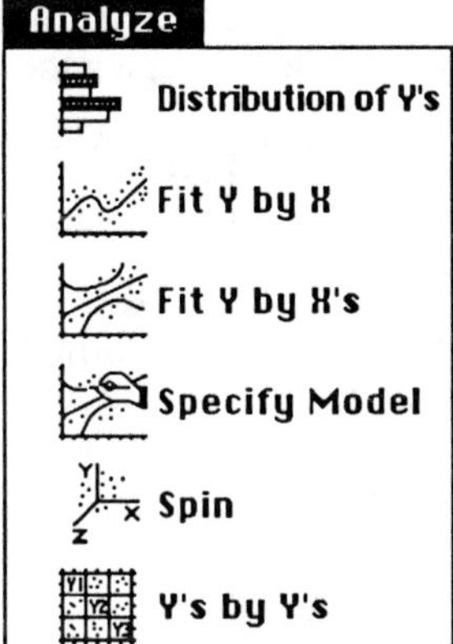

The **Analyze Menu** consists of commands that represent six different statistical platforms. Each platform produces a window that contains graphical displays and supporting text reports. These results can be printed or copied and pasted to other applications. The six platform commands, shown in the menu, process different statistical situations.

Distribution of Y's
describes a distribution of each Y column values using histograms and other graphical and textual reports.

Fit Y by X
describes each pair of X and Y columns. The displays and reports vary depending on the levels of measurement you specify for the X and Y columns, where the level of measurement can be interval, ordinal, or nominal.

Fit Y by X's
fits one model for the Y column and all X columns. It is used for multiple regression, main effects analysis of variance models, analysis of covariance, and analysis of categorical linear models.

Specify Model
displays a dialog that defines complex models automatically, and lets you choose specific model effects and error terms.

Spin
produces a three-dimensional view of data and an approximation to higher dimensions through principal components. The plot is a spinable display of any three numeric columns you choose.

Y's by Y's
describes relationships among Y's with correlations, a scatter plot matrix, and a multivariate outlier plot.

These statistical platforms require that columns be assigned X or Y roles. You can choose roles using the role assignment pop-up menu at the top of each column in the spreadsheet. If the roles have not been assigned previously, you are prompted to assign them by the **Analyze** platform

Distribution of Y's describes a distribution of values with histograms and other graphical and textual reports.

- If a column has interval values, a histogram and quantile box plot are drawn. You can modify the arrangement of the bars using the hand tool from the **Tools** menu.
- For each nominal or ordinal column, a histogram and divided (stacked) bar chart are drawn showing a bar for each level of the ordinal or nominal variable.

The distribution plots are supported with text reports showing selected quantiles and moments of interval values, and counts, proportions, and total log likelihood for nominal or ordinal values.

Fit Y by X studies the relationship of two variables. Each pair of X and Y variables produces plots with accompanying analyses. The type of analysis depends on the data types of the X and Y columns The different analyses are described in the list that follows.

- If both X and Y have interval values, a scatter plot is displayed. Using options available for each plot, you can explore various regression fits for the data and choose the most suitable fit for further analysis. Each fit can be accompanied by tables with supporting statistical analyses and parameter estimates.
- If X is nominal or ordinal and Y is interval, the distribution of Y values is plotted for each discrete value of X. You can use pop-up menu options to see a box plot for each value of X and to compare group means with comparison circles. An accompanying text report shows a one-way analysis of variance table.
- If X has interval values and Y has nominal or ordinal values, then a logistic regression is performed and displayed with a family of logistic probability curves. Accompanying tables show the log likelihood analysis and parameter estimates for each curve.

Note➜ Logistic regressions of ordinal columns are parameterized differently from logistic regressions of nominal columns and therefore produce different results.

- If both X and Y are nominal or ordinal values, a contingency table mosaic chart is shown. A mosaic chart consists of side-by-side divided bars for each level of X, where each bar is divided into proportional segments representing the discrete Y values. Accompanying tables show the log likelihood analysis, cell frequency counts, and cell proportions.

Figure 9.24 shows an example of each type of plot produced by the different combinations of X and Y columns.

Figure 9.24 Analyses from **Fit Y by X**

Interval X
Ordinal or Nominal X
Interval Y
Ordinal or Nominal Y
Regression Lines
One-Way Anova on Means
Logistic Regression
Contingency Table Analysis

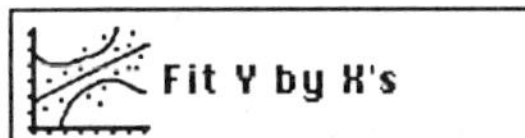

Fit Y by X's performs a fit of one Y variable by all selected X variables. The kind of fit depends on the measurement level of the Y response column.

- If Y has interval values, a least squares fit is done with accompanying leverage plots and an analysis of variance table.
- If Y is nominal, a logistic regression is done by maximum likelihood.
- If Y is ordinal, an ordinal cumulative logistic regression is done by maximum likelihood.

Specify Model displays the dialog shown in **Figure 9.25**. This dialog lets you tailor the analysis using a model specific for your data. The variable selector list in the dialog lists all columns in the current JMP data table. You select columns, assign roles, and choose the model to fit.

Specify Model performs a fit of one Y variable by the X variables selected for the specified model. The kind of fit depends on the measurement level of the Y response column.

- If Y has interval values, a least squares fit is done with accompanying leverage plots and an analysis of variance table.
- If Y is nominal, a logistic regression is done by maximum likelihood.
- If Y is ordinal, an ordinal cumulative logistic regression is done by maximum likelihood.

Figure 9.25 The **Specify Model** Dialog

Model: POPCORN

popcorn
oil amt
batch
yield
trial

» Y » yield
» Wt »
» Freq »

Effects In Model

» Add »
» Cross »
» Nest »
» Surface »
» Factorial »
» Polynomial »
Degree 2

popcorn
oil amt
popcorn*oil amt
batch
popcorn*batch
oil amt*batch
popcorn*oil amt*batch

Remove
Error Term
Run Model
Help

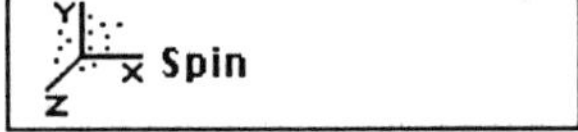

Spin produces a three-dimensional spinable display of values from any numeric columns in the data table as shown in **Figure 9.26**. It can also produce an approximation to higher dimensions through principal components. Three columns spin at a time. However, you can select more than three columns for analysis. A list of selected columns displays to the left of the spin platform. Columns can be added to the list in two ways.

- Designate columns as either X or Y to automatically include them in the list of spin variables.
- Add columns to the list with the **Add Variables** command found in the components pop-up menu of the **Spin** report.

Figure 9.26 Add Variables Dialog and Spin Plot Platform

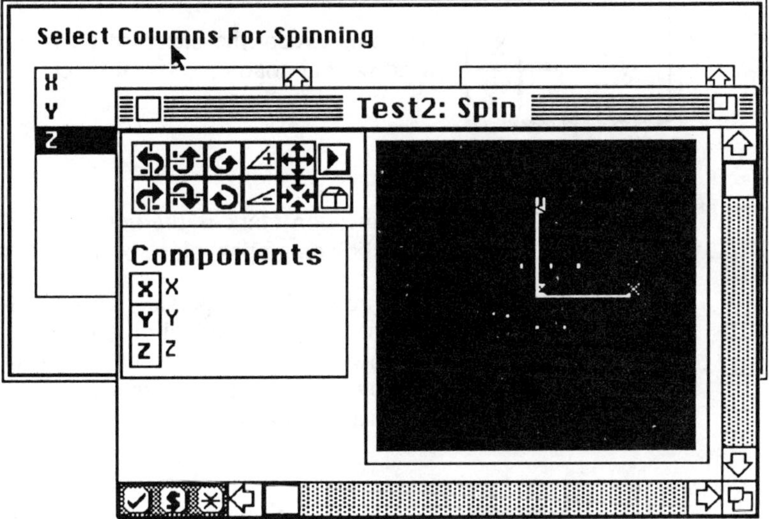

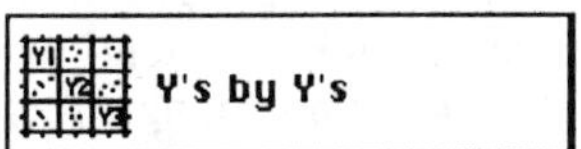

Y's by Y's is used to explore how a number of variables relate to each other. The report produced by this platform is divided into two parts.

The first part of the report examines the relationship between each pair of response (Y) variables showing

- a correlation matrix
- a matrix of scatter plots.

The second part of the report examines the multivariate relationship between all the response variables showing

- the inverse correlation matrix
- a multivariate Mahalanobis distance outlier plot.

The Tools Menu

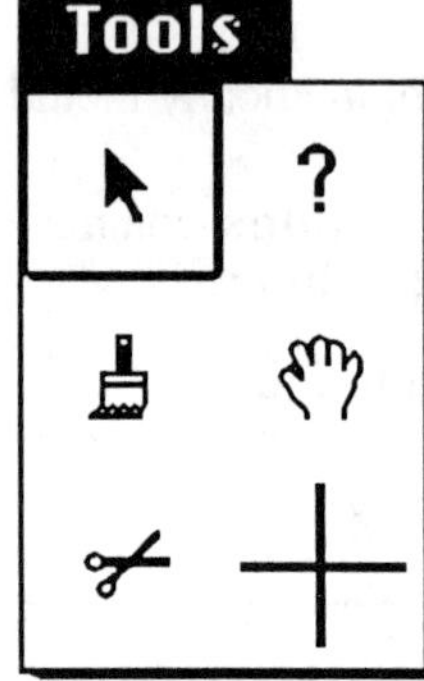

The **Tools** menu is a palette of special tools that determine the effect of mouse actions. The default tool is the arrow cursor. When used in a report window, the arrow reveals text reports, accesses pop-up menus, displays point labels, and can highlight histogram bars.

Other tools cut or copy all or part of a report, access help windows, manipulate graphical displays, and highlight points of interest in a plot.

All tools except the standard arrow cursor function only in report windows but not on the spreadsheet.

The question mark is the help icon. Select the question mark and then click on graphs, plots, or tables to access help windows.

Each help window persists as long as you hold down the mouse button. You can shift-click the question mark to bring up a help window, which stays on the screen until you close it. Some help windows have buttons to reveal further help details and stay on the screen automatically.

The brush tool highlights points in plots. When you drag the mouse a rectangle appears. Move the rectangle over points to highlight them in the plot and in the current data table. Use shift-click to extend the selection. Option-click changes the size of the selection rectangle and also extends the selection.

If the brush tool is in a text report of a report window, it behaves the same as the standard arrow cursor.

The hand tool (or grabber) is for direct manipulation of histograms and spinning plots. You can use the hand to change the number of bars in a histogram or to shift the boundaries of the bars on the axis.

The hand tool is also used on spinning plots to control spin direction and speed. To spin a plot, grab the plot with the hand by holding down the mouse button, and move the hand about. The rotation of the axes tracks the movement of the hand.

Alternatively, if you command-click with the grabber tool, the spin direction depends on the position relative to the origin of the axes. The spin speed is proportional to the distance of the hand from the center.

For more details about using the hand tool in a spinning plot, see Chapter 18, *Analyze: The Spin Command.*

If the hand is in a text report, it behaves the same as the standard arrow cursor.

The scissors are for selecting cut-and-paste territory from a report. Drag the scissors diagonally over any part of the report to define a rectangular area you want to copy to the clipboard. Hold down the option key while clicking the scissors to select blocks of the report window. Use shift-click to extend the selection.

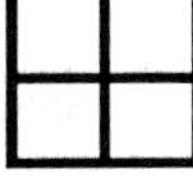

The crosshair is a movable set of axes used to measure points and distances in graphical displays. For example, use the crosshair on a fitted line or curve to identify the response value for any predicted value.

The Window Menu

Window
Redraw
Move To Back
Hide
New Data View
Close *type* Windows
ANIMALS
ANIMALS: Fit Y by X

The **Window** menu helps you organize the windows produced by a JMP session. All windows generated in a session are listed below the dotted line. In the example menu shown to the left, the JMP data table called ANIMALS and a **Fit Y by X** analysis window for the ANIMALS table are both open.

Redraw

The **Redraw** command redraws the active window. It is useful for cleaning up both spreadsheet views and graphical displays that have accumulated stray imperfections resulting from high speed dynamic handling of windows.

Move To Back

The **Move To Back** command moves the active window to the bottom level of windows generated by the current JMP session, leaving the next window in the sequence showing.

Hide

The **Hide** command suppresses the display of the active window but does not close it. To reshow a hidden window, select the window name from the the list beneath the dotted line in the **Window** menu.

New Data View

The **New Data View** command displays a new spreadsheet view of an open data table. The new view is linked to the original view and all corresponding analysis windows. Changes made to a new view reflect on the original view when it is made active.

Close *type* Windows

The **Close *type* Windows** command closes all windows of a given type, where ***type*** is the determined by the active window. For example, if a data table is active, this command is **Close Data Windows** and it closes all data tables when selected.

Chapter 10.
JMP Data Tables

JMP data are organized in memory as rows and columns of a table which is referred to as the *data table*. The columns have names and the rows are numbered. An open data table is kept in memory, and you communicate with it through an active spreadsheet window. You can open as many data tables in a JMP session as memory allows. JMP data tables can be stored in *files* on disk, sometimes called documents.

Commands in the **File**, **Edit**, **Rows**, and **Columns** menus provide data handling operations ranging from standard Macintosh tasks to sophisticated editing and database manipulation. This chapters describes all commands that operate on JMP data tables and includes examples of data table management.

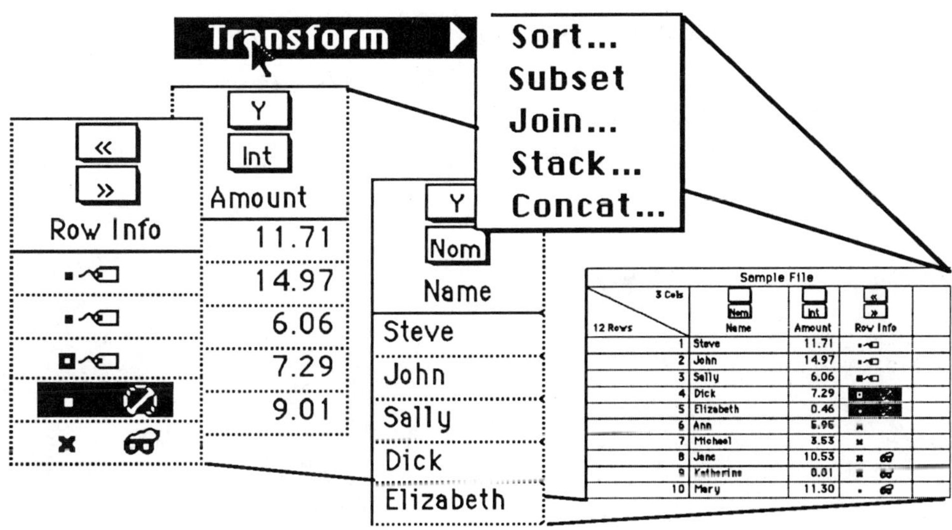

Chapter Contents

Creating a JMP Data Table

To use JMP, you must first store your data in a JMP data table. You create a data table by

- typing data into an empty spreadsheet
- importing data from another application with the **Import** command in the **File** menu
- using the **Paste at End** command from the **Edit** menu.

Once you have data in JMP format, you can use JMP analysis techniques, save information from analyses, generate additional data tables, and export results to other applications. The following sections describe these data table operations.

The New Command

The **New** command in the **File** menu creates a new data table in memory and displays an empty spreadsheet that has one numeric column labelled `column 1`. You can create more columns using the **New Column** command in the **Columns** menu. Once you have created columns, you can type or paste data into the spreadsheet. Alternatively, the **Paste at End** command in the **Edit** menu creates new rows and columns as needed to paste data from the clipboard into the spreadsheet. Use **Save As** from the **File** menu to name the table and to save it as a JMP data file. To open a JMP file, use the **Open** command.

Importing Data

The **Import** command in the **File** menu creates a new data table by reading data from a standard text file, a JMP export file, or a SAS transport file. The dialog shown in **Figure 10.1** displays a modified Macintosh open file dialog with check boxes that let you specify the type of incoming file. To import a text file, you need to specify the end of field and the end of line delimiters by clicking in the appropriate check boxes.

Figure 10.1 The **Import** Dialog

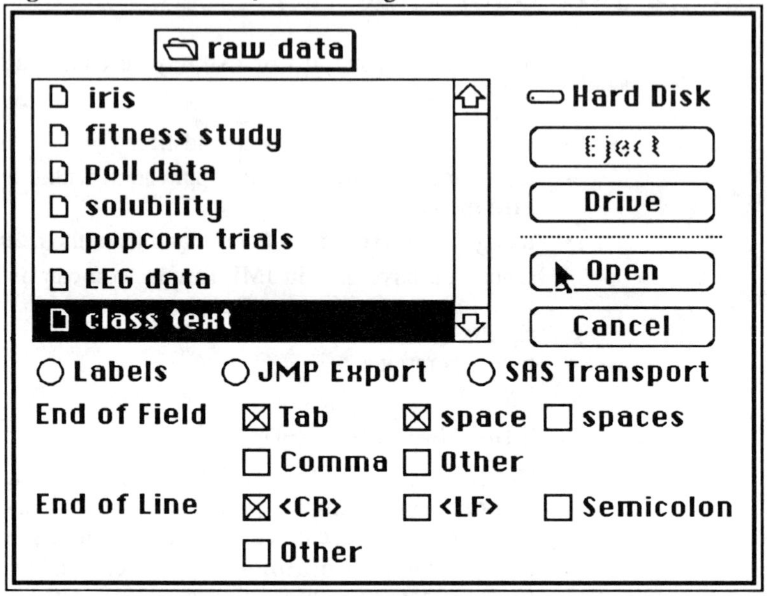

raw data

iris
fitness study
poll data
solubility
popcorn trials
EEG data
class text

Hard Disk
Eject
Drive
Open
Cancel

○ Labels ○ JMP Export ○ SAS Transport

End of Field ☒ Tab ☒ space ☐ spaces
☐ Comma ☐ Other

End of Line ☒ <CR> ☐ <LF> ☐ Semicolon
☐ Other

import example using **Labels** option

JMP automatically creates a data table from a text file of rows and columns. However, JMP reads special file formats specified by radio buttons in the **Import** dialog. For example, the file shown in **Table 10.1** is a text file called CLASS that has column labels as the first line in the file.

Table 10.1 Example of Text File with Labels

```
name   sex age height weight
ALFRED  M  14  69.0  112.5
ALICE   F  13  56.5   84.0
BARBARA F  13  65.3   98.0
CAROL   F  14  62.8  102.5
               .
               .
               .
LOUISE  F  12  56.3   77.0
MARY    F  15  66.5  112.0
PHILIP  M  16  72.0  150.0
ROBERT  M  12  64.8  128.0
RONALD  M  15  67.0  133.0
THOMAS  M  11  57.5   85.0
WILLIAM M  15  66.5  112.0
```

When you select the **Labels** radio button and open the text file above, JMP names the columns of the data table using fields (up to 31 characters) from the first line of the file. The **Import** command creates the JMP data table shown in **Figure 10.2**. The data table is assigned the same name as the text file.

Figure 10.2 **Labels** Option for Text Files with Column Labels

CLASS

	Nom	Nom	Int	Int	Int
	name	**sex**	**age**	**height**	**weight**
1	ALFRED	M	14	69	112.5
2	ALICE	F	13	56.5	84
3	BARBARA	F	13	65.3	98
4	CAROL	F	14	62.8	102.5

The **Import** dialog also reads two other kinds of text files as described below.

When you use the **Save As** dialog and click on the **JMP Export** option, the resulting text file contains header information about the data table and each column. If you need the data in JMP format again, JMP can import this kind of text file more efficiently. See Appendix B, *Questions and Answers* for a discussion of JMP header information.

The **SAS Transport** option in the **Import** dialog creates a data table from a file that is in the SAS transport format. A SAS transport file is created whenever you

- use the SAS System to write an existing SAS data set in transport format
- use the **Save As** command in the **File** menu and click the **SAS Transport Format** radio button before saving the JMP table.

Exporting JMP Files

To export a JMP file, use the **Save As** command in the **File** menu. The dialog shown in **Figure 10.3** has several export options. You write files in standard JMP file format, in standard text format for use in other applications, or as SAS transport files that can be read automatically by the SAS System. **Save as** saves the current data table to a file after prompting you for a name and disk location.

Figure 10.3 The **Save As** Dialog for Exporting JMP Files

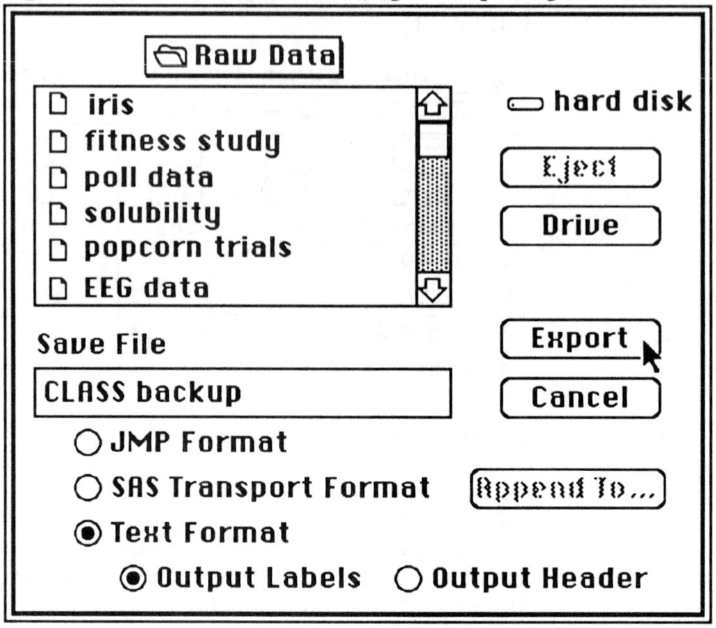

Raw Data

iris
fitness study
poll data
solubility
popcorn trials
EEG data

hard disk

Eject

Drive

Save File

CLASS backup

Export

Cancel

JMP Format

SAS Transport Format

Append To...

Text Format

Output Labels

Output Header

export example using **Labels** option

The **Text Format** option automatically writes the rows and columns of a JMP data table to a rectangular text file. However, a special file format is created when you use the **Output Labels** option. For example, you can create a text file from the CLASS data table (**Figure 10.2** earlier) that contains the column names as fields in the first line, followed by lines of data.

The resulting text file is illustrated shown in **Table 10.2.** The tab characters (represented by right arrows) separate the fields and carriage returns (represented by ¶) delimit the lines. The leading blanks before the numeric values are shown with periods.

Table 10.2 Illustration of Text File Created with **Output Labels** Option

Name➧	Sex➧	Age➧	Weight➧Height¶		
ALFRED.➧	M➧	........14➧		........69➧	.....112.5¶
ALICE..➧	F➧	........13➧		......56.5➧	........84¶
BARBARA➧	F➧	........13➧		......65.3➧	........98¶
CAROL..➧	F➧	........14➧		......62.8➧	.....102.5¶

.
.
.

The Spreadsheet View

This section covers all the cursor forms and describes how to navigate about the spreadsheet to perform cell editing, row and column manipulation, and other spreadsheet tasks. It often refers to a concept called *cell focus.* A cell is focused when it is highlighted or contains the blinking I-beam (see the cursor forms described below). The focused cell responds to cell edit commands.

Cursor Forms

To navigate the spreadsheet, you need to understand how the cursor works. The cursor assumes different forms and performs different actions depending on its location in the spreadsheet. These forms and functions are described below.

Arrow Cursor

The cursor is the standard arrow when it is in the *measurement level box* or the *role assignment box* at the top of a column and when it is in the *Cols* or *Rows* triangular areas in the upper left corner of the spreadsheet.

Cross Cursor

When the cursor is within a column heading or a data cell, it becomes a large cross indicating it is available to select text. When you click the cross cursor, the text is highlighted and can be edited. The cursor then becomes an I-beam. Text within a locked column or a locked data table cannot be edited.

I-beam Cursor

To edit text, position the I-beam within highlighted text. Click or drag the I-beam to mark an insertion point or to select text for replacement. The I-beam deposits a blinking vertical bar to indicate a text insertion point or a highlighted area of text to be replaced. The default selection is the entire word. Use the keyboard to make changes.

Open Cross Cursor

The cursor is a large open cross when you move it into a row or column selection area or into a noneditable area of the spreadsheet. Use the open cross cursor to select a single row or column. Shift-click a beginning and ending range of rows or columns to select an entire array. Use command-click to make a discontiguous selection. See the next section, **Selecting Rows and Columns**, for a more details.

Double Arrow Cursor

The cursor changes to a filled double arrow when placed on a column boundary. This boundary dragger is used to temporarily change the width of a column. However, this does not affect the *column field width* as specified in the **Column Info** dialog.

Navigating the Spreadsheet

The keyboard offers you the ability to navigate efficiently about an active spreadsheet. Keystroke actions are described in this section and summarized in **Table 10.3** at the end of the section.

Keyboard Arrows

← ↑ → ↓

Most Macintosh keyboards include four arrow keys on the lower right half of the keyboard. By pressing one of these keys while a spreadsheet cell is focused, you can move the focus to an adjacent cell. The direction of the arrow determines the direction the focus travels. If there is no adjacent cell to the right or left, the focus moves to the beginning of the next row or to the end of the previous row. When the focus moves to the top of a column, the up arrow key moves it into the column label field. The up arrow key has no further effect on the focus once it reaches a column label, and the down arrow key has no effect once the focus reaches the bottom of a column.

By holding down the shift key, you can modify the effect of an arrow key. Shift-↑ moves the focus to the top-most visible cell within the spreadsheet window, and shift-→ moves the focus to the right-most visible cell in the spreadsheet window. Shift-↓ and shift-← move the focus to the bottom or the far left of the window, respectively. When necessary, the spreadsheet window scrolls to reveal the entire highlighted cell.

The option key also modifies the effect of the arrow keys. If you think of the visible area within the spreadsheet window as a page, then holding down the option key while pressing an arrow key moves the cell focus to an adjacent page. Option-← and option-→ move the focus to the cell in the left-most column in the left or right adjacent page. Option-↑ and option-↓ move the focus to the cell in the top-most row of the page just above or below the visible page.

The left and right arrow keys also affect the position of an I-beam cursor within a spreadsheet cell. If the I-beam indicates an insertion point, then the left and right arrows move the I-beam left or right one character at a time within the cell.

The tab, shift-tab, return, shift-return, enter, and shift-enter keys can also be used to move the focus. Tab and Shift-tab move the focus one adjacent cell the the right or left. Return and shift-return move the focus one cell up or down. Enter and shift-enter move the focus to the first cell in the left or right adjacent column.

Table 10.3 Keyboard Commands and Actions

Keystroke	Moves focus...
→	one adjacent cell to the right (or moves I-beam cursor one character to the right)
←	one adjacent cell to the left (or moves I-beam cursor one character to the left)
↑	up one cell in the same column
↓	down one cell in the same column
Shift-→	to the right-most visible cell in the spreadsheet
Shift-←	to the left-most visible cell in the spreadsheet
Shift-↑	to the top-most visible cell in the spreadsheet
Shift-↓	to the bottom-most visible cell in the spreadsheet
Option-→	to cell in left-most column in right adjacent page
Option-←	to cell in left-most column in left adjacent page
Option-↑	to cell in top-most row in page just above visible page
Option-↓	to cell in top-most row in page just below visible page
Tab	one adjacent cell to the right
Shift-tab	one adjacent cell to the left
Return	down one cell in the same column
Shift-return	up one cell in the same column
Enter	to the first cell in the next row
Shift-enter	to the first cell in the previous row

Selecting Rows and Columns

The selection of rows and columns in a JMP spreadsheet is done by highlighting them as shown in **Figure 10.4**. To highlight a row, click on the space that contains the row number. To highlight a column, click on the background area above the column name.

Figure 10.4 Select Rows and Columns

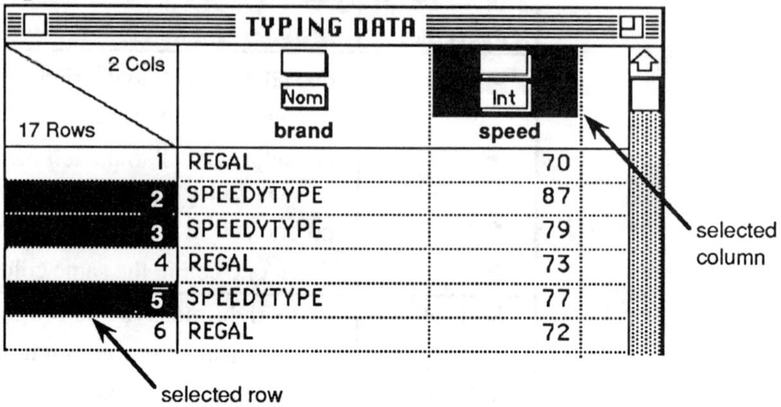

TYPING DATA

2 Cols
17 Rows

	brand (Nom)	speed (Int)
1	REGAL	70
2	SPEEDYTYPE	87
3	SPEEDYTYPE	79
4	REGAL	73
5	SPEEDYTYPE	77
6	REGAL	72

selected column

selected row

To extend the selection of rows or columns, drag across the array or shift-click the first and last row or column range. To make a discontinuous selection, use command-click to make a discontinuous selection. To select a subset of the spreadsheet, drag the mouse to form a rectangular box. The corresponding rows and columns become highlighted as shown below.

Figure 10.5 Selecting a Subset of the Spreadsheet

BIG CLASS

5 Cols
40 Rows

drag mouse to select a rectangular subset

BIG CLASS

5 Cols
40 Rows

	name (Nom)	age (Ord)	sex (Nom)	height (Int)
1	KATIE	12	F	59
2	LOUISE	12	F	61
3	JANE	12	F	55
4	JACLYN	12	F	66
5	LILLIE	12	F	52
6	TIM	12	M	60

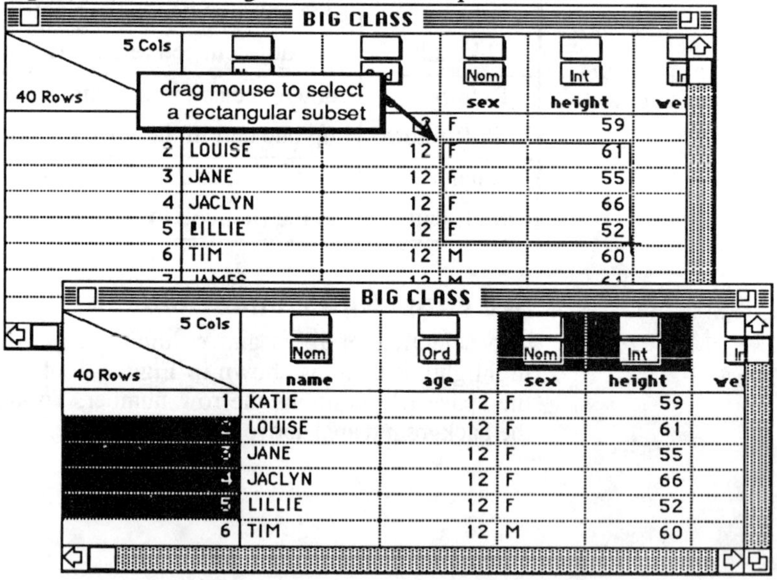

You can also select rows by using the **Select** command from the **Rows** menu. **Select All Rows** selects the entire JMP data table. Selecting all rows is useful for

- assigning a special marker or color to all points in a graphical display
- adding new columns to the data table with the **Paste at End** command in the **Edit** menu.

You can select rows with the excluded, hidden, or labelled row state characteristic using the **Select** submenu in the **Rows** menu.

Select Excluded
selects all excluded rows regardless of their current selection status. Selected rows that are not excluded are deselected.

Select Hidden
selects all hidden rows regardless of their current selection status. Selected rows that are not hidden are deselected.

Select Labelled
selects all labelled rows regardless of their current selection status. Selected rows that are not labelled are deselected.

Row states are discussed in detail later in this chapter.

Locating Selected Rows

Because JMP commands and graphical displays make it possible for you to select a large number of discontiguous rows, there are two special commands to help you locate selected rows in the spreadsheet. The **Locate Next** and **Locate Previous** commands in the **Rows** menu search for the next closest selected row below or above the *current row*, respectively. If there are no selected rows in the spreadsheet, these two commands are dimmed.

current row

The starting point of a locate next or previous search is always the *current row*. The current row is defined in any of the following ways.

- When you open a JMP data table, the first row is the current row.
- The most recently edited row is the current row by default.
- You can change the current row by option-clicking anywhere within a row.
- A row found by **Locate Next** or **Locate Previous** is the current row.

Whenever you locate a selected row, the spreadsheet window scrolls to make that row visible, if necessary. The selection area of the row you locate also blinks briefly to distinguish it from any other selected rows in view.

You can use either of the locate commands repeatedly to step through the selected rows in the spreadsheet. If there are no more selected rows between the current row and the bottom of the spreadsheet, the **Locate Next** command generates a beep that alerts you, and the location of the current row does not change. Likewise, if there are no more selected rows above the current row, the **Locate Previous** command alerts you.

Command-click to deselect rows or columns. Click in the triangular *Cols* or *Rows* area in the upper left corner of the spreadsheet to deselect all rows or columns, respectively.

Note ➜ **Invert Selection** is a **Rows** menu command that deselects all currently selected rows and selects all unselected rows. This menu item is useful when analyzing exclusive subsets of rows separately.

Adding and Deleting Rows

To add rows to a JMP spreadsheet, use the **Add Rows** command from the **Rows** menu or double-click in the triangular *Rows* area in the upper left corner of the spreadsheet. The dialog shown in **Figure 10.6** prompts you to enter the number of rows you want to add. If a column is computed using the calculator, then its new cells are automatically filled with values. Otherwise, the cells have missing values.

You can also add new blank rows by clicking anywhere in the data area beyond the last non blank row.

Figure 10.6 The **Add Rows** Dialog

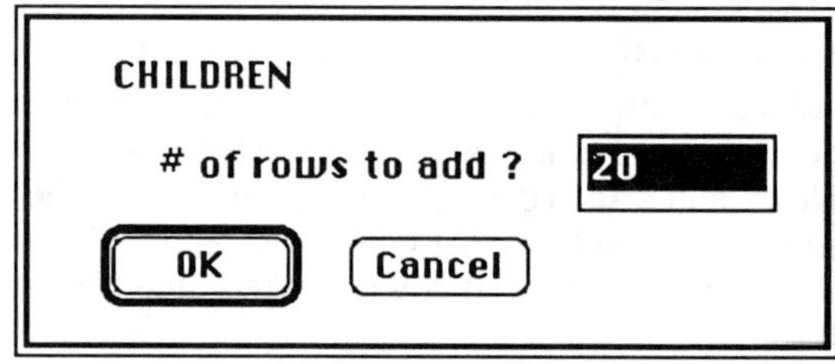

To delete rows from the spreadsheet, select the rows you want to delete and choose the **Delete Rows** command from the **Rows** menu. If you mistakenly delete rows, immediately select the **Undo Delete Rows** command from the **Edit** menu.

Adding and Deleting Columns

To add column to a JMP spreadsheet, use the **New Column** command from the **Columns** menu or double-click in the triangular *Cols* area in the upper left corner of the spreadsheet. The dialog, shown in **Figure 10.7**, asks you to name the new column and provide column characteristics.

Optionally, you can specify a formula to compute column values. You can check the lock box to protect any column from being changed. To unlock a column, check the lock box again. A column is automatically locked when you click on **Use Formula** or when the data type is **Row State**. For more details on new column dialog options, refer to the **Columns** Menu section in Chapter 9, *The Menu Bar*.

Figure 10.7 The Column Definition Panel

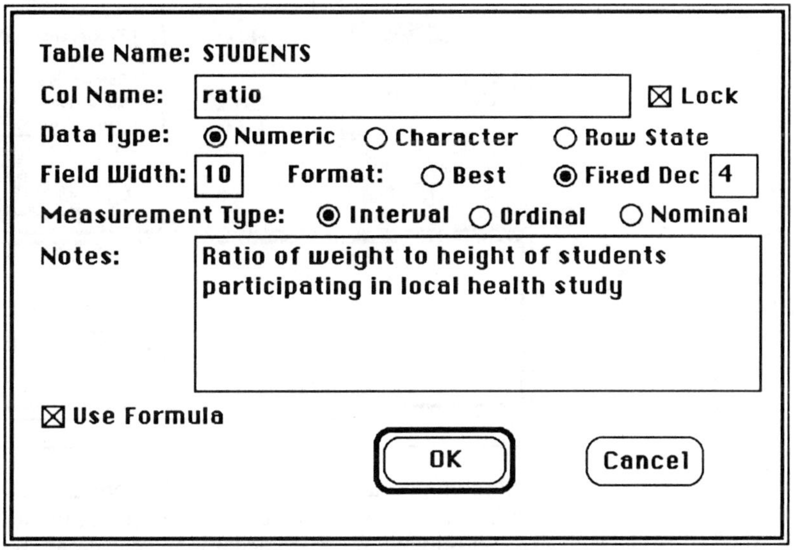

Table Name: STUDENTS

Col Name: ratio ☒ Lock

Data Type: ◉ Numeric ○ Character ○ Row State

Field Width: 10 Format: ○ Best ◉ Fixed Dec 4

Measurement Type: ◉ Interval ○ Ordinal ○ Nominal

Notes: Ratio of weight to height of students participating in local health study

☒ Use Formula

OK Cancel

If you click the **Use Formula** box, the calculator window shown in **Figure 10.7** is displayed. You can use this window to construct a formula to compute values for the new column. The formula can include existing columns, functions, and constants, and it can use conditional logic. When the **Use Formula** box is checked, the new column is automatically locked and cannot be edited on the spreadsheet.

Figure 10.8 The Calculator Window

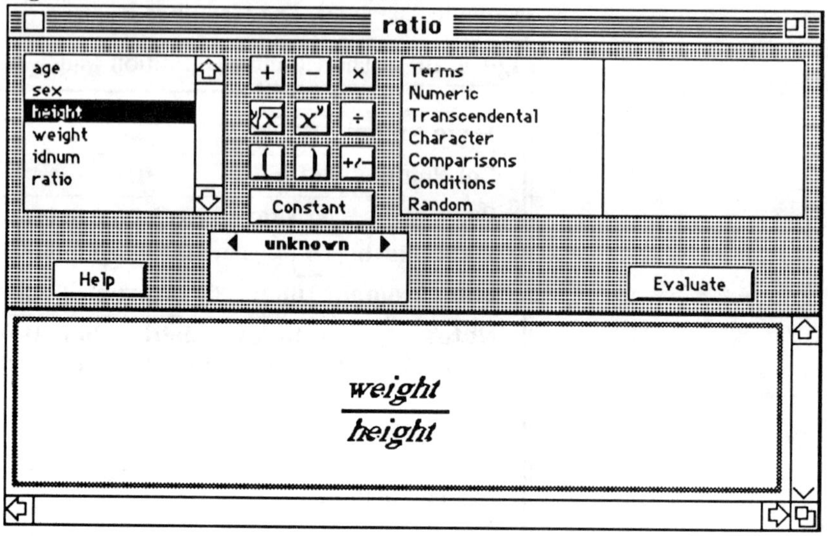

Columns can also be added to the spreadsheet by using save commands within the statistical platforms. These commands are available in the save ($) pop-up menu, which accompanies each analysis window.

To delete columns from the spreadsheet, select the columns you want to delete and choose the **Delete Columns** command from the **Columns** menu. If you mistakenly delete columns, immediately select the **Undo Delete Columns** command from the **Edit** menu.

Row States

Row states are characteristics associated with a row. They are used to distinguish subsets of your data, exclude data from analyses, and customize the appearance of graphical displays. Row state commands are located in the **Rows** menu and affect only highlighted rows. After row states are assigned, they can be saved permanently with the data table in a special row state column. The six row states are defined in the following list.

Exclude/Include is a toggle that excludes selected rows from statistical analyses. For example, you may want to eliminate known outliers from a regression analysis. To do this, select the rows that contain extreme values and choose **Exclude/Include** from the **Rows** menu. Data remain excluded until **Exclude/Include** is chosen again for selected rows.

Excluded data are not automatically hidden in plots even though they are excluded from calculations in text reports and graphical displays.

Hide/Unhide is a toggle that suppresses the display of points in all scatter plots. To hide data, select rows in the spreadsheet and choose **Hide/Unhide** from the **Rows** menu. For example, you can exlude points from analysis and then hide those same points in scatterplots. The data remain hidden until **Hide/Unhide** is chosen again for hidden rows.

Hidden points are not automatically excluded from statistical computations that affect text reports and graphical displays even though they are not displayed in the plots.

Label/Unlabel is a toggle that labels points on all scatter plots. To label points, select the rows containing the points and use the **Label/Unlabel** command. This row state remains in effect until **Label/Unlabel** is chosen again for the selected rows.

By default, the row number is used as the label value on scatter plots. However, you can assign the values of any column to be labels which identify points in plots.

Colors

Colors lets you assign any colors to highlighted rows (on color monitors). The points in scatterplots and spinning plots display using the color you select from the **Colors** submenu. Each active color displays next to its row number in the spreadsheet.

Markers assigns a character to replace the default dot in scatterplots and spinning plots. There are eight markers available in **Markers** submenu. Each active marker displays next to its row number in the spreadsheet.

6

Selection status is not assigned using the **Rows** menu. However, it is a row state characteristic that can be saved in a row state column. For details about how to select rows, see **Selecting Rows and Columns** earlier in this chapter. In the spreadsheet, the row number of a selected row is highlighted.

Saving row states

To save row state information, create a new **Row State** column. You can assign row state information into the new column two ways.

- Use the calculator to assign row state information based on a formula.
- Select the **Copy from Row State** command in the >> pop-up menu at the top of the row state column.

To activate a row state column, Select the **Copy to Row State** command in the << pop-up menu at the top of the row state column. The active row states appear in the row number space as shown in **Figure 10.9**.

Figure 10.9 Row State Column Buttons

BIG CLASS

3 Cols / 40 Rows		weight (Int)	height (Int)	markers
	1	95	59	
	2	123	61	
	3	74	55	
	4	145	66	
	5	64	52	
	6	84	60	
	7	128	61	
	8	79	51	
	9	112	60	
	10	107	61	

copy column data to active row state

copy active row state to column

active row state information

Cutting and Pasting

Edit menu commands have different effects depending on the current spreadsheet selection. These commands either affect selected rows and columns or a focused cell. The **Edit** commands operate according to the following rules.

- If there is a focused cell, all **Edit** commands apply only to the cell.
- The **Edit** commands affect all values in selected rows if no columns are selected. Likewise, they affect all values in selected columns (except the header field) if no rows are selected.
- If both rows and columns are selected, **Edit** commands affect the subset of cells defined by their intersection.
- If there is no focused cell or selected region, all **Edit** commands are dimmed except **Paste at End**, which is described in the following list.

The spreadsheet supports the following **Edit** menu commands.

Copy copies selected fields (either data or a column heading) from the active data table to the clipboard.

Cut copies selected fields (either data or a column heading) from the active data table to the clipboard and replaces them with missing values. **Cut** is equivalent to **Copy**, then **Clear**.

Clear replaces selected fields (either data or a column heading) in the active data table with missing values but does not copy them to the clipboard.

Paste copies data from the clipboard into a selected area of a JMP data table.

Undo cancels the effect of the most recent reversible command. If **Undo** is available, it displays in the **Edit** menu as **Undo *command***, where command is the reversible command. Most destructive spreadsheet operations (such as cutting, pasting and deleting) are reversible. However, **Undo** is dimmed when it is unavailable. Once you have selected **Undo**, the command changes to **Redo *command.***

Paste at End pastes data from the clipboard into selected rows or selected columns in the active spreadsheet (but not both). If both rows and columns are selected, the **Paste at End** command is dimmed. While the **Paste at End** command reads only data in text format from the clipboard, it can paste data into character, numeric columns, and row state columns as long as these corresponding columns match in data type (numeric, character or row state). However, it cannot paste other data types not supported by JMP.

Paste at End copies character strings from the clipboard, pasting each string which is separated by a tab character into a separate spreadsheet cell. Likewise, each return character in the clipboard represents the end of a spreadsheet row. Any cells which have been copied from JMP are automatically stored in this format.

If one or more rows are selected when you choose **Paste at End**, the command adds columns to the right of existing columns. The command pastes the number of rows from the clipboard which correspond to the number of selected rows in the spreadsheet. All other rows in the clipboard are ignored, and the cells corresponding to other rows in new columns fill with missing values.

If one or more columns are selected when you choose **Paste at End**, the command adds rows to the bottom of existing rows. The command pastes the number of columns from the clipboard which correspond to the number of selected columns in the spreadsheet. All other columns in the clipboard are ignored, and the cells corresponding to other columns in new rows fill with missing values.

It you select **Paste at End** for rows in a spreadsheet that contain row state columns, an error message is generated. An error message is also generated if you **Paste at End** to numeric columns when the clipboard contains non numeric character data.

If there are no selected rows or columns, **Paste at End** behaves as if all existing columns are selected. Therefore, you can use this command to import data into a new spreadsheet because it creates the rows and columns needed for the data in the clipboard.

Printing and Journaling

The **Print** command in the **File** menu displays the standard Macintosh print dialog. You can print the entire spreadsheet or only a subset of data defined by selected rows and columns.

The **Journal** command from the **File** menu lets you save the spreadsheet in a journal file. You can journal the entire spreadsheet or only a subset of data defined by selected rows and columns.

Note➡ A column containing row state data cannot be saved in a journal file.

Transforming JMP Data Tables

The following sections cover the **Transform** command in the **Edit** Menu. The **Transform** command creates a new data table by modifying the active data table or by combining several tables.

Figure 10.10 The **Transform** Command Submenu

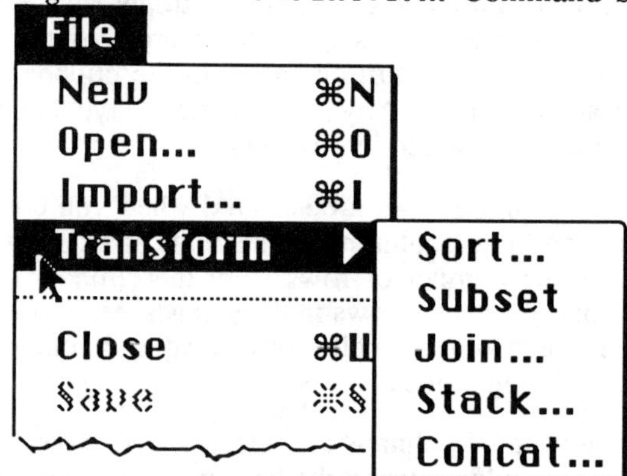

Sort... sorts a JMP data table by one or more columns.

Subset creates a new data table that is a subset, which consists of selected rows and columns from the active spreadsheet.

Join... creates a new data table by merging (joining) two tables side-by-side.

Stack... creates a new data table by stacking specified columns from the active data table into a single new column.

Concat... creates a new data table from two or more open data tables by combining the tables end-to-end.

Sorting a Data Table

The **Sort** command displays a dialog (shown in **Figure 10.11**) for specifying columns as sort fields. To sort a JMP data table, select sort fields from the **Columns** list and add them to **Sort By** list with the **Add** button.

Alternatively, double-click on column names to add them to the **Sort By** list. You can remove sort fields by selecting them in the **Sort By** list and clicking on the **Remove** button.

ascending

descending

The columns you add to the **Sort By** list establish the order of precedence for sorting. The first column is the major sort field, and each successive column in the list sorts minor to the previous column. The icon to the left of the sort column indicates whether the values sort in ascending or descending order. Values sort in ascending order by default. You can toggle the sort between ascending and descending for any selected column in the **Sort By** list with the **a...Z/Z...a** button.

Click on the **Sort** button when you have completed the dialog.

Figure 10.11 The Sort Dialog

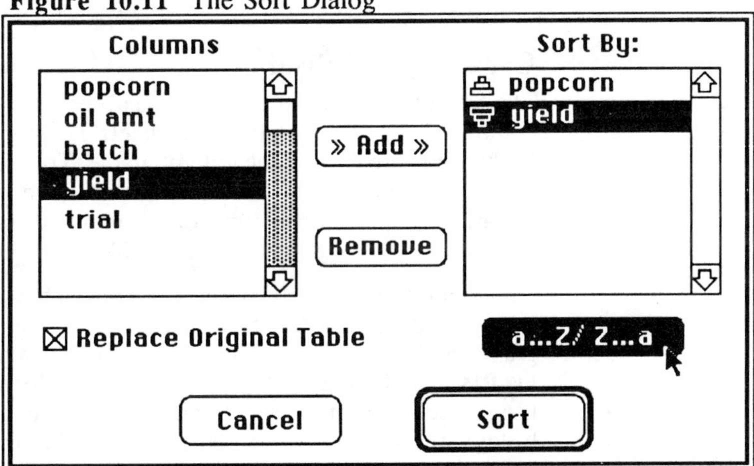

Figure 10.12 shows the POPCORN data table in its original form on the left and sorted by **popcorn** and **yield** (within **popcorn**) on the right. Both fields are sorted in ascending order. In this example, the **Sort** command created a new data table. However, you can check the **Replace Original Table** option to overwrite the original data table with the sorted version.

Figure 10.12 Unsorted and Sorted POPCORN Data Table

POPCORN

	popcorn (Nom)	yield (Int)
6	gourmet	12.1
7	plain	10.6
8	gourmet	18.0
9	plain	8.8
10	gourmet	8.2
11	plain	8.8

Untitled

	popcorn (Nom)	yield (Int)
6	gourmet	9.2
7	gourmet	8.6
8	gourmet	8.2
9	plain	10.6
10	plain	10.4
11	plain	10.1

Creating a Subset

The **Subset** command produces a subset consisting of selected rows and columns from the active spreadsheet. To select a column, click in the background area above the column name. To select a row, click on the row number in the spreadsheet or highlight the corresponding value in a graphical display. If no rows or columns are selected, then the **Subset** command reproduces the entire data table.

All graphical displays that represent a data table are linked to each other and the corresponding spreadsheet. Therefore, highlighting bars or points in graphical displays selects the corresponding spreadsheet rows. For example, the **Distribution of Y's** command in the **Analyze** menu is frequently used to select a range of column values. **Figure 10.13** illustrates selected histogram bars and the corresponding selected rows in the spreadsheet.

Figure 10.13 Highlighting Columns with a Histogram

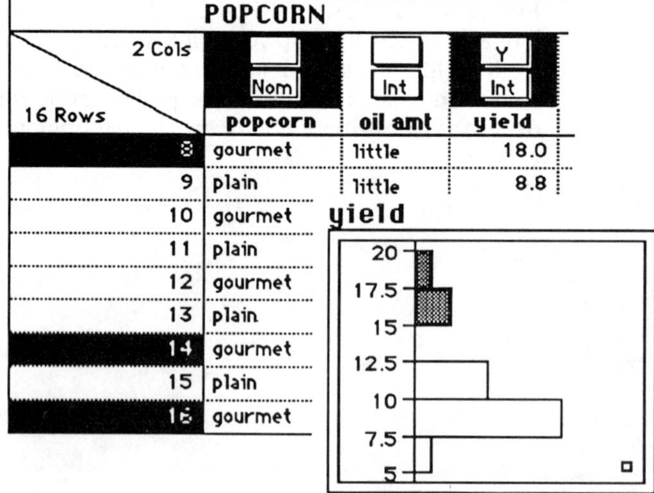

After selecting the rows and columns above, the **Subset** command creates the new untitled data table shown below.

Figure 10.14 Subset of POPCORN Data Table

Untitled

5 Cols / 3 Rows	popcorn (Nom)	yield (Int)
1	gourmet	18.0
2	gourmet	15.9
3	gourmet	16.0

Joining Data Tables

The **Join** command creates a new data table by merging (joining) two open data tables side-by-side. Tables can be joined

- by row number
- by matching the values in one or more columns that exist in both data tables
- in a Cartesian fashion where all values in a column of one data table are merged with all values in a column of another table.

join by row number

The simplest join combines tables by row number. For example, suppose you want to combine the eight rows from each data table shown in **Figure 10.15** into a single data table. (This example uses the TRIAL1 and TRIAL2 data from the SAMPLE DATA folder. The **batch** and **oil amt** columns are hidden in some of the figures that follow.)

Figure 10.15 The TRIAL1 and TRIAL2 Data Tables

TRIAL1

	popcorn (Nom)	yield (Int)
1	plain	8.2
2	gourmet	8.6
3	plain	10.4
4	gourmet	9.2
5	plain	9.9
6	gourmet	12.1
7	plain	10.6
8	gourmet	18.0
9		

TRIAL2

	popcorn (Nom)	yield (Int)
1	plain	8.8
2	gourmet	8.2
3	plain	8.8
4	gourmet	9.8
5	plain	0.1
6	gourmet	5.9
7	plain	7.4
8	gourmet	6.0
9		

When you select the **Join** command, the dialog shown in **Figure 10.16** displays the name of the active table next to the word **Join**. You select an open data table to join **With** from the selector list in the **Join** dialog. Optionally, specify the name of the new data table. By default, the new data table name is the concatenation of the original table names, separated by a plus sign.

Figure 10.16 The Join Dialog

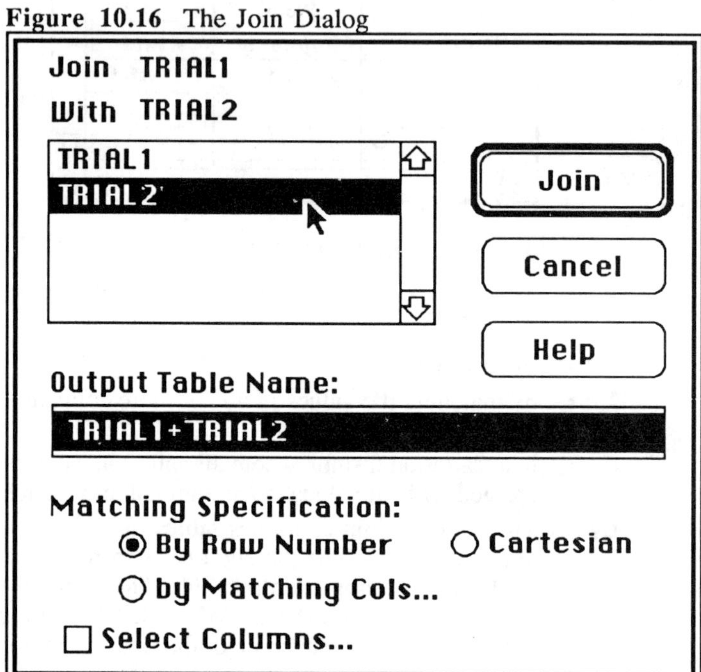

After you select a **With** table, choose a matching specification from the bottom part of the dialog as shown in **Figure 10.16**. The default matching option is **By Row Number**. If you match by row number and click on **Join**, JMP creates the new data table shown in **Figure 10.17.** The two original tables are joined side-by-side, and the new table has all columns from both tables. If a column name is unique, it is written directly to the new table. Columns with the same names in both tables are not overwritten. Instead, both columns are included in the new table. Each is named by appending the original column name to its data table name.

Figure 10.17 Result of Join by Row Number

TRIAL1+TRIAL2

	popcorn:TRIAL1 (Nom)	yield :TRIAL1 (Int)	popcorn:TRIAL2 (Nom)	yield :TRIAL2 (Int)
1	plain	8.2	plain	8.8
2	gourmet	8.6	gourmet	8.2
3	plain	10.4	plain	8.8
4	gourmet	9.2	gourmet	9.8
5	plain	9.9	plain	10.1
6	gourmet	12.1	gourmet	15.9
7	plain	10.6	plain	7.4
8	gourmet	18.0	gourmet	16.0
9				

select a subset of columns

If you don't want to join all columns from the original data tables, the **Select Cols** option in the **Join** dialog allows you to select a subset of columns. In the example above, the data in two popcorn columns are identical. If you click the **Select Cols** box in the **Join** dialog before joining these tables, the dialog shown in **Figure 10.18** lets you select the columns you want to include in the new data table.

Figure 10.18 The Select Columns for Join Dialog

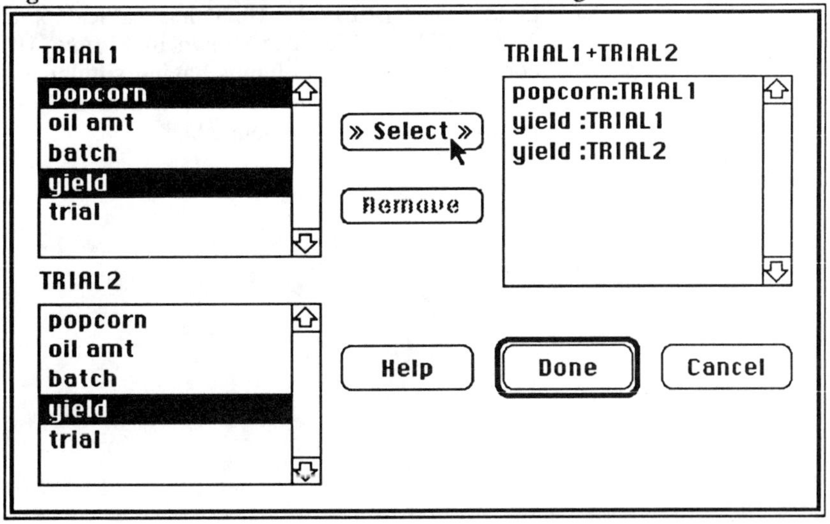

The new table shown in **Figure 10.19** contains only the selected columns. You can also delete and rearrange columns, use data table editing features to rename the columns, and select the **Save As** command from the **File** menu to save the table with a new name.

Figure 10.19 Joined Table with Column Selection

TRIAL1+TRIAL2

	Nom popcorn:TRIAL1	Int yield :TRIAL1	Int yield :TRIAL2
1	plain	8.2	8.8
2	gourmet	8.6	8.2
3	plain	10.4	8.8
4	gourmet	9.2	9.8
5	plain	9.9	10.1
6	gourmet	12.1	15.9
7	plain	10.6	7.4
8	gourmet	18.0	16.0
9			

join by matching columns

Often you need to join two tables that have unequal numbers of rows. This means columns that contain similar information do not have all matching values. For example, suppose the popcorn trials experiment is complete for the first trial but partially complete for the second trial. The TRIAL1 data table shown in **Figure 10.20** has values for 8 experimental conditions. The LITTLE table, also shown in **Figure 10.20**, only has values for 4 conditions. Further, columns having similar information in the two tables do not have the same names (**oil amt** and **oil**) and are not in the same relative positions in the data tables.

Figure 10.20 Joined Tables with Column Selection

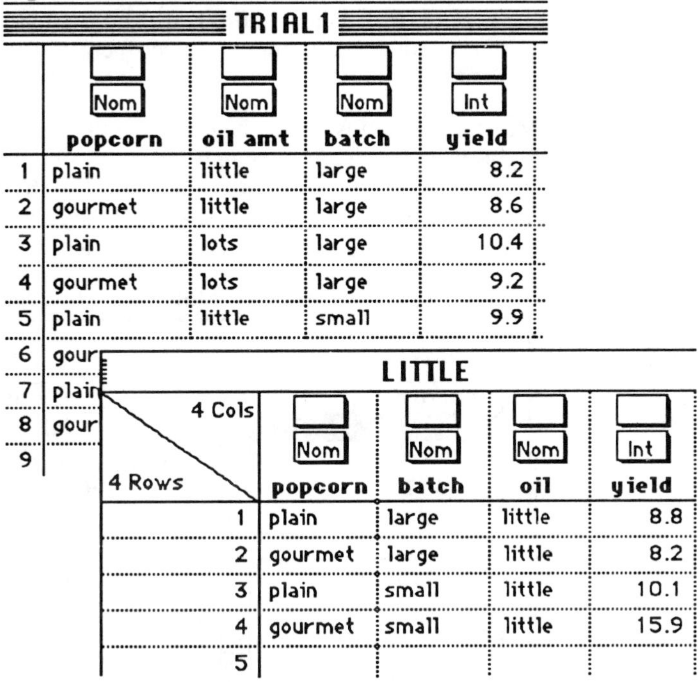

TRIAL1

	popcorn (Nom)	oil amt (Nom)	batch (Nom)	yield (Int)
1	plain	little	large	8.2
2	gourmet	little	large	8.6
3	plain	lots	large	10.4
4	gourmet	lots	large	9.2
5	plain	little	small	9.9
6	gour			
7	plai			
8	gour			
9				

LITTLE

4 Cols / 4 Rows

	popcorn (Nom)	batch (Nom)	oil (Nom)	yield (Int)
1	plain	large	little	8.8
2	gourmet	large	little	8.2
3	plain	small	little	10.1
4	gourmet	small	little	15.9
5				

To join these tables correctly, you must match the values for **popcorn** in both tables, **oil amt** in the TRIAL1 table with **oil** in the LITTLE table, and **batch** in both tables. The **By Matching Cols** option at the bottom of the **Join** dialog lets you specify columns whose values must match in order to complete a join.

Figure 10.21 Join By Matching Cols Option

Output Table Name:

TRIAL1+LITTLE

Matching Specification:

○ By Row Number ○ Cartesian

◉ By Matching Cols...

☐ Select Columns...

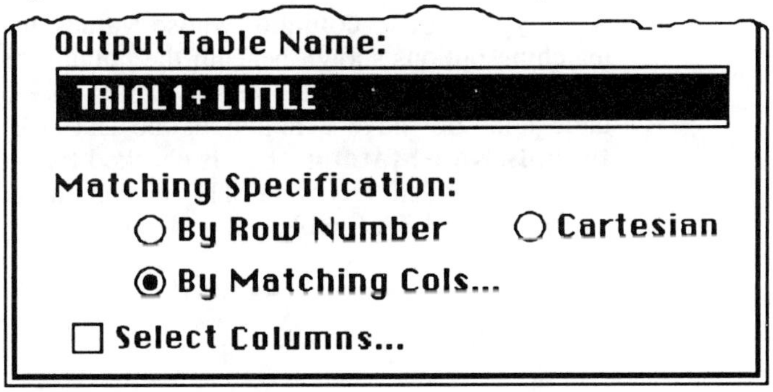

When you click on the **By Matching Cols** radio button, a second dialog shown in **Figure 10.22** prompts you to select columns whose values must match in both tables for rows to be joined. After selecting a column from each list, click on the **Match** button. This displays the columns in the lower boxes. Select additional pairs of columns as needed. If you want to remove matched columns, select them in the lower boxes and click on the **Remove** button.

Matching columns do not have to have the same names and do not have to be in the same relative column position in both tables. When you click on the **Done** button, the first column in the lefthand list is paired with the column in the righthand list. Likewise the second columns are paired, and so on. Rows are joined only if values match for all the column pairs.

Figure 10.22 The Specify Matching Columns Dialog

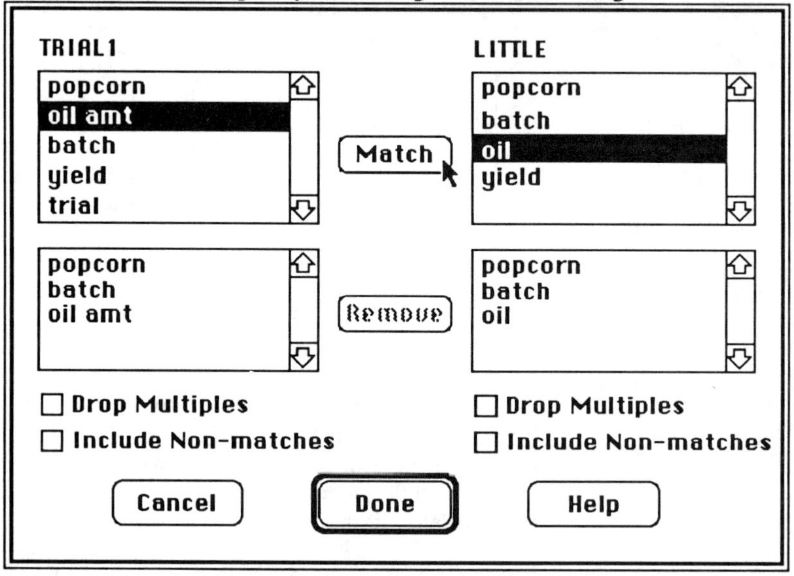

After you choose columns whose values must match, the additional matching options shown beneath the column lists display for both tables. In the data tables above (see the tables in **Figure 10.20**) the TRIAL1 table does not have a match in the LITTLE table for each row. The **Include Non-Matches** box is checked for the TRIAL1 table so that the new table, called TRIAL1+LITTLE has all rows in TRIAL1 even if no match in the LITTLE table is found.

Figure 10.23 Choose Matching Options

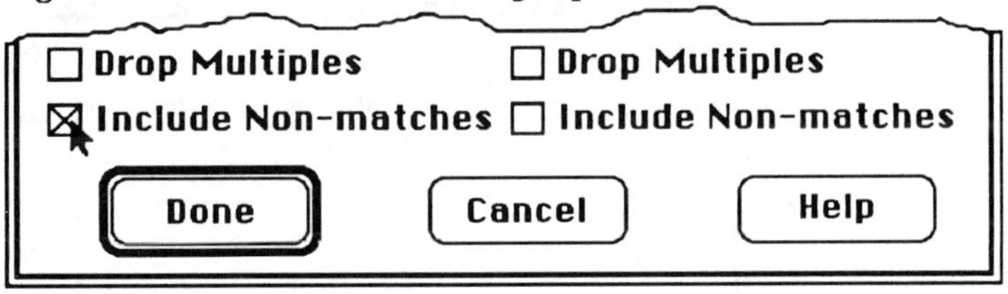

Drop Multiples
Include Non-matches
Drop Multiples
Include Non-matches
Done
Cancel
Help

After this dialog is complete, click the **Done** button to display the **Join** dialog again. This lets you use the **Select Columns** option.

Figure 10.24 Choose the Select Columns option

Output Table Name:
TRIAL1+LITTLE
Matching Specification:
By Row Number
Cartesian
By Matching Cols...
Select Columns...

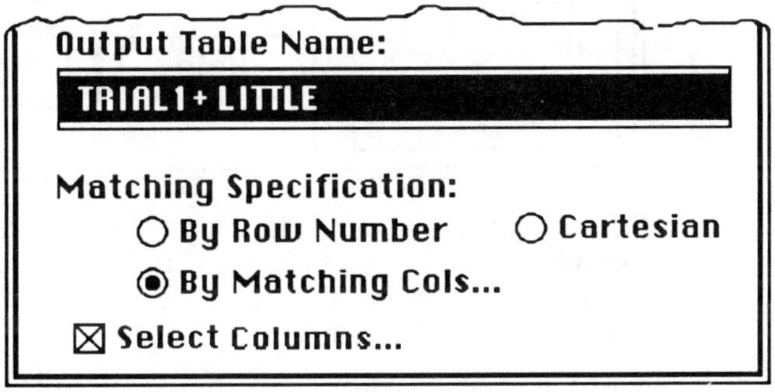

When you click on the **Select Columns** check box, you see the dialog shown in **Figure 10.25**. After selecting all the columns you want from both tables, click on the **Select** button. If a column name is the same in both tables, the name in the new table is the column name appended to the data table name separated by a colon. If a name (such as oil amt) occurs in only one table, it is written directly to the new table. The new column names are listed in the box under the new table name. To remove a column from this list, select it and click the **Remove** button.

Click on **Done** when you are ready to complete the join. This displays the the **Join** dialog once more.

Figure 10.25 The Select Columns Dialog

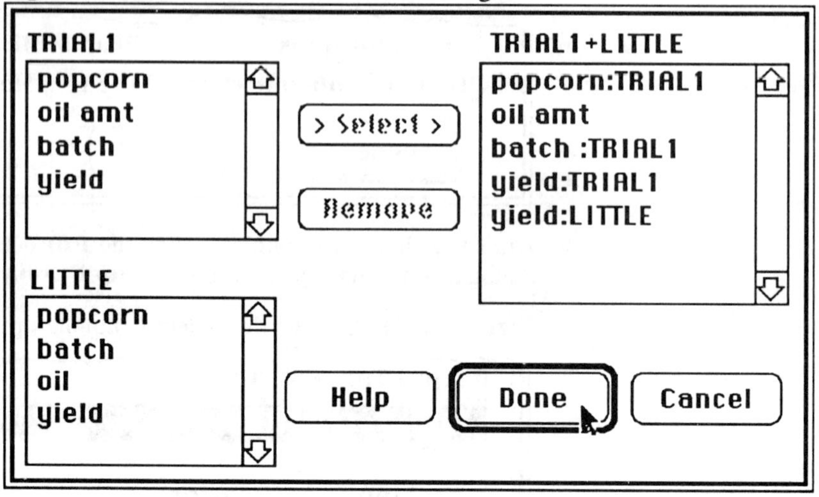

TRIAL1
popcorn
oil amt
batch
yield

> Select >

Remove

TRIAL1+LITTLE
popcorn:TRIAL1
oil amt
batch :TRIAL1
yield:TRIAL1
yield:LITTLE

LITTLE
popcorn
batch
oil
yield

Help
Done
Cancel

When you click on the **Join** button, JMP creates the data table shown in **Figure 10.26.** The **yield** column from the LITTLE data table contains missing values for rows found in the TRIAL1 data table without matching rows in the LITTLE data table.

Notice that the new table is now sorted by the matching columns. The sort is performed by JMP and is necessary for the matching to take place properly. You do not need to sort tables previously to join them with matched columns.

Figure 10.26 Joined Table with Columns

TRIAL1+LITTLE

	popcorn...	oil amt	batch ...	yield :...	yield :LITTLE
	Nom	Nom	Nom	Int	Int
1	gourmet	little	large	8.6	8.2
2	gourmet	little	small	12.1	15.9
3	gourmet	lots	large	9.2	•
4	gourmet	lots	small	18.0	•
5	plain	little	large	8.2	8.8
6	plain	little	small	9.9	10.1
7	plain	lots	large	10.4	•
8	plain	lots	small	10.6	•

Additional join options are described below.

Drop Multiples
You can specify **Drop multiples** in either or both of the data tables being joined. If you drop multiples in both tables, only the first match found is written to the new table. If you specify this option on one table, the first match value is joined with all matches in the other table.

Include Non-Matches
When you **Include non-matches** from a data table, each row from that data table is included in the new data table even if there is no matching value. Because there is no matching row in the other data table, cells for columns from the other data table have missing values for that row. You can specify this option for either or both data tables being joined.

If you do not select any options, a Cartesian join is performed within each group of matching column values.

Cartesian join

If you choose the Cartesian join, each row in the **Join** data table is joined with each row in the **With** data table. To illustrate this, suppose you want to construct the experimental design for the POPCORN data table used throughout the transform examples (see **Figure 10.32**). You can begin with the three small tables shown in **Figure 10.27**. Each table contains two rows and one column. The values are the experimental categories for the popcorn yield trials.

Figure 10.27 JMP Tables for a Cartesian Join

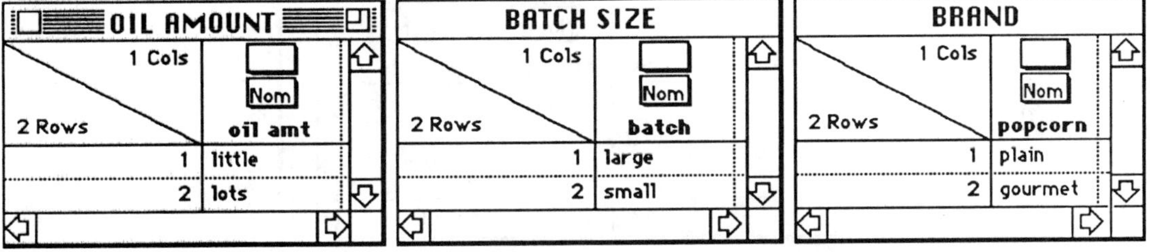

The **Join** command joins the active data table with any other open data table you choose. In this example, you must use the **Join** command twice. The first join (shown in **Figure 10.28**) combines the OIL AMOUNT data table with the BATCH SIZE data table using the Cartesian option.

Figure 10.28 Cartesian Join Dialog

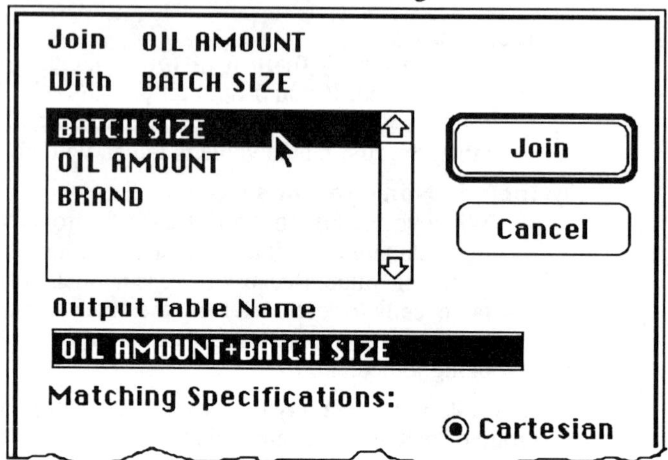

The joined data table is shown in **Figure 10.29**. The new table has each column from the original tables. Each value in the OIL AMOUNT table is paired with each value in the BATCH SIZE table, resulting in a new table with four rows.

Figure 10.29 Cartesian Join Example

OIL AMOUNT+BATCH SIZE

2 Cols / 4 Rows	oil amt (Nom)	batch (Nom)
1	little	large
2	little	small
3	lots	large
4	lots	small

The next step is to make the BRAND table active and to select the **Join** command again. The new table, OIL AMOUNT+BATCH SIZE, is now in the list of open data tables and is selected as the **With** table. Type POPCORN as the **Output Table Name**, and click on **Join** with the **Cartesian** option to produce the final data table.

Figure 10.30 Cartesian Join Dialog

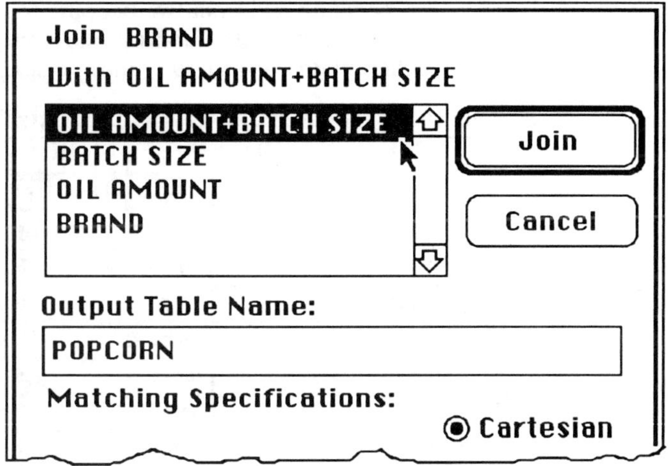

This final data table has a row for each experiment condition. The new POPCORN data table is ready for recording the results of the popping trials.

Figure 10.31 Final Table from Cartesian Join Example

POPCORN

3 Cols / 8 Rows	popcorn (Nom)	oil amt (Nom)	batch (Nom)
1	plain	little	large
2	plain	little	small
3	plain	lots	large
4	plain	lots	small
5	gourmet	little	large
6	gourmet	little	small
7	gourmet	lots	large
8	gourmet	lots	small
9			

Note➜ The number of rows produced by a Cartesian join is the product of the number of rows in the two original tables. Remember, a data table must fit in memory.

Stacking Columns

The **Stack** command creates a new data table from the active table by stacking specified columns into a single new column.

In the example below, a JMP data table called POPCORN TRIALS has two columns that contain yield from two popping trials conducted under various conditions. To complete the appropriate analysis of the data, the two columns, called **yield1** and **yield2**, must be *stacked* into a single column.

Figure 10.32 The POPCORN TRIALS Data Table

	popcorn (Nom)	oil amt (Nom)	batch (Nom)	yield1 (Int)	yield2 (Int)
1	plain	little	large	8.2	8.8
2	gourmet	little	large	8.6	8.2
3	plain	lots	large	10.4	8.8
4	gourmet	lots	large	9.2	9.8
5	plain	little	small	9.9	10.1
6	gourmet	little	small	12.1	15.9
7	plain	lots	small	10.6	7.4
8	gourmet	lots	small	18.0	16.0

The **Stack** command from the **Transform** submenu displays the dialog shown in **Figure 10.33.** To stack columns, select their names from the **Columns** selector list and click on the **Add** button. The column names appear in the **Stacking Columns** list.

This example dialog specifies that the two columns, **yield1** and **yield2**, be stacked into a single column called **yield.** You can enter any name for the new stacked column in the **Stacked Column Name** box. If you do not enter a name for the new stacked column, it is named _Stacked_ by default. If you delete the default name and do not specify a new name, the new stacked column is not written to the data table.

The **Stack** command creates a *type* column to identify each row in the new table. The value in the type column is the name of the column in the original table that contained the stacked value. In the example below, the type column is called **trial** and is assigned either the literal value 'yield1' or 'yield2'. If you do not enter a name in the **Type Column Name** box, the new type column is named _Type_ by default. If you delete the default name and do not specify a new name, the new *type* column is not written to the data table.

You can enter a name for the data table in the **Output Table** box as shown below. This example uses the default name, **Untitled.**

Figure 10.33 The Stack Dialog

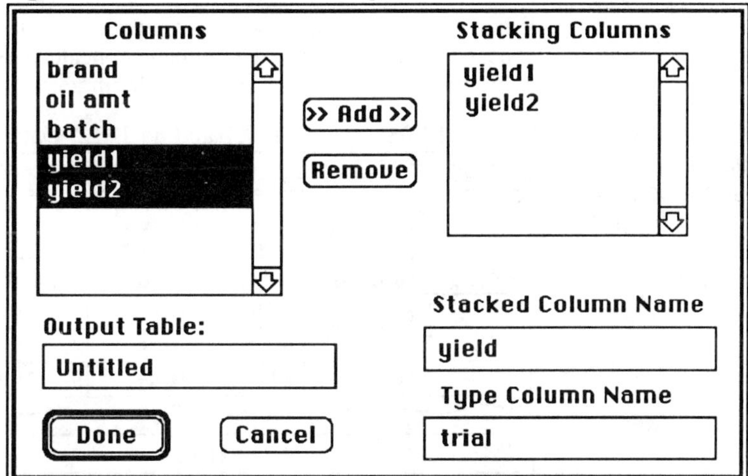

After you complete the dialog, click on **Done**. In this example, the stack operation creates the new untitled data table shown in **Figure 10.34**. This new table has twice as many rows as the original table because two columns are stacked. The values of **yield** are the experimental results, and the values in **trial** column indicate which trial each row represents.

Figure 10.34 Example of a Stacked Column

	Nom	Nom	Nom	Int	Nom
	popcorn	**oil amt**	**batch**	**yield**	**trial**
5	plain	lots	large	10.4	yield1
6	plain	lots	large	8.8	yield2
7	gourmet	lots	large	9.2	yield1
8	gourmet	lots	large	9.8	yield2
9	plain	little	small	9.9	yield1
10	plain	little	small	10.1	yield2
11	gourmet	little	small	12.1	yield1
12	gourmet	little	small	15.9	yield2

Note➜ You can stack as many columns as needed. However, the number of rows in the new stacked table is the product of the number of stacked columns and the number of rows in the original data table. Be sure you don't create a data table too large for available memory.

Concatenating Data Tables

When two or more data tables are appended end-to-end, they are said to be concatenated. The **Concat** command concatenates data tables and creates one column in the new table for each column name in the original tables.

concatenating tables with the same column names

Figure 10.35 The TRIAL1 and TRIAL2 Data Tables

TRIAL1 (5 Cols, 8 Rows)

	popcorn (Nom)	oil amt (Nom)	batch (Nom)	yield (Int)	trial (Int)
1	plain	little	large	8.2	1
2	gourmet	little	large	8.6	1
3	plain	lots	large	10.4	1

TRIAL2 (5 Cols, 8 Rows)

	yield (Int)	trial (Int)	popcorn (Nom)	oil amt (Nom)	batch (Nom)
1	8.8	2	plain	little	large
2	8.2	2	gourmet	little	large
3	8.8	2	plain	lots	large

Suppose you want to concatenate two data tables with the same column names as shown in **Figure 10.35**. The column names do not have to be ordered the same in each table. When you select the **Concat** command from the **Transform** submenu, the dialog shown in **Figure 10.36** lists all the open JMP data tables. To concatenate the data tables above, select TRIAL1 and TRIAL2 from the list on the left and click on the **Add** button.

Figure 10.36 The **Concat** dialog

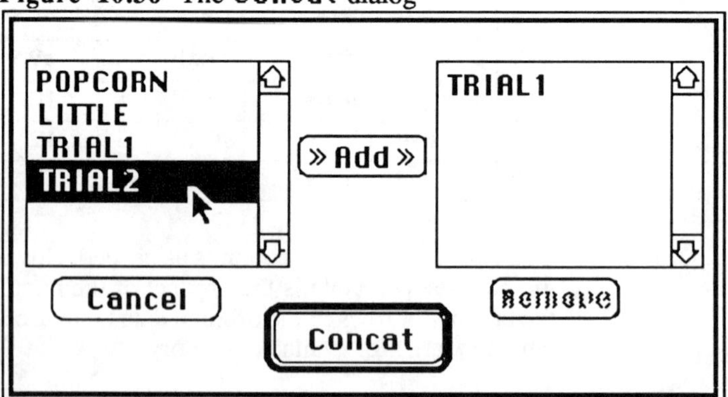

When you click on the **Concat** button, the data tables are combined into a new untitled data table as shown in **Figure 10.37**. All the rows from the first data table are followed by all of the rows from the second data table.

Figure 10.37 Concatenation of the TRIAL1 and TRIAL2 Data Tables

Untitled

5 Cols / 16 Rows	popcorn (Nom)	oil amt (Nom)	batch (Nom)	yield (Int)	trial (Int)
1	plain	little	large	8.2	1
2	gourmet	little	large	8.6	1
3	plain	lots	large	10.4	1
4	gourmet	lots	large	9.2	1
5	plain	little	small	9.9	1
6	gourmet	little	small	12.1	1
7	plain	lots	small	10.6	1
8	gourmet	lots	small	18.0	1
9	plain	little	large	8.8	2
10	gourmet	little	large	8.2	2

concatenating tables with different column names

A data table always contains a column for each column name in the original data tables. For example, the tables shown in **Figure 10.38** each have a column called **trial** and **batch**, but the new table has only one column for each of these. However, even though **yield1** and **yield2** contain similar information, both are included in the new table.

Figure 10.38 Data tables with Different Column Names

TRIAL1

batch (Nom)	yield1 (Int)	trial (Int)
large	8.2	1
large	8.6	1
large	10.4	1

TRIAL2

batch (Nom)	yield2 (Int)	trial (Int)
large	8.8	2
large	8.2	2
large	8.8	2

The concatenated data table shown in **Figure 10.39** shows columns **yield1** and **yield2**. These columns have missing values for rows originating from the table in which the column did not exist.

Figure 10.39 Concatenation When Column Names are Different

Untitled

	popcorn (Nom)	oil amt (Nom)	batch (Nom)	yield1 (Int)	trial (Int)	yield2 (Int)
1	plain	little	large	8.2	1	•
2	gourmet	little	large	8.6	1	•
3	plain	lots	large	10.4	1	•
4	gourmet	lots	large	9.2	1	•
5	plain	little	small	9.9	1	•
6	gourmet	little	small	12.1	1	•
7	plain	lots	small	10.6	1	•
8	gourmet	lots	small	18.0	1	•
9	plain	little	large	•	2	8.8
10	gourmet	little	large	•	2	8.2
11	plain	lots	large	•	2	8.8
12	gourmet	lots	large	•	2	9.8
13	plain	little	small	•	2	10.1
14	gourmet	little	small	•	2	15.9
15	plain	lots	small	•	2	7.4
16	gourmet	lots	small	•	2	16.0

Note➜ You can concatenate as many tables as needed. However, the number of rows in the new table is the total number of rows in all the tables. Be sure you don't create a data table too large for available memory.

Chapter 11.
Using the Calculator

The JMP calculator is a powerful tool for building formulas that calculate column values. JMP formulas can use information from existing columns in the data table, built-in JMP functions, and constants. Formulas can be simple assignments of numeric, character, or row state constants or complex evaluations based on conditional clauses.

When you create a formula for a column, that formula becomes an integral part of the data table. The formula is stored as part of a column's information when you save the data table, and it is retrieved when you reopen the data table. You can examine or change a column's formula at any time by opening the calculator window.

A column whose values are computed by a formula is both linked and locked. It is linked to (or dependent on) all other columns that are part of its formula. Its values are automatically recomputed whenever you edit the values in these columns. It is also locked so that it cannot be edited, which would invalidate its formula.

The calculator window operates like a pocket calculator with buttons, displays, and an extensive list of easy-to-use features. In this chapter, there is a reference section with examples for all calculator features. Additional sections describe how to construct formulas, how to correct mistakes, and how to interpret alert and error messages.

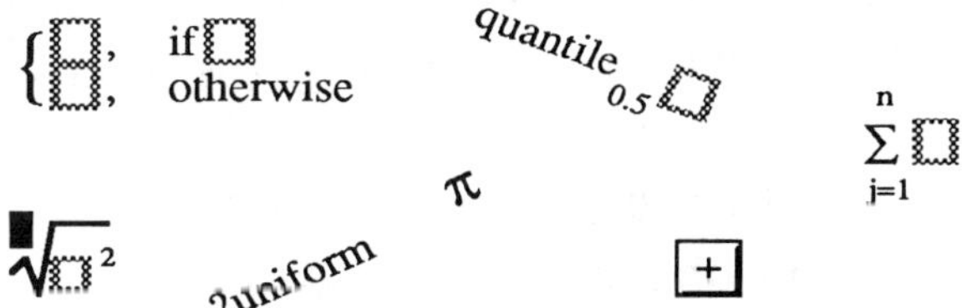

Chapter Contents

A Quick Example

The following example is designed to give you a quick look at the basic features of the calculator. If you prefer a hands on approach to learning software, you can review this example and then explore the calculator on your own. The keypad and function browser definitions in this chapter serve as a reference when needed.

For this example, open the JMP file called STUDENTS (in the SAMPLE DATA folder) which contains a column called **weight**. To begin, create a new column to contain standardized weight values by selecting **New Column** from the **Columns** menu as shown in **Figure 11.1**.

Figure 11.1 The Columns Menu

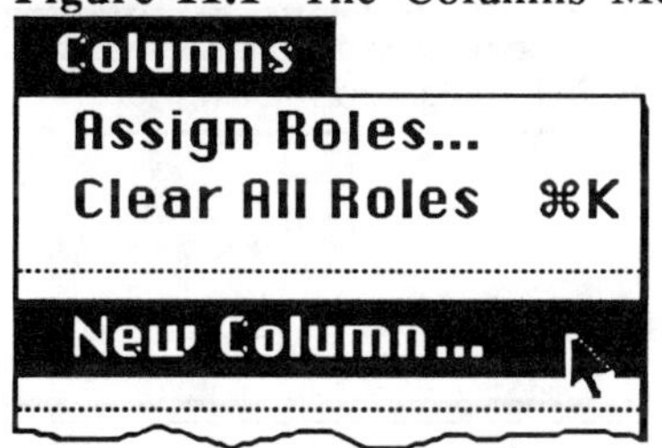

The dialog shown in **Figure 11.2** lets you set the new column's characteristics. Type the new name, **Std. Weight**, in the **Col Name** area and click on **Use Formula**. By default, the other column characteristics are set for a numeric interval variable that is correct for this example.

Figure 11.2 The New Column Dialog

Table Name: Untitled
Col Name: Std. Weight ☒ Lock
Data Type: ◉ Numeric ○ Character ○ Row State
Field Width: 10 Format: ◉ Best ○ Fixed Dec
Measurement Type: ◉ Interval ○ Ordinal ○ Nominal
Notes:
☒ Use Formula
OK Cancel

When you click on the **OK** button in the **New Column** dialog, the calculator window opens as shown in **Figure 11.3.**

Figure 11.3 The Calculator Window

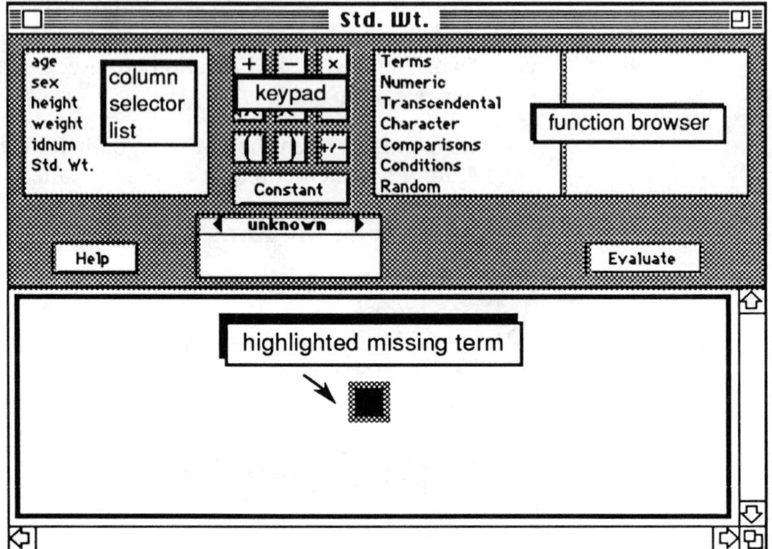

Next, enter the formula that standardizes the weight values by following the steps below.

1. While the initial missing term is highlighted, click on the column named weight in the column selector list.

   weight

2. Click on the minus button in the keypad.

   weight – ■

3. While the new missing term is highlighted, click on weight again.

   weight–*weight*

4. Click in the function browser topics list and scroll down the left side to locate Statistical functions. Click on this topic to see a list of statistical functions on the right half of the function browser. Click on Mean in this list.

   weight–$\overline{weight}$

5. Click on the right parenthesis in the keypad to highlight the entire expression.

$\overline{weight-weight}$

Note➡ Step #5 is important because the currently highlighted expression becomes the argument for the next function you choose. The argument is grouped with that function such that the function cannot be highlighted separately.

6. While the entire expression is highlighted, click on the division button in the keypad.

$$\frac{(weight-\overline{weight})}{\blacksquare}$$

7. Choose weight again from the column selector list.

$$\frac{(weight-\overline{weight})}{weight}$$

8. While weight is still highlighted in the denominator, choose Std. Deviation from the right half of the function browser.

$$\frac{(weight-\overline{weight})}{\text{std } weight}$$

You have now entered your first formula. Close the calculator window to see the column fill with values. If you change any of the weight values, the calculated Std. Wt. values change automatically.

What if you made a mistake?

If you make a mistake entering a formula, first try choosing **Undo** from the **Edit** menu. **Undo** negates the effect of the last command. There are other editing commands available to help you modify formulas including **Cut**, **Copy**, and **Paste**, and you can use the delete key to remove selected expressions.

If you need to cut or delete a function, you can save the arguments that are grouped with it. Use the **Copy** command to temporarily copy arguments to the clipboard or scrapbook, or delete all but one argument and then *peel* the function from it. To peel, first highlight the function with the mouse and then type option-delete. If you prefer to use menus, you can choose **Clear** from the **Edit** menu in place of pressing the delete key.

The Calculator Window

You can open the calculator window by clicking on **Use Formula** and **OK** in a **New Column** dialog or by clicking in the formula display field of a **Column Info** dialog. The calculator window displays as shown in **Figure 11.4**.

The two main areas of the calculator window are called the *calculator panel* and the *formula display*.

- The calculator panel is composed of buttons, selector lists, and a constant entry field.
- The formula display is an editing area you use to construct and modify formulas.

Figure 11.4 The Calculator Window

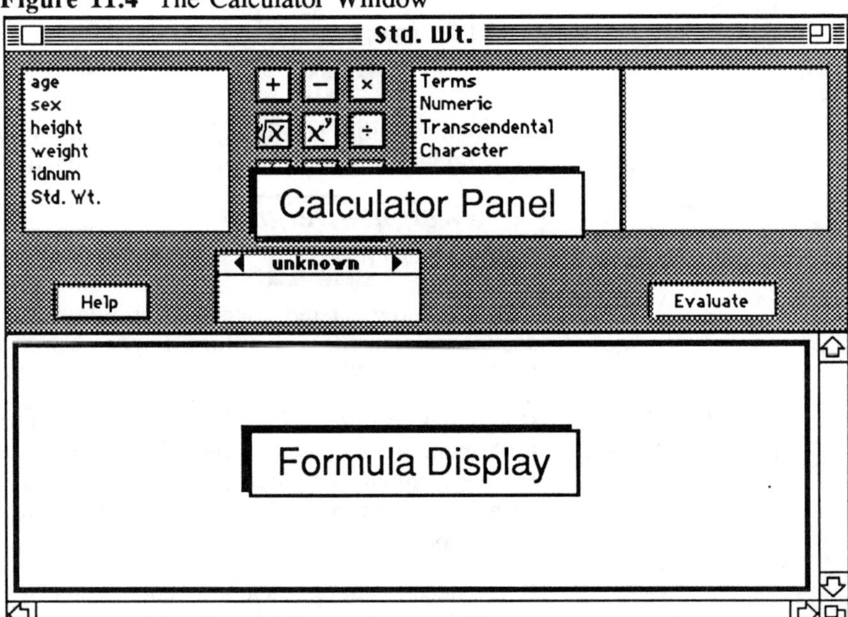

Calculator Terminology

The following list is a glossary of terminology used throughout this chapter. Each italicized word within a definition is also defined.

argument
: A constant or variable *expression* (including mathematical operands) that is operated on by a *function*.

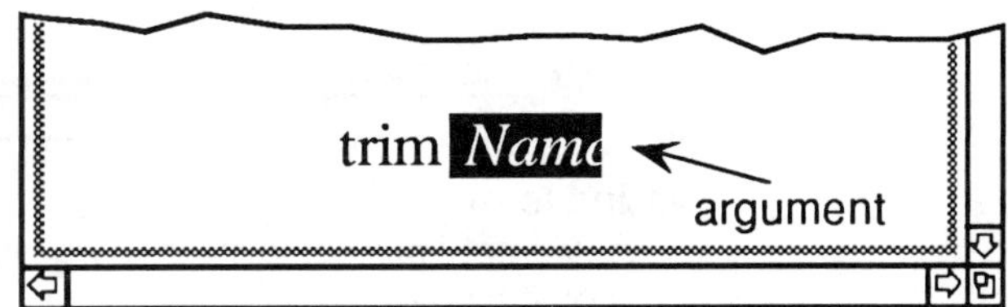

term
: An indivisible part of an expression. Constants and variables are terms.

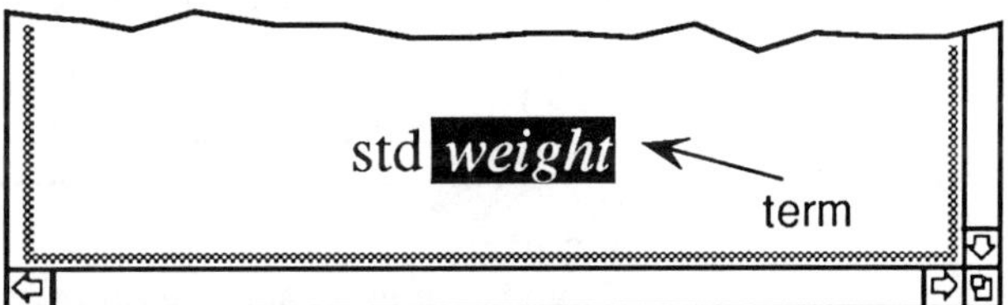

expression
: A formula or any part of it that can be highlighted as a single unit, including *terms*, *missing terms*, and *functions* grouped with their *arguments*.

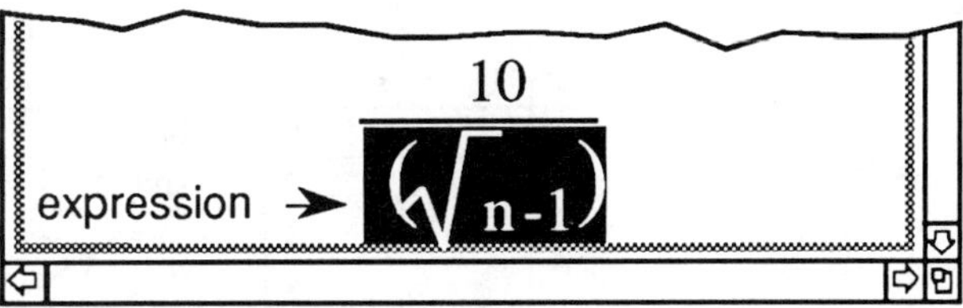

function
: A mathematical or logical operation that performs a specific action on one or more *arguments*. A function always produces a unique result. Functions include most items in the function browser and all keypad operators.

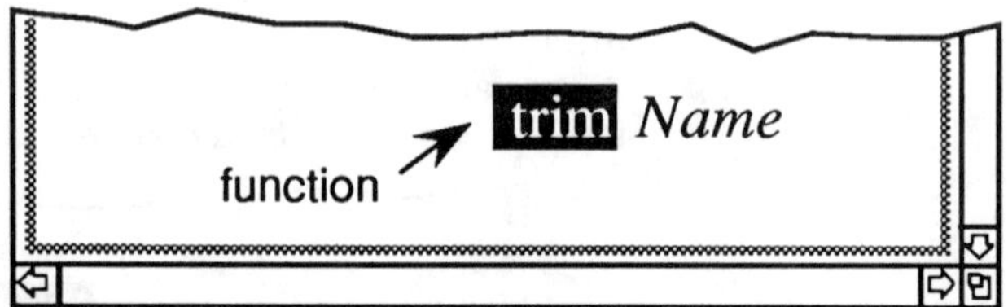

missing term
: Any empty placeholder for an *expression*, represented by the small box shown below.

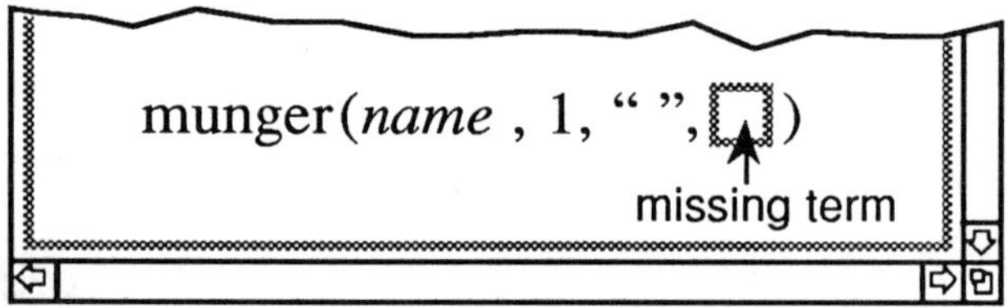

missing value
: Excluded or null data consisting of the missing value mark (•), numeric infinity (∞, -∞), or null character strings. Numeric infinity usually results from invalid computations such as dividing by zero.

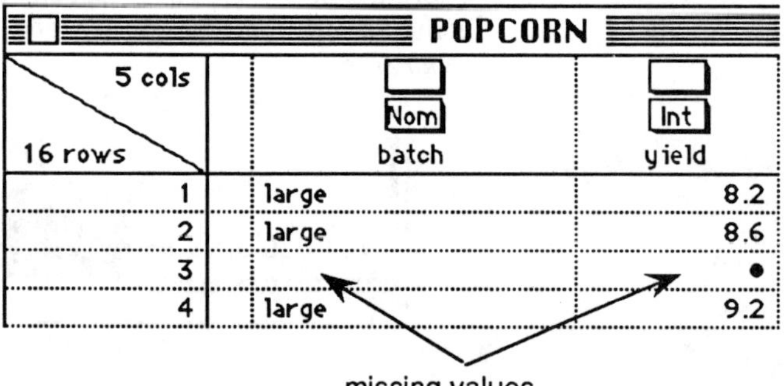

The Calculator Panel

The calculator panel is composed of buttons and selection lists as shown in **Figure 11.5**. Each component of the calculator is described in the following paragraphs.

Figure 11.5 The Calculator Panel

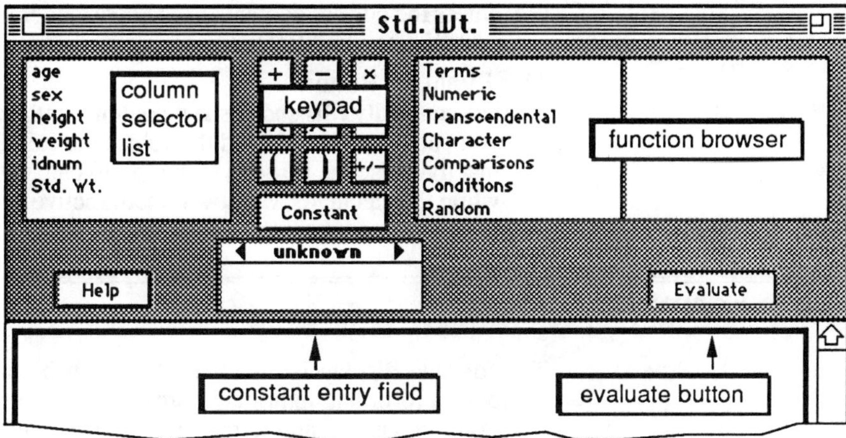

The Column Selector List
: displays all columns from the current data table. To choose a column, highlight an expression in the formula and click on the appropriate column name.

The Keypad
: is a set of buttons for commonly used calculator functions.

The Function Browser
: groups collections of calculator functions and features in lists organized by topic. To enter a function in a formula, highlight an expression and click on any item in one of the function browser topics.

The Help Button
: displays a help window that provides information to help you use the calculator.

The Constant Entry Field
: is an editable field used to enter any type of constant. The bar above this field identifies the current data type, and the triangles on either side of the bar scroll to other data types. You can automatically switch the data type to character by preceding your entry with a quotation mark.

To enter a constant, highlight an expression in the formula and scroll to the appropriate data type. The correct data type often displays automatically based on the data type of the highlighted expression in the formula display. After entering the constant, press the **Constant** button or type the return key to enter the value in the formula.

If you specify a constant as row state data, the default row state constant (the black dot) appears in the constant entry field. This field then accesses a pop-up menu containing options for all the row states. You can use the pop-up menu to modify the row state constant.

The Constant Button
: enters the current contents of the constant entry field in the formula, replacing the highlighted expression.

The Evaluate Button
: automatically fills spreadsheet columns with calculated values whenever you change a formula and then close the calculator window or make it inactive. Use the **Evaluate** button to calculate a column's values while the calculator window remains active.

The Formula Display

The formula display is the area where you build and view a formula. To compose a formula, highlight expressions in the formula display and apply functions, constants, and terms from the calculator panel. Functions always operate upon highlighted expressions; constants and terms always replace highlighted expressions.

In the formula display, arguments are always grouped with functions. Highlight a function using the mouse to verify which expressions serve as its arguments. These groupings also indicate how order of precedence rules apply and show which arguments will be deleted if you delete a function.

Keypad Functions

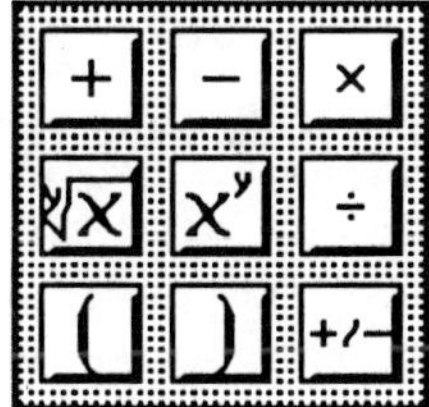

The keypad is composed of common operators (referred to as functions in all calculator documentation). Enter a keypad function by highlighting an expression in the formula display and clicking on the appropriate keypad button.

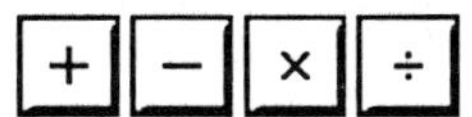

The four arithmetic functions shown at left work as they normally do on a pocket calculator. The example below shows the addition function.

column 1 + *column 2*

The radical function calculates the specified root of a radicand. It has an implied index of 2 (a square root), which is not displayed. The index area is highlighted so that you can enter a different value as shown below.

The exponential function raises a given value to a specified power. It has an exponent of two by default. The power is also highlighted and can be changed to another value as shown below.

The left parenthesis key applies parentheses around a highlighted expression. The applied parentheses are a special function that controls the order in which an expression is evaluated. The example below includes parentheses to force the addition of *column 1* and *column 2* before multiplication.

column 1 • (*column 1* + *column 2*)

)

The right parenthesis key highlights an expression, starting with the current expression and expanding outward to the next parentheses each time the key is clicked. If no parentheses exist further outward, the right parentheses highlights the entire formula.

In the following example, *column 1* is highlighted and then the right parenthesis is clicked twice.

$$\left(\frac{column\ 1 \cdot (column\ 1 + column\ 2)}{10}\right)$$

$$\left(\frac{column\ 1 \cdot (column\ 1 + column\ 2)}{10}\right)$$

$$\left(\frac{column\ 1 \cdot (column\ 1 + column\ 2)}{10}\right)$$

+/-

The unary function inverts the sign of its argument. Apply the function to variable expressions or use it to enter negative constants as shown below.

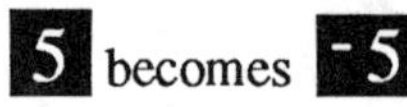

Function Browser Definitions

Terms

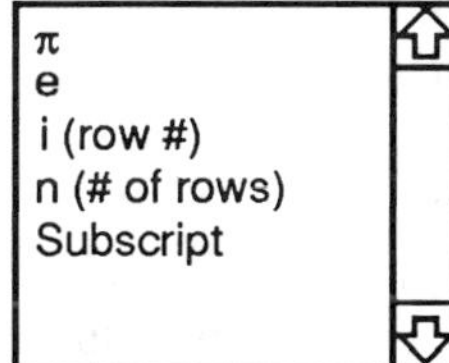

Terms

The Terms selector list consists of terms commonly used in formulas. To select from this list, highlight an expression and click on a term. Each term is described below.

π is the numeric constant 3.14159265....

e is the numeric constant 2.718... that is used as the base of natural logarithms.

i(row#) is the current row number when an expression is evaluated for that row. You can incorporate i in any expression, including those used as column name subscripts. The default subscript of a column name is i unless otherwise specified.

n (# of rows) is the total number of rows in the active data table.

Subscript identifies a specific row in a column. Highlight a column name in the formula display and click on subscript to display the column's default subscript i. The i is highlighted and can be changed to any numeric expression. Subscripts which evaluate to nonexistent row numbers produce missing values. Any column name without a subscript is interpreted as $Column_i$.

The formula in **Figure 11.6** uses subscripts to calculate the difference between each pair of values from the column named **count**. When i is 1, the computation produces a missing value.

Figure 11.6 Example of Using Subscripts

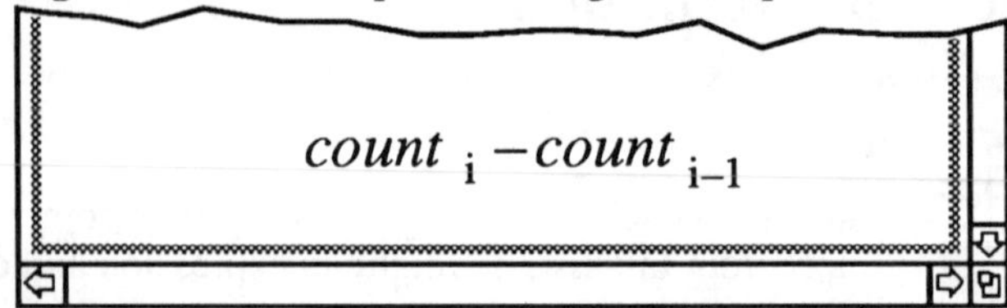

The formula in **Figure 11.7** calculates a column called **fib**. It shows the use of subscripts to recursively calculate the *Fibonacci* series (each value is the sum of the two preceding values in the calculated column). A recursive formula is one that includes the name of the calculated column, subscripted such that it only references evaluated rows (rows 1 through i-1).

Figure 11.7 Example of Recursion

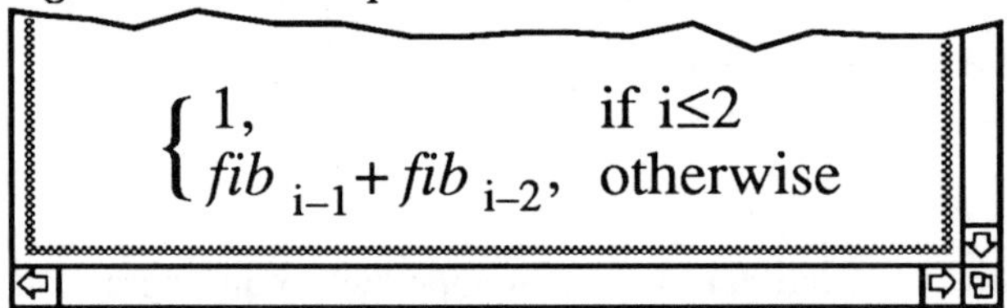

The calculation of the Fibonacci series shown above includes a conditional expression and a comparison. See the sections called **Conditions** and **Comparisons** for more information about these function browser topics.

Numeric

Floor
Ceiling
Modulo
Absolute Value

Numeric Functions

Numeric functions require numeric values as arguments and calculate numeric values as results. Each numeric function is described below.

$\lfloor \square \rfloor$

The Floor function returns the largest integer less than or equal to its argument.

For example,

$$\lfloor 2.7 \rfloor$$

results in 2.

$\lceil \square \rceil$

The Ceiling function returns the smallest integer greater than or equal to its argument.

For example,

$$\lceil 2.3 \rceil$$

results in 3.

$\square \bmod \square$

The Modulo function returns the the remainder when the second argument is divided into the first.

For example,

$$6 \bmod 5$$

results in 1.

$|\square|$

The Absolute Value function returns a positive number of the same magnitude as the value of its argument.

For example, both

$$|5|_{\text{and}}|-5|$$

result in 5.

Transcendental

Sine
Cosine
Tangent
Natural Log
Common Log

Transcendental Functions

The calculator supports logarithmic functions for any base and selected trigonometric functions. Each is described below.

$\sin \square$

The Sine function calculates the sine of an argument given in radians.

The expression, $\sin 0$, evaluates as 0.

$\cos \square$

The Cosine function calculates the cosine of an argument given in radians.

The expression, $\cos 0$, evaluates as 1.

$\tan \square$

The Tangent function calculates the tangent of an argument given in radians.

The expression, $\tan\left(\frac{\pi}{4}\right)$, evaluates as 1.

$\ln \square$

The Natural Log function calculates the natural logarithm (base e) of its argument.

The expression, $\ln e$, evaluates as 1.

$\log_{10} \square$

The Common Log function calculates a common logarithm (base 10). You can change the default log base from 10 to another value using the constant field. Both log arguments can be numeric expressions.

The expression, $\log_2 32$, evaluates as 5.

Character

- Munger
- Concatenate
- Trim
- Length
- Num to Char
- Char to Num

Character Functions

Character functions accept character arguments or return character strings or both. The character functions are described below.

munger

Munger can compute new character strings from existing strings by inserting or deleting characters. It can also produce substrings, calculate indices, and perform other tasks depending on how you specify its arguments. Munger's four arguments are the *search string*, the *offset*, the *find value*, and the *replace string* as shown in **Figure 11.8**.

Figure 11.8 The Munger Function

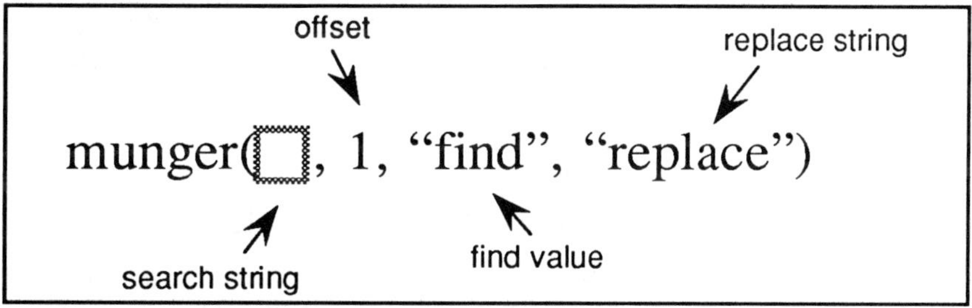

Search String
: is an expression that evaluates to a character constant. Munger applies the other three arguments to this string to compute a result.

Offset
: is an integer representing the starting search position in the search string. If the offset is greater than the search string's length, Munger uses the string's length as the offset.

Find Value
: is a character string or an integer. Use a character string as search criterion, or use a positive integer to represent the number of consecutive characters to find. If you specify a negative integer as the Find Value, Munger searches through to the end of the string.

Note ➡ The Munger function treats upper and lower case letters as different characters.

Replace String
: The replace string can be a string or a missing term. If it is a string, Munger replaces the search criterion with the Replace String to form the result. If it is a missing term, Munger calculates either a substring

or an index depending on whether the Find Value is an integer or a string, respectively.

The following are examples of Munger formulas and their results.

The formula

munger(*name* , 1, “ ”, “, ”)

produces a string consisting of the *name* string with a comma-space pair replacing the first space in the string. For example the name, ‘Smith John’ evaluates as ‘Smith, John’.

The formula

munger(*name* , 1, “,”, “”)

produces a string consisting of the *name* string with the first comma deleted. For example, ‘Smith, John’ evaluates as ‘Smith John’.

The formula

munger(*course* , 5, -1, “”)

produces a new string from the *course* string, excluding all characters from the fifth position to the end of the string. For example, ‘Stat 102’ evaluates as ‘Stat’.

The formula

munger(*course* , 6, 3, □)

produces a three character string corresponding to three character substring that starts at the sixth position of the *course* string. For example, ‘Stat 102’ evaluates as ‘102’.

The formula

munger(*course* , 1, “ ”, □)

produces the position of the first blank in each string from the *course* column. For example, the blank in ‘History 111’ has an index of 8.

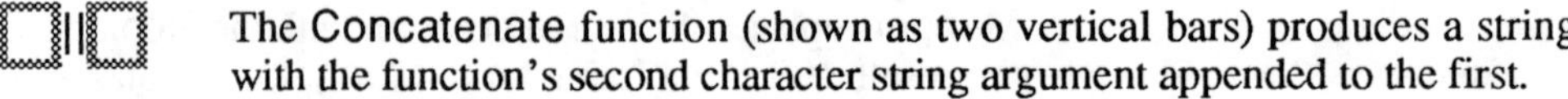

The Concatenate function (shown as two vertical bars) produces a string with the function’s second character string argument appended to the first.

For example, "Dr. "‖*name* produces a new string consisting of the title "Dr." followed by a space and the contents of the *name* string.

trim ▢

The Trim function produces a new character string from its argument, removing any trailing blanks.

For example, trim "John   " evaluates as 'John'.

length ▢

The Length function calculates the length of its argument.

For example, length "Elizabeth" evaluates as 9.

char ▢

The Num to Char function produces a character string that corresponds to the digits in its numeric argument.

For example, char 1.123 evaluates as '1.123'.

num ▢

The Char to Num function produces a numeric value that corresponds to its character string argument when the character string consists of numbers only. All other character data produce a missing value.

For example, num "1.123" evaluates as 1.123.

Comparisons

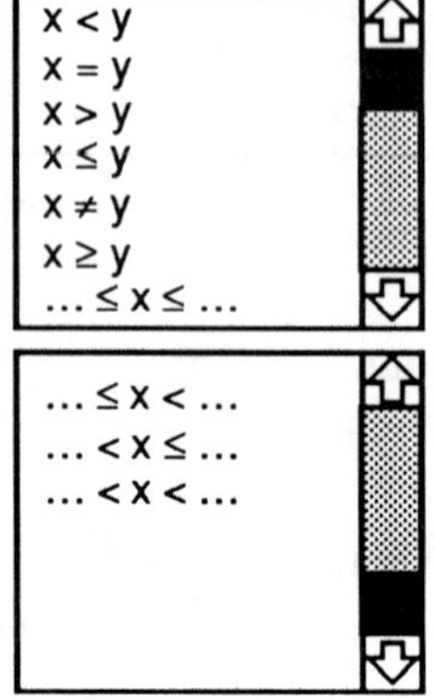

Comparisons

Comparisons comprise functions that include relational symbols comparing the ranks of their arguments. Each relationship evaluates as *true* or *false* based on the comparison of numeric magnitudes or character rankings.

Comparisons are most useful when you include them in conditional expressions, but they can also stand alone as numeric expressions. A true relationship evaluates as 1 and a false relationship evaluates as 0.

The relational symbols are

- < (less than)
- = (equal to)
- > (greater than)
- ≤ (less than or equal to)
- ≥ (greater than or equal to)
- ≠ (not equal to)

A Comparison's arguments can be any expressions. However, all arguments being compared must be of the same data type. A missing term can replace any data type and is considered less than all other values including numeric negative infinity.

The calculator uses the international utilities package when comparing character strings. These resources contain different rankings for each international character set and take diacritical marks into consideration.

Often, you can express two comparisons as one range comparison. A range comparison has three arguments and evaluates as true whenever the specified variable falls within the designated interval. For example, the expression

$$100 < weight < 150$$

evaluates as true whenever a value in the weight column falls between 100 and 150. This comparison is a more efficient way of expressing

$$(100 < weight) \text{ and } (weight < 150)$$

The function browser shows relational symbols comparing *x's* to *y's*. To enter a comparison into a formula, highlight an expression in the formula and click on the appropriate relationship from the comparison list. The comparison appears in the formula display, and your highlighted expression replaces the *x* in a comparison or the lower bound (the first ellipsis) in a range comparison. Missing terms appear as place holders for the other arguments.

Conditions

New Condition
Add Clause
Delete Clause
Not
And
Or

Conditions

Items in the Conditions list let you build a sequence of clauses paired with *result expressions* as shown in **Figure 11.9**. Constructing a sequence of clauses is the way you *conditionally* assign values to cells in a calculated column. The calculator searches down from the top of the sequence for the first true if-clause and evaluates the corresponding result expression. Subsequent true if-clauses are ignored. For maximum efficiency, it is best to list the most frequently evaluated if-clause/result pairs first in the sequence. When no if-clause is true, the calculator evaluates the result expression that accompanies the *otherwise* clause.

Figure 11.9 Conditional Expression with 3 if-clauses

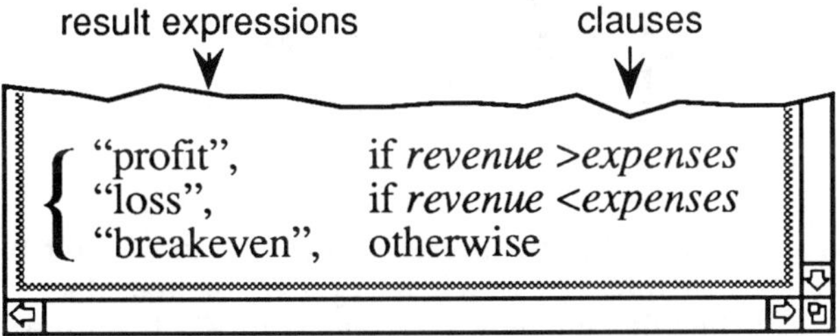

An if-clause is usually a comparison. However, any expression that evaluates as a numeric value can be an if-clause.

Note ➡ By definition, expressions that evaluate as zero or missing are *false*. All other numeric expressions are *true*. All false expressions evaluate as 0 and all true expressions evaluate as 1 when used as numeric expressions.

All result expressions in a conditional expression must evaluate to the same data type. A missing term matches any data type. When a missing term is evaluated, it produces a missing value.

To construct a conditional expression, first use New Condition. Then use the other features in the conditions list to edit and expand the conditional expression. Each feature is described below.

New Condition

When you highlight an expression and click on New Condition, the calculator creates a new conditional expression with one if-clause and one otherwise clause as shown in **Figure 11.10**. The highlighted expression becomes the first result expression.

Figure 11.10 A New Conditional Expression

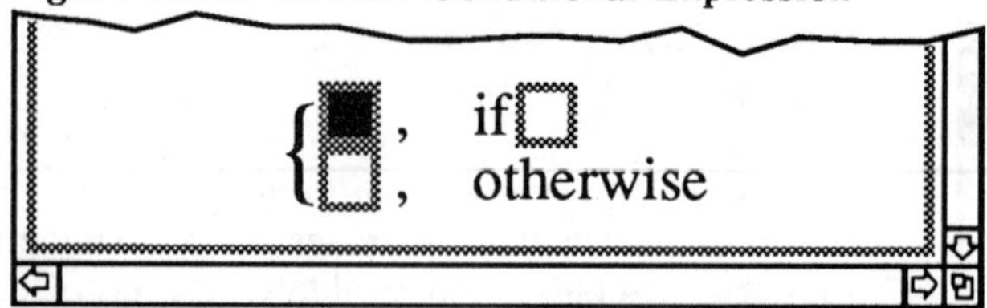

To enter a conditional expression, fill the three missing terms with expressions.

For example, to calculate *count* as a percentage of *total* when *total* is not 0, enter the conditional expression

$$\begin{cases} \frac{count}{total} \cdot 100, & \text{if } total \neq 0 \\ 0, & \text{otherwise} \end{cases}.$$

Add Clause

Add Clause inserts a new if-clause/result pair immediately before an existing clause when any part of that clause or its result expression is highlighted. However, when you Add Clause, the highlighted area cannot extend beyond a single clause or result expression.

When you click on Add Clause, the expression

$$\begin{cases} \frac{count}{total} \cdot 100, & \text{if } total \neq 0 \\ 0, & \text{otherwise} \end{cases}$$

becomes

$$\begin{cases} \frac{count}{total} \cdot 100, & \text{if } total \neq 0 \\ \square, & \text{if } \square \\ 0, & \text{otherwise} \end{cases}.$$

Delete Clause

Delete Clause deletes an if-clause/result pair when any part of the clause or its result expression is highlighted. However, when you use Delete Clause, the highlighted area cannot extend beyond a single clause or result expression.

When you click on Delete Clause, the expression

$$\begin{cases} \frac{count}{total} \cdot 100, & \text{if } total \neq 0 \\ \blacksquare, & \text{if } count > total \\ 0, & \text{otherwise} \end{cases}$$

becomes

$$\begin{cases} \frac{count}{total} \cdot 100, & \text{if } total \neq 0 \\ \boxed{0} & \text{otherwise} \end{cases}$$

not $\square$

The Not function is usually part of a conditional expression. When its argument is false, the Not function evaluates as true. Otherwise, the Not function evaluates as false.

When you apply the Not function, use parentheses where necessary to avoid ambiguity. For example, not $(weight = 64)$ can evaluate as either true or false, but $(\text{not } weight) = 64$ is always false. Expressions such as not $(weight = 64)$ can also be entered as $weight \neq 64$.

$\square$ **and** $\square$

The And function is usually part of a conditional expression. It evaluates as true when both of its arguments are true. Otherwise it evaluates as false. See **Figure 11.11**.

The formula

$$\begin{cases} \text{“driver”}, & \text{if } licensed = \text{“yes” and } age > 21 \\ \text{“passenger”}, & \text{otherwise} \end{cases}$$

labels participants as drivers only if both comparisons are true.

☐ or ☐

The Or function is usually part of a conditional expression. It evaluates as true when either of its arguments is true. If both of its arguments are false, then the Or expression evaluates as false. See **Figure 11.11**.

The formula

$$\begin{cases} \text{“cabin 1”}, & \text{if } \textit{instructor} = \text{“yes” or } \textit{age} > 21 \\ \text{“cabin 2”}, & \text{otherwise} \end{cases}$$

assigns instructors and all participants over 21 to cabin 1.

Figure 11.11 Evaluations of And and Or Expressions

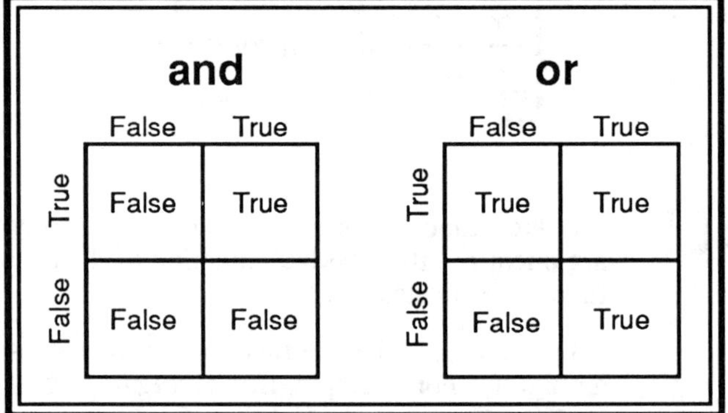

Random

Uniform
Normal
Shuffle

Random Number Functions

Random number functions generate real numbers by effectively 'rolling the dice' within the constraints of the specified distribution. The three Random number functions in JMP are Uniform, Normal, and Shuffle. Each time you press the **Evaluate** button in the calculator window, these functions produce a new set of random numbers. The Random functions are described below.

?uniform

Uniform generates random numbers uniformly between 0 and 1. This means that any number between 0 and 1 is as likely to be generated as any other. The result is an approximately even distribution. **Figure 11.12** shows a histogram of 2000 ?uniform random numbers.

Figure 11.12 Uniform Distribution of Random Numbers

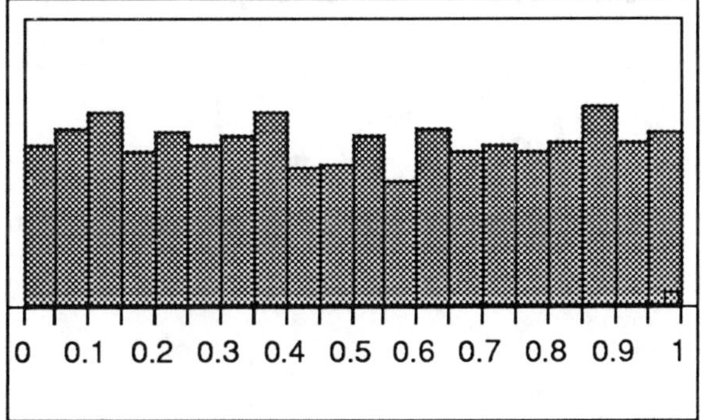

?normal

Normal generates random numbers that approximate a normal distribution with a mean of 0 and variance of 1. **Figure 11.13** shows a histogram of 2000 ?normal random numbers.

Figure 11.13 Normal Distribution of Random Numbers

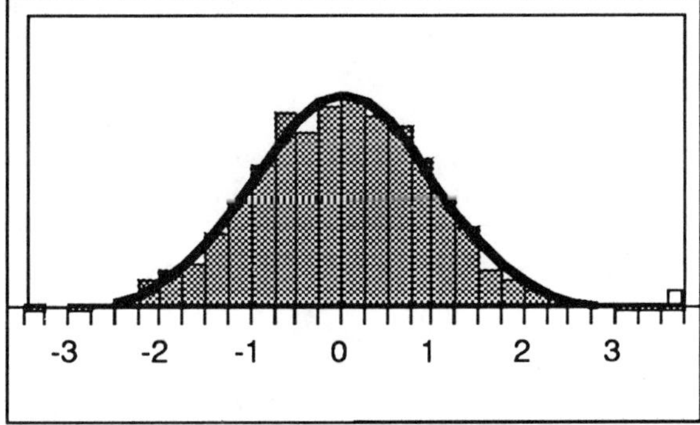

The formula

?normal•5+30

generates a random normal variable with a mean of 30 and a standard deviation of 5. The constant values can be changed to specify a different mean and standard deviation.

?shuffle

When Shuffle is used as a subscript, it returns a value selected at random from the column that serves as its argument. Each value from the original column is assigned only once as Shuffle's result. If Shuffle is not used as a column subscript, it selects a row number at random from the current data table. Each row number is selected only once.

To identify a 50% random sample, use the following formula.

$$\begin{cases} X_{\text{?shuffle}}, & \text{if } i \le \frac{n}{2} \\ \square, & \text{otherwise} \end{cases}.$$

This formula chooses half the values (n/2) from the column X and assigns them to the first half of the rows in the computed column. The remaining rows of the computed column fill with missing values.

Statistical

# of non-missing
Quantile
Mean
Std Deviation
Sum
Product
Index

Statistical Functions

The calculator evaluates Statistical Functions differently from other functions. Most functions evaluate data for the current row only. However, all statistical functions require a set of values upon which to operate.

Note ➡ The Sum and Product functions always evaluate for an explicit range of column values. All other statistical functions *always* evaluate for i = 1 to n non-missing values *on every row*.

Except for the # of Non-missing function, statistical functions apply only to numeric data. The calculator excludes missing numeric values from its statistical calculations.

Each statistical function is described below.

N ☐ The # of Non-missing function counts the number of *non*-missing values in the column you specify. **Figure 11.14** shows how missing values are represented on the spreadsheet for each type of data.

A missing numeric value is either a missing value mark (•) or the result of an invalid operation (such as division by zero) that returns positive or negative infinity (∞,-∞). Missing character values are null character strings.

All cells of a row are considered missing values whenever the exclude row state is active for that row (active row states display to the left of row numbers). The calculator interprets other missing values according to their data types. In row state columns, the excluded row state characteristic is treated as missing values when used in formulas.

Figure 11.14 Spreadsheet Representation of Missing Values

MISSINGS

5 Cols / 40 Rows	sex (Nom)	ht (Int)	wt (Int)	ratio (Int)	markers
1	F	59	95	1.61	▪
2	F	•	123	•	▪
3	F	55	•	•	▪
4		0	45	∞	▪ ⊘
5	F	52	64	1.23	▪
6	M	60	84	1.40	▪

missing character value — missing numeric values — missing row state value

quantile$_p$ ☐

The Quantile function computes the value at which a specific percentage of the values is less than or equal to that value. For example, the value calculated as the 50% quantile, also called the median, is greater than or equal to less than the data and below half the data.

The Quantile function's subscript, p, represents the quantile percentage divided by 100. The 25% quantile, also called the lower quartile, corresponds to $p = .25$, and the 75% quantile, called the upper quartile, corresponds to $p = .75$. The default value of p is .5 (the median).

The calculator computes a quantile for a column of N non-missing values by first assigning each value a rank. The rank is determined by arranging the values in descending order. The subscripts, $y_1, y_2, \ldots, y_N$, represent the ranks of each value.

The pth quantile value is calculated by rank using the formula $p \cdot (i/(N+1))$ where p is the quantile, i is the ith rank, and N is the total number of non-missing values. If the result is an integer, then the value corresponding to that rank is the quantile value. If the result is not an integer, then the value is interpolated by assigning the integer part of the result to I, the fractional part to f, and applying the formula

$$q_p = (1-f)y_I + (f)y_{I+1} \quad .$$

For example, suppose a column has values 2, 4, 6, 8, 10, 12, 14, 16, 18, and 20. The 50% quantile is calculated as

$$.5 * (10+1) = 5.5 \quad .$$

Because the result is fractional, the 50% quantile value is interpolated as

$$(1-.5) * 12 + (.5) * 10 = (.5)12 + (.5)10 = 6 + 5 = 11 \quad .$$

The following are example Quantile formulas.

- $\text{quantile}_1 \, age$ calculates the maximum *age*.
- $\text{quantile}_{0.75} \, age$ calculates the upper quartile *age*.
- $\text{quantile}_{0.5} \, age$ calculates the median *age*.
- $\text{quantile}_{0.25} \, age$ calculates the lower quartile *age*.
- $\text{quantile}_0 \, age$ calculates the minimum *age*.

The formula

$$\text{quantile}_{\left(\frac{(i-1)}{(n-1)}\right)} age$$

lists the values of age in ascending order in the calculated column.

The formula

$$\text{quantile}_{\left(\frac{(n-i)}{(n-1)}\right)} age$$

lists the values of age in descending order in the calculated column.

The Mean function calculates the mean (or arithmetic average) of the numeric values identified by its argument. The mean of a variable, y, is denoted $\overline{y}$ and is computed

$$\overline{y} = \frac{\sum_{i=1}^{n} y_i}{N} \quad .$$

The formula, $\overline{age}$, calculates the average of all non-missing values in the *age* column.

std ☐

The Std. Deviation function measures the spread around the mean of the distribution identified by its argument. In the normal distribution, about 68% of the distribution is within one standard deviation of the mean, 95% of the distribution is within two standard deviations of a mean and 99 percent of the distribution is within three standard deviations of the mean. The standard deviation of a variable, y, is usually denoted s and computed

$$s = \sqrt{s^2}, \text{ where } s^2 = \frac{\sum_{i=1}^{N} (y - \bar{y})^2}{N-1} .$$

The formula, std *height*, calculates the standard deviation of the non-missing values in the column called height.

$\sum_{j=1}^{n}$ ☐

The Sum function uses the summation notation shown at left. To calculate a sum, replace the missing term with an expression containing the index variable j. Sum repeatedly evaluates the expression for j = 1, j = 2, through j = n and then adds the results together to determine the final function result.

You can replace n, the number of rows in the active spreadsheet, and the index constant, 1, with any constant value appropriate for your formula.

For example, the following expression

$$\sum_{j=1}^{i} revenue_j$$

computes the total of all revenue values for row 1 through the current row number, filling the calculated column with the cumulative totals of the *revenue* column.

Note ➜ See Index for an explanation of nesting summations.

$\prod_{j=1}^{n}$ ☐

The Product function uses the notation shown at left. To calculate a product, replace the missing term with an expression containing the index variable j. Product repeatedly evaluates the expression for j = 1, j = 2, through j = n and multiplies the results together to determine the final function result.

You can replace n, the number of rows in the active spreadsheet, and the index constant, 1, with any constant value appropriate for your formula.

For example, the expression

$$\prod_{j=1}^{i} j$$

calculates i! (each row number's factorial).

Note ➔ See Index below for an explanation of nesting products.

Index

Index works with Sum and Product, usually as a column Subscript (see Terms). The calculator matches a new index to the first summation or product index variable it finds while searching outward from the highlighted expression. Each time you choose Index while one expression remains highlighted, the calculator matches the index to the next furthest out index variable.

Sometimes, it is useful to designate an expression containing a summation or product as the argument of a summation or product function. This is called nesting. The calculator automatically assigns a unique index variable for each Σ and Π, assigning j as the left-most index and reassigning the existing indices as necessary. All corresponding index variables within the formula also adjust as necessary.

The expression

$$\sum_{j=1}^{n} \sum_{k=1}^{n} \blacksquare$$

becomes the following expression when index is applied.

$$\sum_{j=1}^{n} \sum_{k=1}^{n} k$$

Apply index again while k is highlighted to change k to j.

$$\sum_{j=1}^{n} \sum_{k=1}^{n} j$$

Since there are no further index variables, applying index again changes j back to the first variable, k.

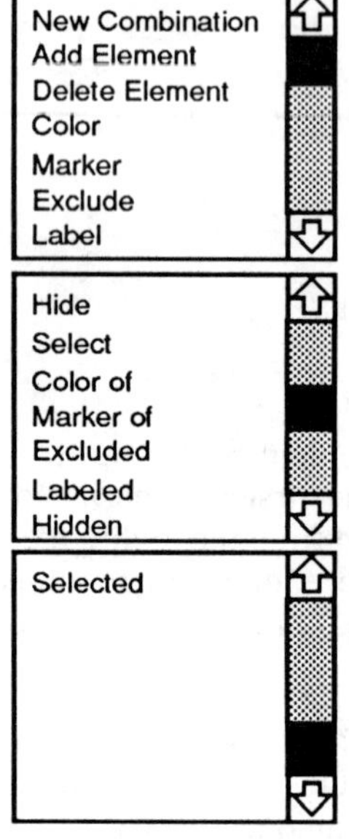

Row State

Formulas process Row State data just as they process character and numeric data. Use row state information to distinguish subsets of your data and to exclude data from analyses.

The six row states are Select, Hide, Exclude, Label, Color, and Marker. Each characteristic is described below.

- Select rows to identify data applicable to JMP commands.
- Hide rows to hide data in report displays.
- Exclude rows to eliminate data from analysis calculations and displays.
- Label rows to identify points in plots.
- Use one or more of the 15 JMP colors to mark points in report displays.
- Use one or more of the 8 JMP markers to mark points in report displays.

A row can be assigned any combination of row states. A list of multiple row states is referred to as a row state combination. Row state functions and terms are described below.

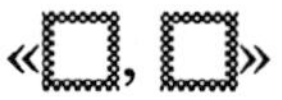

New Combination generates a row state combination with two missing terms. Replace each element with a row state term or expression that evaluates to a row state.

The formula

«select (i mod 2), label ((i+1) mod 2)»

alternately labels or selects each row in the calculated row state column. The select and label functions are defined later in this section.

Note ➡ If you include conflicting row states in a combination, the results are unpredictable.

Add Element

Add Element works with New Combination. If you need more than two elements, choose Add Element to add a missing term to your row state combination.

The expression

is a row state combination expanded to hold 3 row state expressions.

Delete Element

Delete Element removes the highlighted expression from an existing row state combination.

To remove the highlighted expression in

«color i, marker i, **label i**» ,

click on Delete Element. The new formula is

«color i, marker i» .

The color, marker and label functions are defined later in this section.

color □

The Color function returns the color from the JMP Color table (see **Table 11.1** below) that corresponds to its integer argument. JMP colors are numbered 0 through 15, but larger integers map to color indices using the formula *Index mod 16.* Both 0 and 15 map to black.

Table 11.1 JMP Color and Marker Tables

0, 15 Black	1 Grey	2 White
3 Red	4 Green	5 Blue
6 Brown	7 Blue-Green	8 Purple
9 Yellow	10 Cyan	11 Magenta
12 Yellow-Green	13 Light Blue	14 Fuchsia

0 •	1 +
2 ✖	3 □
4 ◆	5 ▲
6 Y	7 Z

The formula

color i

assigns all the row state colors in a repeating sequence to the calculated row state column.

marker □

The Marker function returns the marker from the JMP Marker table (see **Table 11.1** above) that corresponds to its integer argument. JMP markers are numbered 0 through 7, but larger integers map to marker indices using the formula *Index mod 8*.

The formula

marker i

assigns all the row state markers in a repeating sequence to the calculated row state column.

Exclude, Label, Hide, and Select

The Exclude, Label, Hide, and Select functions accept either 0 or 1 as an argument. Enter 1 to activate one of these row states or 0 to deactivate a row state.

For example,

exclude (i mod 2)

excludes all odd numbered rows in the calculated row state column.

colorOf ☐

ColorOf accepts an expression that evaluates to a number or constant and returns a color using the JMP Color table (see **Table 11.1** above). The default color index is 0 (black).

For example, colorOf [▪] is 0.

markerOf ☐

MarkerOf accepts an expression that evaluates to a number or constant and returns a marker in the JMP marker table (see **Table 11.1** above). The default marker value is 0 (the dot).

For example, markerOf [◆] is 4.

Excluded, Labeled, Hidden, and Selected

Excluded, Labeled, Hidden, and Selected accept any row state expression or constant and return a 1 if the row state is active or a 0 if the row state is inactive. These characteristics are inactive by default.

The formula

exclude (excluded *status 1* or hidden *status 1*)

excludes all rows in the row state column *status 2* if they are excluded or hidden in the row state column *status 1*.

Editing

Revert
Rotate
Reduce
Enlarge

Editing

Editing features are designed to help you build formulas with ease. Each feature is described below.

Revert discards all changes made to the formula after the data table was last saved.

Rotate swaps the first argument in a highlighted expression to the last argument's position, moving all other arguments in the expression to the left.

The expression $A + B + C$ becomes $C + A + B$ when rotated once.

Reduce decreases the type sizes in the formula display. Using Reduce, you can view a large formula without scrolling. You can also use Reduce to adjust a formula's size before copying it.

The formula

$$10.75 + \left(\begin{cases} 0.525, & \text{if } \textit{oil amt} = \text{"lots"} \\ -0.525, & \text{if } \textit{oil amt} = \text{"little"} \\ \bullet, & \text{otherwise} \end{cases} \right) + \left(\begin{cases} 1.75, & \text{if } \textit{batch} = \text{"small"} \\ -1.75, & \text{if } \textit{batch} = \text{"large"} \\ \bullet, & \text{otherwise} \end{cases} \right)$$

has been reduced several times to fit in the space above.

Enlarge increases the type sizes in the formula display. Using Enlarge, you can get a better view of the formula. You can also use Enlarge to adjust a formula's size before copying it.

The formula display in **Figure 11.15** shows part of a formula enlarged for a better view. To view more of any large formula, stretch the calculator window using the box in the lower right corner of the window or scroll through the formula using the horizontal scroll bar.

Figure 11.15 Enlarged View in Formula Display

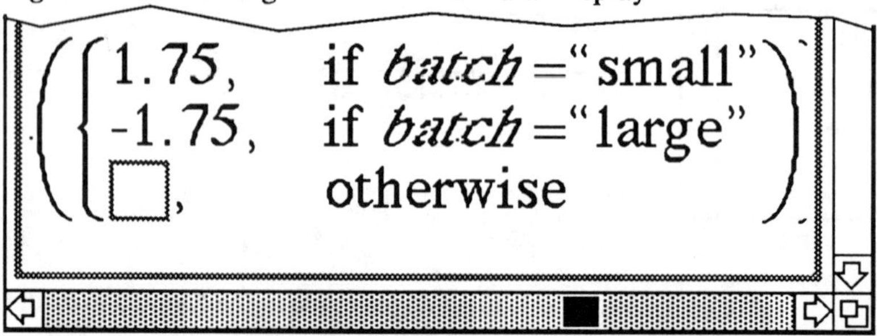

Note➡ The initial type style and size of characters in the formula display depends on the selection of fonts available in your Macintosh system. The calculator first searches for the Times font and then the New York font. If neither font is available, formulas display in your system's default application font (usually Geneva). The calculator automatically chooses the largest type size available that is smaller than 24 point.

Working with Formulas

Computational Order of Precedence

As you build a formula, keep in mind that all functions have an order of precedence defined by levels 1 through 6 as described in **Table 11.2.** Expressions with a high order of precedence are evaluated before those at lower levels. When precedences are equal, an expression is evaluated from left to right. You can use parentheses to override other precedence rules when necessary because any expression within parentheses is always evaluated first.

Terms have no order of precedence because they cannot be evaluated further.

Table 11.2 Precedence Order of Functions

Level 1:	Parentheses
Level 2:	Functions: |x|†, Ceiling†, Char to Num, Color, ColorOf, Log, Conditions†, Cos, Exclude, Excluded, Floor†, Hide, Hidden, Label, Labeled, Length, Marker, MarkerOf, Mean†, Munger, N, ln, Num to Char, Power, Π, Quantile, Root, Row State Combinations†, Select, Selected, Sin, Std. Deviation, Σ, Tan, Trim
Level 3:	*, ÷, mod
Level 4:	+, -
Level 5:	Comparisons: <, ≤, =, ≠, >, ≥, ≤ x ≤, ≤ x <, < x ≤, < x <
Level 6:	Logical Functions: And, Not, Or

† When one of these functions has an expression as its argument, the argument has a higher order of precedence as it would if enclosed in parentheses.

Building a Formula in Order of Precedence

Since functions have a high order of precedence and are grouped with their corresponding arguments, it is best to build a formula starting with any expression that serves as an argument. It is a good idea to create expressions working from highest to lowest order of precedence when possible. If you need parentheses, be sure to click on the left parenthesis in the keypad before entering the expression to be enclosed.

For example, given a data table with the columns *A, B* and *C*, use the steps below to compose the expression *A•(B+C)*. Note that this expression is not the same as *A+B•C*, which evaluates as *A+(B•C)*.

1. While the first missing term is highlighted ■, click on column A in the column selector list.

2. Click on the multiplication button in the keypad.

3. While the new missing term is highlighted, click on column B in the column selector list.

   A •B

4. While column B is highlighted, click on the left parenthesis in the keypad to apply parentheses.

   A •(B)

5. While column B is still highlighted, click on addition button in the keypad.

   A •(B +■*)*

6. While the new missing term is highlighted, click on column C in the column selector list.

Because order of precedence determines which arguments are affected by each function, it also affects the grouping of expressions. Highlight functions in the formula to verify how the order of precedence rules have been applied.

Efficiency

Usually, it is not necessary to structure formulas with efficient evaluation in mind. Most formulas evaluate almost instantaneously regardless of their structure. This is because statistical functions and constant expressions are evaluated only once when a column's values are calculated.

Constant Expressions

Once JMP has evaluated a formula, you can highlight a constant expression to see its value in the constant entry field.

Focus

When you click the mouse in the constant entry field or in a selector list on the calculator panel, the keyboard *focus* shifts to that item. For example, if you click on Character in the function browser, the first character function, Munger, automatically highlights, indicating that it is focused. Likewise, if you click in the column selector list, a column name is focused. If you click in the constant entry field or begin typing a constant value, then that constant value is focused. A focused item in the calculator panel automatically replaces the highlighted expression in the formula display if you press the return key.

When you click in the formula display area, the grey border around the formula changes to black, indicating that no item in the calculator panel is currently focused. Keystrokes and editing commands affect the formula instead.

Cutting and Pasting

You can cut or copy an entire formula or any expression and paste it into another formula display. If you paste a formula in the scrapbook, it is saved both as PICT and Expr (expression format) data. You can copy expression format data into any formula display, and you can paste PICT format information into other Macintosh applications. When you copy an expression from one data table to another, the column names in the formula change to match column names in the new spreadsheet. Each column name is reassigned to the column that matches the original column's position in the original spreadsheet.

If this produces undesirable results, you can replace all references to a column throughout a formula by holding down the option key while you change one instance of the name. The other instances of the name change automatically.

Changing a Formula

If you need to change a formula, double click on the corresponding column's selection area to bring up its **Column Info** dialog. To reopen the calculator window, click on the picture of the formula in this dialog.

When you change a formula, note that deleting a function also deletes its arguments. However, you can save complicated expressions that serve as arguments. Use the **Copy** command to temporarily copy arguments to the clipboard or scrapbook, or delete all but one argument and then *peel* the function from the remaining argument as shown in **Figure 11.16**.

To peel a function from an argument, first highlight the function with the mouse and then type option-delete. If you prefer to use menus, you can choose **Clear** from the **Edit** menu in place of pressing the delete key.

Figure 11.16 Peeling and Argument

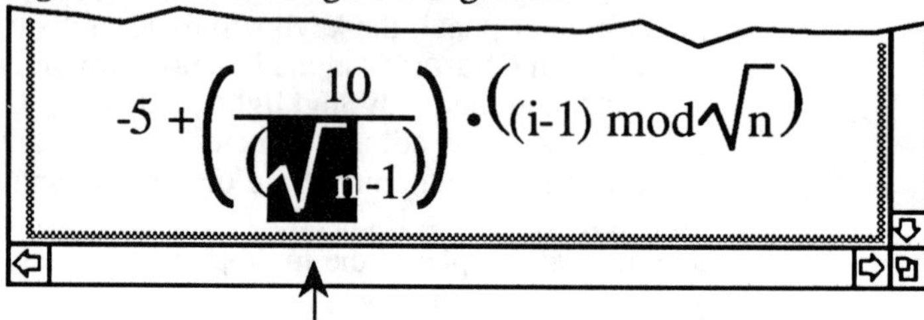

to replace $\sqrt{n}$ with n , highlight $\sqrt{n}$ and type option-delete.

After you complete the changes above, the new values fill the column automatically when you click on the **Evaluate** button or close the calculator window.

Changing Column Values

Once you have created a formula, you can change the column values that are part of your formula. The calculator automatically recalculates the values in the dependent column.

Caution and Error Messages

Error and warning messages alert you that there is an error in your formula and warn you whenever you choose a command or use a JMP feature which can have unforeseen effects on calculated data. A beep often serves as the first warning of a problem, followed by a message if the problem persists. Calculator messages include both *Caution Alerts* and *Stop Alerts* as described below.

Caution Alerts allow you to continue a specific course of action but warn you of possible problems. They typically give you a choice of response buttons.

Stop Alerts detect errors that must be corrected before processing can continue. They often inform you that an invalid action has been ignored. Stop Alerts also appear each time you attempt to close the calculator window or click in another window while an invalid formula is in the formula display. When this happens, click on the **OK** button and correct or delete the invalid formula.

All messages that relate to the calculator window are explained below.

Caution Alerts

The formula for "Column 2" depends on "Column 1." Do you want to delete "Column 1" anyway?

This message appears when you choose to delete a column in the data table that affects another column's values. In this example, the formula for *Column 2* includes a reference to column *Column 1*. If *Column 1* is deleted, all references to *Column 1* in the *Column 2* formula change to missing terms which then calculate as missing values. If other columns depend on *Column 1*, this message may appear up to three times after which there is no further warning. Click on **Cancel** to keep *Column 1* and the dependent data in *Column 2*.

Are you sure you want to revert to the last formula saved?

This message appears when you click on Revert in the Editing topic list of the function browser. Click **OK** to discard all changes made to your formula since the last time you saved the data table. Click on **No** to keep the current formula.

Stop Alerts

The data type of the selected expression is *{numeric, character, row state}*, but it should be *{numeric, character, row state}*. Please change it to the correct type.

The highlighted expression evaluates to a data type that is invalid in either the context of the expression or the column type. Replace the highlighted expression with an expression that evaluates to the required data type as specified in the message. This message can appear any time the calculator attempts to evaluate a formula or you reopen the data table.

You must select a piece of the formula to manipulate first.

No expression in the formula display is highlighted. A function or feature of the calculator panel can affect a specific expression in the formula display only when that expression is highlighted.

You must choose a column from the column list first.

No column name is highlighted in the formula display. Some functions, such as subscript, require that the currently highlighted expression be a column name.

To add a clause, select the clause that you want to precede. To delete a clause, select it first.

No result expression has been highlighted in the conditional expression. To add a clause to a conditional expression, first highlight all or part of the result expression that is to follow the new clause. Likewise, highlight all or part of a result expression in a clause before attempting to delete the clause.

There will always be an "otherwise" clause. You cannot delete it unless you delete the entire condition.

The otherwise clause is highlighted for deletion. However, the result expression within an otherwise clause is always executed when no if-clause is true and therefore cannot be deleted. Delete the entire conditional expression instead.

You can only have one non-missing argument in the selected expression. Cut or clear all of the arguments but one and try again.

There is more than one argument in the highlighted expression. If you need to delete a function but want to save one of its arguments, you can *peel* the function from the argument. To do this, change all the function's other arguments to missing terms by highlighting each argument and pressing the delete key. Peel the function by highlighting it in the formula display and typing option-delete. The above message appears if you attempt to peel a function from more than one non-missing argument.

The selected column references a row that has not been evaluated yet. The subscript must be between 1 and i-1.

A row has been referenced that is not yet evaluated. The calculator allows recursive formulas as long as they only reference a row number between 1 and i-1. Referencing any other row number is invalid. See the function browser definition of Subscript in the Terms topic list for more information about recursive formulas.

The selected column already has a subscript. To delete a subscript, just choose the column again.

Subscript (from the Terms list) has been chosen, but the highlighted column name is already subscripted. If you need to remove the subscript, highlight the column name in the formula display and choose the same column again from the column selector list. This replaces the subscripted column name with the same column name unsubscripted.

To add an element to the combination, select the element that you want to precede. To delete an element, select it first.

No row state element has been highlighted in the row state combination. To add a row state element to a combination, first highlight the element that is to follow the new element. Likewise, highlight an element before attempting to delete it.

There will always be at least one element in a row state combination. You cannot delete it unless you delete the entire combination.

The last element of a row state combination is highlighted for deletion. Delete the entire row state combination instead.

You can only use an index with the Sum (Σ) and Product (Π) functions.

Index has been chosen outside the scope of a summation or product function. Apply Index only to summations and products.

Computing Grid Coordinates

The following paragraphs guide you through the process of computing grid coordinates using the JMP calculator window. You can use the same technique to derive formulas and fill columns with grid and response data needed for your statistical analyses.

Figure 11.17 Diagram of Cartesian Grid Coordinates

X	Y	X	Y	X	Y	X	Y	X	Y	X	Y
-5	-5	-3	-5	-1	-5	1	-5	3	-5	5	-5
-5	-3	-3	-3	-1	-3	1	-3	3	-3	5	-3
-5	-1	-3	-1	-1	-1	1	-1	3	-1	5	-1
-5	1	-3	1	-1	1	1	1	3	1	5	1
-5	3	-3	3	-1	3	1	3	3	3	5	3
-5	5	-3	5	-1	5	1	5	3	5	5	5

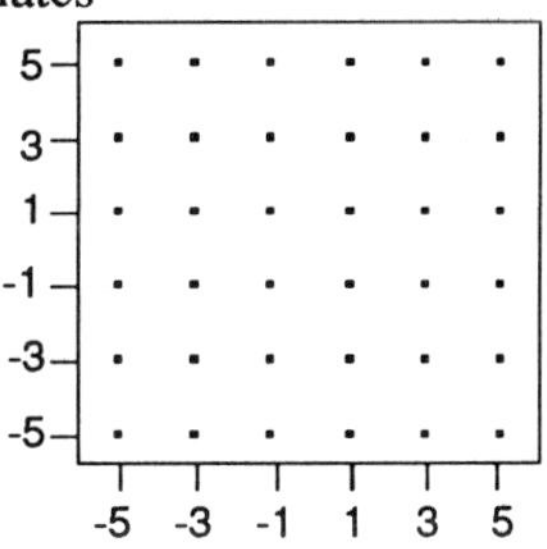

In **Figure 11.17**, the grid is a perfect square, symmetrical about both the x and y axes with points evenly spaced and one corner located at (5, 5). The following paragraphs describe how to build a similar grid with formulas that allow the grid's density to vary according to the number of rows you choose.

Creating X and Y Formulas

Begin by creating a new data table. Double click on the row selection area of the spreadsheet and enter the number of rows to add such that the total number of rows is a perfect square. In our example, there are 441 rows forming a 21 by 21 grid as shown in **Figure 11.18**.

Figure 11.18 21 by 21 Grid

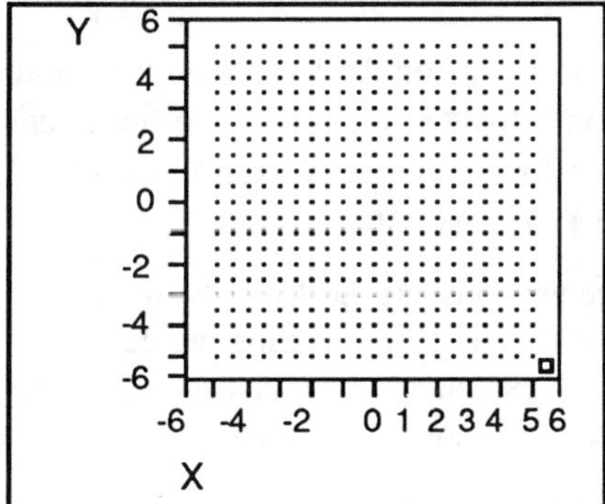

Next double click the mouse in the first column's selection area to access the **Column Info** dialog. Change the column's name to 'X' and click on *Use Formula.* The column's data type is numeric by default. Click OK. The calculator window opens ready for you to enter your first formula in the formula display as shown in **Figure 11.19.**

Figure 11.19 The Formula Display area of a New Calculator Window

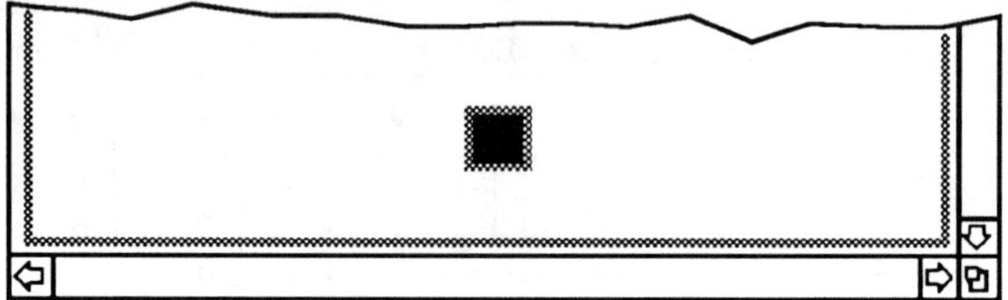

The formula for column X must fill each of the first $\sqrt{n}$ cells with $^{-}5$. Each successive value also repeats $\sqrt{n}$ times, with each group of values increasing by a uniform increment until the final value is 5. This pattern forms a square when matched with the values you will specify for the Y column.

You can think of the values of X as a function of the current row number i. The value in each row equals -5 added to the product of (i-1) and a constant that is determined by the number of rows. Choose (i-1) instead of i so that the expression evaluates to 0 in row #1.

This expression displays in calculator format as

$$-5+\square\bullet(i-1) \quad .$$

Begin entering the expression using the following steps.

1. Type the number 5 and press <Return>.
2. Click the mouse on the +/- button in the keypad to change 5 to $^{-}5$.
3. Click on (or type) the addition button (+) followed by the product button (∗).
4. Click the mouse on the left parenthesis in the keypad.
5. Click on Terms in the function browser and choose i.
6. Click on (or type) the subtraction button (–).
7. Type 1 and press <Return>.

To calculate the constant for the given number of rows, divide the difference between the starting and ending values ($5 - {}^{-}5 = 10$) by the number of additional values you need. Since each value repeats $\sqrt{n}$ times, there are $\sqrt{n}$ values in the X column. You need $\sqrt{n}$-1 additional values to fill the column.

By entering the constant as a variable expression, you can modify the number of rows without changing the formula. The variable expression is.

$$-5+\left(\frac{10}{\left(\sqrt{n}-1\right)}\right)\cdot(i-1)$$

Enter this expression into the existing formula as follows.

8. Highlight the missing term and click on the left parenthesis.
9. Type the number 10 and press <Return>.
10. Type the slash mark (/) or click on the division button (÷) in the keypad.
11. Click the mouse on the left parenthesis.
12. Click on Terms in the function browser and choose n.
13. Click the mouse on the radical button in the keypad. The index of the radical is 2 by default, but the 2 is not displayed. However, the area that represents the index is highlighted so that you can enter a different value. The default index of 2 is correct for this formula.
14. Click on the subtraction button.
15. Type 1 and press <Return>.

This expression is also used in the formula for column Y. Highlight the entire expression and choose **Copy** from the edit menu. Paste the expression into the Macintosh scrapbook for use later.

Finally, force each number to repeat $\sqrt{n}$ times by dividing $\sqrt{n}$ into the row number expression. By taking the floor of each result, the expression converts the first $\sqrt{n}$ row numbers to 0, the next group to 1 and so forth.

Click on the calculator window to make it active again and then complete the formula as described below.

16. Click on the minus sign in the expression (i-1) to highlight the expression.
17. Click on the division button.
18. Click on Terms in the function browser and choose n.
19. Click the mouse on the radical button in the keypad.
20. Click on the right parentheses to highlight the entire row number expression.
21. Click on Numeric functions in the function browser and choose Floor.

The final formula is

$$-5+\left(\frac{10}{\left(\sqrt{n}-1\right)}\right)\cdot\left[\left(\frac{(i-1)}{\sqrt{n}}\right)\right].$$

Close the calculator window to see column X fill with the correct values.

Next, double click in the column selection area to insert another column in the data table. Name the column 'Y' and click on *Use Formula* and *OK*. Column Y also starts with ⁻5 and increases in uniform increments through 5. However, the entire pattern repeats $\sqrt{n}$ times instead of each individual value.

To construct the Y formula, use the expression saved in the scrapbook from column X and modify the row number expression. The expression saved in the scrapbook is

$$-5+\left(\frac{10}{\left(\sqrt{n}-1\right)}\right)\cdot(i-1).$$

Modify (i-1) using the Modulo function so that each series of values, -5 through 5, repeats. Divide the row number expression by $\sqrt{n}$ to produce a repeating series of numbers from 0 to ($\sqrt{n}$-1).

Follow the steps below to enter the formula.

1. Copy the expression you saved from column X into the new calculator window.
2. Click on the minus sign in the expression (i-1) to highlight the expression.
3. Click on Numeric functions in the function browser and choose Modulo.
4. Click on Terms in the function browser and choose n.
5. Click the mouse on the radical button in the keypad.

The final formula is

$$-5+\left(\frac{10}{\left(\sqrt{n}-1\right)}\right)\cdot\left((i-1) \bmod \sqrt{n}\right).$$

Close the calculator window to see column Y fill with the correct values.

Figure 11.20 shows a spreadsheet that contains 441 rows. The data vary, depending on the number of rows you specify in the spreadsheet.

Figure 11.20 Grid Data Table Spreadsheet

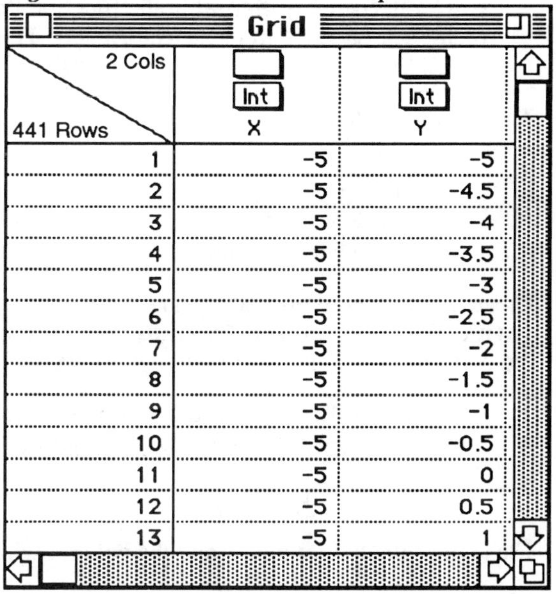

Grid

2 Cols / 441 Rows	X (Int)	Y (Int)
1	-5	-5
2	-5	-4.5
3	-5	-4
4	-5	-3.5
5	-5	-3
6	-5	-2.5
7	-5	-2
8	-5	-1.5
9	-5	-1
10	-5	-0.5
11	-5	0
12	-5	0.5
13	-5	1

Select the **Fit Y by X** command from the **Analyze** menu to view the grid.

The Cowboy Hat Data Table

The Cowboy Hat 3-dimensional image is a set of points you can plot and rotate in 3-space using the JMP Spin Analysis. To create the cowboy hat, you must form a square 2-dimensional grid symmetric about X and Y and then project it onto a z-axis using the formula

$$z = \sin\sqrt{x^2+y^2} \; .$$

To create the cowboy hat data table, start by creating the grid described in the previous section. Then create another column, 'Z', and specify its formula as follows.

1. Select column X from the column selector list.
2. Click the mouse on the exponent button in the keypad.
3. Click the mouse on the addition button in the keypad.
4. Select column Y from the column selector list.
5. Click the mouse on the exponent button in the keypad.
6. Click on the right parenthesis in the keypad to highlight the entire expression.

7. Click the mouse on the radical button in the keypad.
8. Click on the right parenthesis in the keypad to highlight the entire expression.
9. Click on Transcendental in the function browser's topics list and then click on Sine in the selector list of Transcendental functions to the right.

The final formula is

$$\sin \sqrt{(X^2+Y^2)} \; .$$

Close the Z calculator window.

Create a fourth column, setting its data type as **Row State** in the **Column Info** dialog. Specify **Use Formula** in the same manner as before. Click **OK** and follow the steps below to color the cowboy hat data points.

1. Click on Terms in the function browser and click on i.
2. Click on Row State in the function browser and choose Color.

   The formula is color i .

Close the calculator window and the column will fill with row state colors.

Figure 11.21 shows the spreadsheet view of the cowboy hat data table. The data vary depending on the number of rows. For best results choose a number of rows, n, such that 225 < n < 625.

Figure 11.21 Spreadsheet View of the Cowboy Hat Data Table

COWBOY HAT

4 Cols / 441 Rows	X (Int)	Y (Int)	Z (Int)	Colors
▪ 19	-5	4	.11965157...	▪
▪ 20	-5	4.5	.42921793...	▪
▪ 21	-5	5	.70886129...	▪
▪ 22	-4.5	-5	.42921793...	▪
▪ 23	-4.5	-4.5	.08068791...	▪
▪ 24	-4.5	-4	-.25938757...	▪
▪ 25	-4.5	-3.5	-.54995318...	▪
▪ 26	-4.5	-3	-.76745272...	▪
▪ 27	-4.5	-2.5	-.90669040...	▪
▪ 28	-4.5	-2	-.97760363...	▪

To see the data spin, follow the steps described in the last section of Chapter 1, called *Take a Quick Spin.*

Keyboard Shortcuts

You can use the mouse or the keyboard shortcuts shown in **Table 11.3** to access calculator features.

Table 11.3 Keyboard Shortcuts

Replace a highlighted expression with the function or argument shown below by typing the corresponding key command.

?normal .... <Option>-?
?uniform .... ?
i (Term) .... @
Index .... [
Missing Element .... <Delete>
n (Term) .... #
π (Term) .... <Option>–p

Apply one of the functions listed below to a highlighted expression by typing the corresponding key command.

((in the keypad) .... (
) (in the keypad) ....)
x (in the keypad) .... *
+ (in the keypad) .... +
– (in the keypad) .... -
÷ (in the keypad) .... /
And .... &
Combination, New .... <Option>–\
Condition, New .... {
Mean .... _
Not .... !
Or .... |
Power .... ^
Quantile .... %
Radical .... <Option>–v
Std Deviation .... <Option>–j
x < y .... <
x = y .... =
x > y .... >
x ≠ y .... <Option>–=
x ≤ y .... <Option>–,
x ≥ y .... <Option>–.
± .... <Option>–+
Σ .... <Option>–w
Π .... <Shift–Option>–p

Press the <Enter> key to enter a focused item from the calculator panel in a formula. For more information about calculator focus, see the section called **Working with Formulas** in this chapter.

Type <Option-r> to change the data type in the calculator's constant entry field to row state.

Use <Shift> with keyboard shortcuts that includes a shift-key character.

The arrow keys operate as keyboard shortcuts in the formula display for extending and reducing the current selection (the up and down arrows) or changing the current selection to another argument within an expression (the left and right arrows). The left and right arrows search in a loop moving in the indicated direction each time the key is pressed.

You can also use the up and down arrow keys to traverse the column selector list and the function browser.

Chapter 12.
JMP Report Windows

This chapter describes features common to all report windows in JMP. These include all the standard controls in Macintosh windows along with some special additional features.

To make the most of each analysis, all report windows have button controls, pop-up menus, and help windows. In addition, you can select from six special tools that change the function of the mouse. This puts report features literally at your fingertips.

You can customize the appearance of report windows by changing type styles, adding colors and labels to plots, and more. Once you've explored the report windows on your monitor, you can print them directly from JMP. If you want to include analyses in your own comprehensive report, you can journal report windows to word processing files or copy any part of a report window into another application.

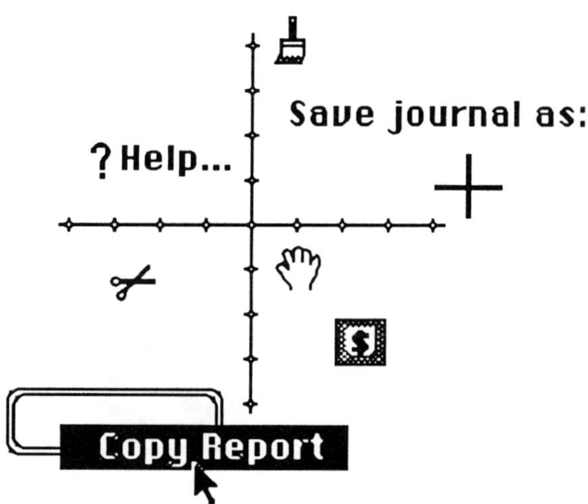

Chapter Contents

Report Windows

All analysis platforms produce report windows. These windows contain both text reports and graphical displays. They are designed to be convenient in several ways.

- Although report windows consist of complex plots and calculations, they appear on the screen almost instantaneously most of the time.
- Individual plots can be resized, and text reports can be concealed to optimize the use of space on the monitor.
- Report windows always print so that each individual plot and text report appear completely on one page if possible.
- Many areas of report windows are linked to HELP documents.

All or part of any report window can be copied to other Macintosh applications, journaled to a special file, or printed directly from JMP.

Standard Macintosh Window Features

JMP report windows include all standard Macintosh interfaces including scroll bars, a resize box, a zoom box, and a close box. Report windows also include other special buttons and pop-up menus as shown in **Figure 12.1**.

Figure 12.1 Report Window

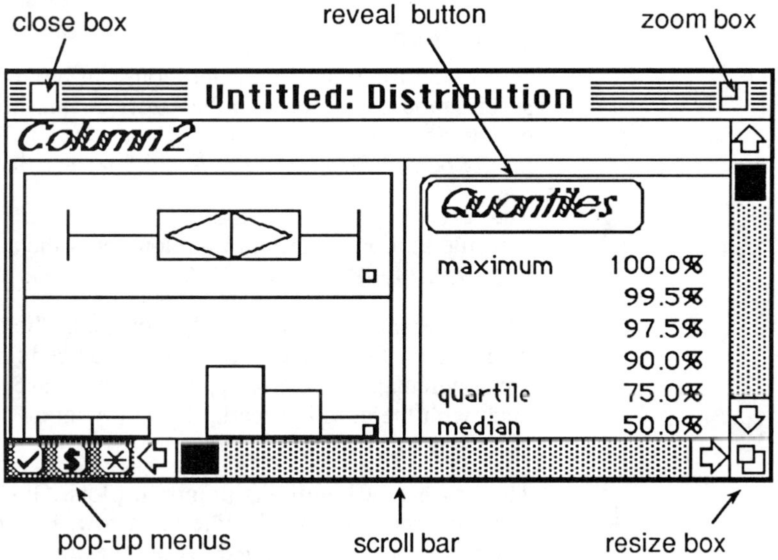

Pop-up Menu Icons

There are three special icons to the left of the horizontal scroll bar in all JMP report windows. These icons identify the three pop-up menus described below.

The check mark identifies options which are different for each report window. For example, in the **Distribution of Y's** analysis, this menu has options for fitting a normal curve to each histogram, for switching displays from vertical to horizontal orientation, and more.

The dollar sign identifies a pop-up menu for saving data generated by the current analysis. The new data are saved as additional columns in the data table that corresponds to the report. These commands differ for each analysis platform. For example, on two platforms you can save predicted values and on another you can save ranks.

The asterisk accesses a pop-up menu of options common to all data reports. These options are described below.

Help brings up a help window for the current platform. The help window has buttons you can use to bring up further help.

Reveal All opens all text reports in the active report window.

Conceal All closes all the reveal buttons, concealing all text reports in the active report window.

JMP Tools

To expand the usability of the mouse in report windows, you can choose special mouse cursors from the **Tools** menu. Mouse actions have different effects depending on which tool you choose. The six tools are described briefly below. Chapter 9, *The Menu Bar*, describes these tools in detail.

The arrow is the default tool and the standard text cursor. It is the most versatile tool. For more information about the arrow, see the section called **Highlighting Points in Plots** later in this chapter.

The question mark accesses help windows. Click and hold the mouse button on an area within a report window if you need further information. The help window remains on the screen until you release the button. When you shift-click with the question mark, the help window remains on the screen until you close it.

The brush tool highlights points in plots. Outside of a graphical display area, the brush tool works like the arrow. For more information about the brush, see the section called **Using the Brush** later in this chapter.

The hand tool (or *grabber*) directly manipulates histograms and spinning plots. On all other parts of report displays, it works like the arrow.

The scissors tool selects an area in the report window to copy.

The cross tool measures the position of points on a plot against their axes. Drag the mouse around on a plot to see a set of axes appear that follow the movement of the mouse.

Reveal Buttons

Buttons with double round-rectangle borders are called *reveal* buttons. Click on a reveal button to alternately reveal and conceal its corresponding text report as shown in **Figure 12.2.**

Figure 12.2 Concealed Text Report

Summary of Fit

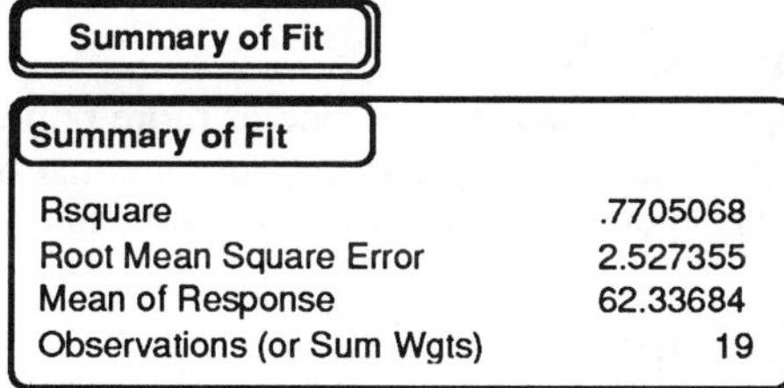

Summary of Fit

Rsquare	.7705068
Root Mean Square Error	2.527355
Mean of Response	62.33684
Observations (or Sum Wgts)	19

Sometimes reports are nested in multiple layers as shown in **Figure 12.3**. You can either click to open each layer individually or shift-click to open all layers. Shift-clicking again conceals all the layers of nested reports.

Figure 12.3 Nested Text Reports

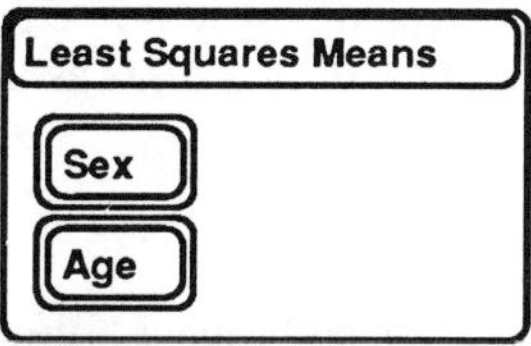

Least Squares Means

Sex

Level	Least Sq Mean	Std Error	Mean
F	61.407410941	1.1408765027	60.5889
M	64.131838422	1.0026927810	63.9100

Age

Level	Least Sq Mean	Std Error	Mean
11	54.4	2.1159568468	54.4000
12	59.167557251	1.3458885343	59.4400
13	61.887404580	1.7440790361	61.4333
14	64.9	1.4962074351	64.9000
15	65.625	1.4962074351	65.6250
16	70.637786259	3.0768834307	72.0000

You can also reveal or conceal all text reports using the **Reveal All** and **Conceal All** commands listed in the asterisk pop-up menu shown in **Figure 12.4.** This menu is located to the left of the horizontal scroll bar in each report window.

Figure 12.4 Asterisk Pop-up Menu

Pop-up Menus

Many report windows include one or more pop-up menus (accessed by the icon shown at left). These pop-up menus list different options and commands depending on the analysis platform.

Report Fonts and Sizes

One of the ways you can customize the appearance of reports is to change text styles using the **Preferences** command in the **File** menu. The Preferences dialog is shown in **Figure 12.5**. This dialog has two pop-up menus called **Font** and **Size**. The **Font** menu lists different type faces, and the **Size** menu lists both point sizes and styles. Hold down the Shift key while selecting from the **Font** menu to see each font displayed. You can change the fonts for *text*, *headings*, and *titles* as described below.

- *Text* refers to the body of reports.
- *Headings* refers to column headings.
- *Titles* refers to window titles and reveal buttons.

To simultaneously apply a type preference to all of the above, hold down the option key while selecting from the **Font** or **Size** menu.

Figure 12.5 The **Preferences** Dialog

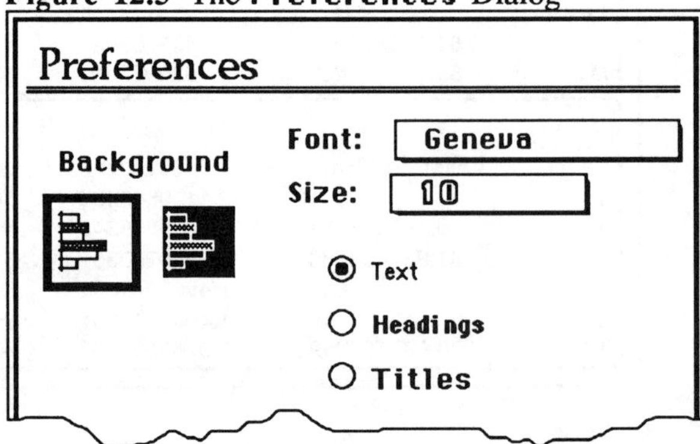

Graphical Displays

Graphical displays are plots of data within JMP report windows. Special options are available to manipulate these plots in each analysis window.

Resizing Graphs

In addition to the standard window resize box, there is a small box in the lower right corner of all JMP plots. When you click and drag in this box, you adjust the height and width of all plots in a frame independently of other frames in the same report window (see **Figure 12.6** below). It functions the same as the resize box for an entire window. However, while it adjusts widths of plots independently, it adjusts the depths of all plots in one frame as a single unit.

Figure 12.6 Plot Frame Resize Boxes

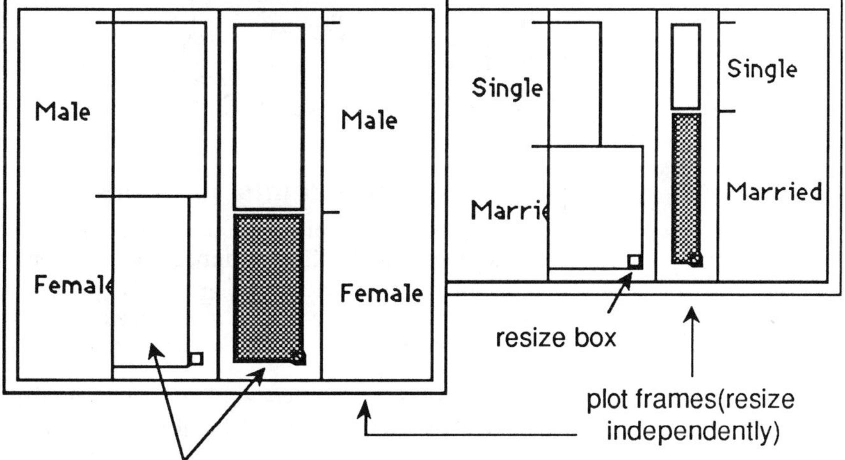

When you use this special resize box, the option and shift keys have special effects as described below.

- Adjusting a plot's size with option-drag resizes the graph in 8-pixel increments. This makes it easier to line up several graphs.
- Adjusting the plot's size with shift-drag makes the resized plot frame square, rather than rectangular.
- Adjusting the plot's size with the option and shift keys pressed adjusts it in 8-pixel increments and makes the plot frame square.

Changing the Plot Background

Graphs can have either white or black backgrounds. To set the background, choose the **Preferences** command from the **File** menu. The background control is shown in **Figure 12.7**. Note that some plots are not affected by this setting. For example, the spin plot has its own background setting.

Figure 12.7 The Plot Background Control

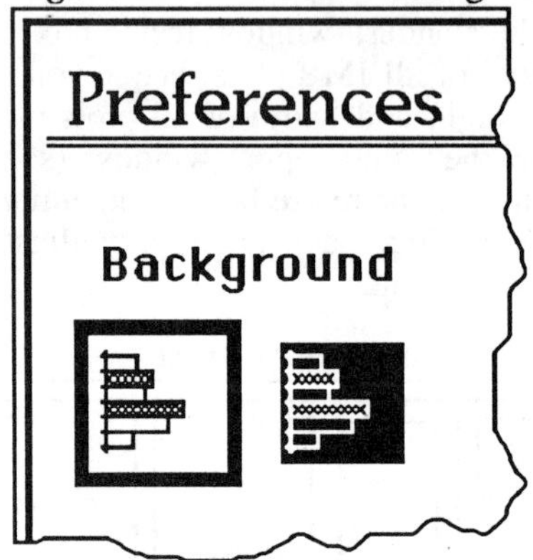

Highlighting Points in Plots

Most graphical displays are plots of columns from the data table. Points in these plots have different appearances depending on the active row state characteristics. These include select (highlight), exclude, hide, label, marker and color. There are 7 JMP markers and 15 JMP colors (available on color monitors). These row states are described in Chapter 10, *The Data Table*.

Label is a special row state that assigns a label to one or more rows in the data table. To designate a label for every row, assign the role of *label* to one column in the spreadsheet using the column's role assignment box. Row numbers are the default labels when there is no label column.

To apply a row state, select rows in any representation of the data table and choose the row state you want from the **Rows** menu. The data table and plots automatically show the effect of the row state.

You can save row states permanently by creating a new column and setting its type as **Row State**. The two boxes at the top of the new row state column are pop-up menus containing the commands **Copy to RowState** and **Copy from RowState**. **Copy to RowState** makes a row state column's information active. **Copy from RowState** copies the active row state information into the row state column.

You can select rows in graphical displays using either the standard arrow cursor or the brush tool as described in the following paragraphs.

Using the Arrow Cursor

Clicking the arrow cursor on a point in a plot highlight (selects) the point and the corresponding row in the current data table. If the corresponding row is not already labelled, its label appears on the display for as long as you hold down the mouse button.

When you click the arrow in a display, previously selected data are no longer selected. You can either select a new point or no points depending on where you click. To extend a selection rather than replace it, hold down the shift key while you click on a new point.

Using the Brush Tool

Brushing points in scatterplots was introduced by Becker and Cleveland in *Technometrics* (1987, p. 127-142).

The brush tool in JMP highlights a group of points delineated by a rectangle. As you drag the brush over a plot, points that fall within the rectangle are temporarily highlighted. Since the markers of highlighted points are larger except in spin plots, the brush tool is effectively a magnifying lens as it passes over points. To select points using the brush tool, release the mouse button while the chosen points are within the rectangle.

You can shift-drag the brush tool to extend the current selection. The new selection includes the previous selection and all points that pass within the rectangle while the shift key is pressed. The extended selection areas need not be contiguous.

You can also option-drag the brush tool to resize its rectangular selection area independently for each plot. You can use this feature to drag a rectangle around the area containing all the points you want to select. Sometime this is referred to as *marquee-selection*. You can also drag out a thin vertical or horizontal rectangle to produce a slicing tool that can traverse and highlight across either axis.

If you hold down the command key and drag the brush tool, the selection status of points within the rectangle continuously inverts, making the points blink on all representations of the data.

If you command-option-drag *(push)* with the brush tool and then let go, the rectangle bounces around in the frame on its own. Its direction and speed are determined by the speed and direction you push (or drag) the mouse. Click the mouse button in the report window to stop the rectangle from bouncing.

Help Windows

Help windows contain documents to answer questions you may encounter while using JMP. Sometimes they include pictures and examples, but they always include textual explanations. Many help windows also include buttons to access further help.

The file called JMP Help, which is distributed with JMP, contains the help documents. Be sure to keep it in the same folder as JMP.

You can access help windows almost anywhere in JMP. The paragraphs that follow describe different ways to access help.

Help from the Apple Menu

If you have general questions or want to see a list of help topics, choose **About JMP** from the Apple menu. This brings up the first help document with buttons for further help as shown in **Figure 12.8**. All the help documents are arranged hierarchically with buttons for accessing more detailed information when applicable.

Figure 12.8 About JMP Dialog Box

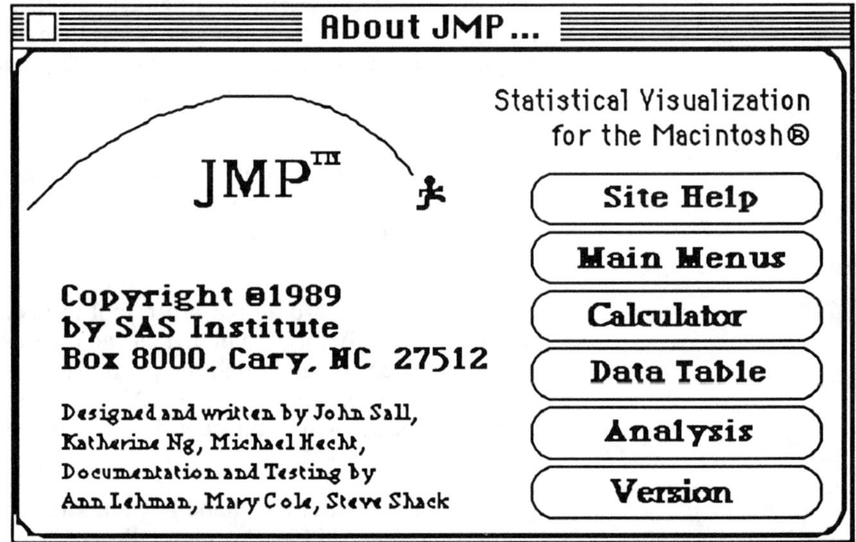

Help from Buttons in Dialogs

Many dialog windows in JMP include a **Help** button. For example, there are help buttons on the **Specify Model** dialog, the calculator window, the **Preferences** dialog, and on data manipulation dialogs.

Help from a Platform Window

If you want help using an analysis platform, select **Help** from the asterisk pop-up menu (shown below) located to the left of the horizontal scroll bar.

Figure 12.9 The Help Command in a Platform Window

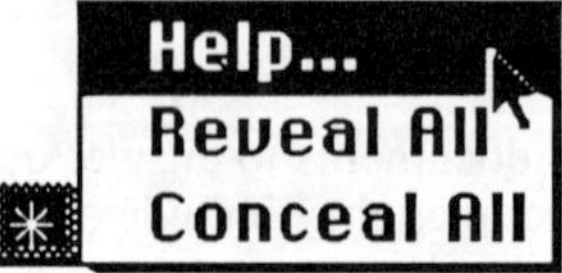

For example, the **Help** command for the **Distribution of Y's** platform is shown in **Figure 12.10** along with a second layer of help for a specific component of the platform.

Figure 12.10 Example of a Help Window

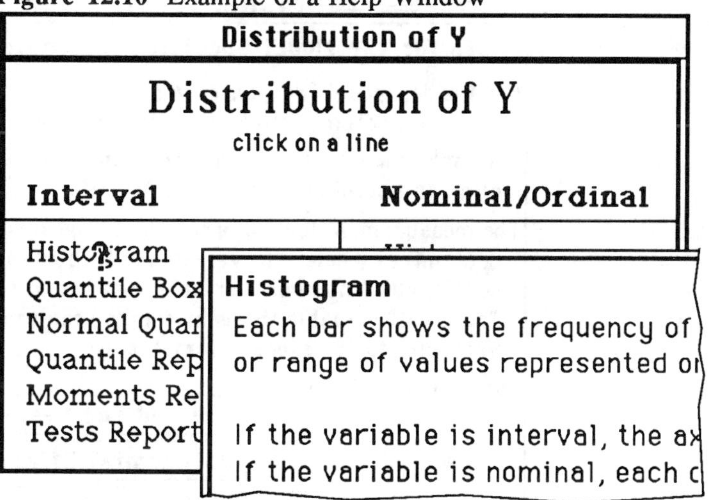

Help by Clicking on an Item

When you need help concerning a specific item in a JMP window, select the question mark from the **Tools** menu and click on the item. For example, when you click on **Std Error** as shown in **Figure 12.11**, a help window displays and persists as long as you hold the mouse button down. Shift-click to keep a window on the screen until you close it with the close box.

Figure 12.11 Clicking on an Item for Help

Parameter Estimates

Term	Estimate	Std Error	t Ratio	Prob>|t|
Intercept	42.570142	2.67988	15.89	0.0000
Height	.19761499	.026157	7.55	0.0000

std error

Standard Error

The standard error is a measure of the variability of the main parameter estimate It is an estimate of the standard deviation of the distribution of the estimate. It is used to construct t-tests and confidence intervals concerning the parameter.

If a help window contains buttons for further help as shown in **Figure 12.12**, it automatically remains on the screen.

Figure 12.12 Help Window with Additional Help Buttons

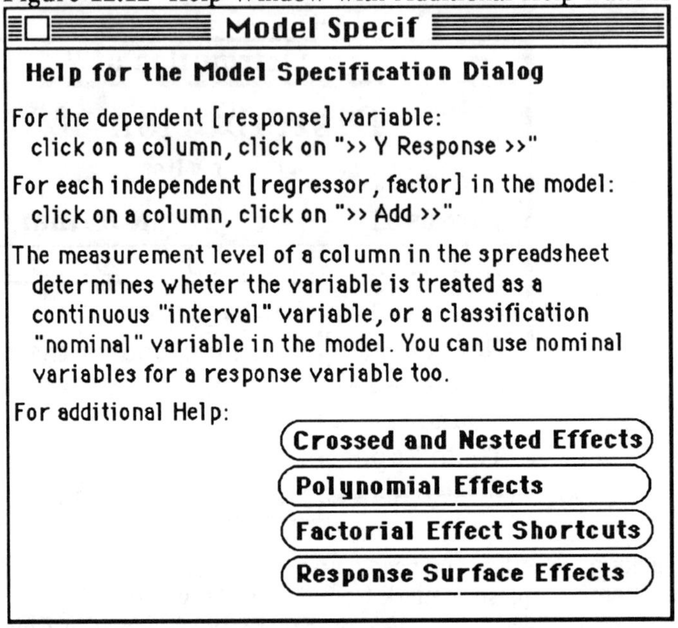

Model Specif

Help for the Model Specification Dialog

For the dependent [response] variable:
click on a column, click on ">> Y Response >>"

For each independent [regressor, factor] in the model:
click on a column, click on ">> Add >>"

The measurement level of a column in the spreadsheet determines wheter the variable is treated as a continuous "interval" variable, or a classification "nominal" variable in the model. You can use nominal variables for a response variable too.

For additional Help:

Crossed and Nested Effects

Polynomial Effects

Factorial Effect Shortcuts

Response Surface Effects

Copying from JMP Report Windows

You can use the standard Macintosh copy and paste operations to move data and graphical displays from JMP to other applications. However, although the **Edit** menu includes both **Cut** and **Copy** commands, they perform the same tasks in document windows. **Cut** copies all or part of the active report window into the clipboard but does not delete the image from JMP. There is also a special command called **Copy Report** (discussed below) which copies all text from an active report window but no plots.

Copying Graphical Displays

When you copy from a report window, the image is stored in the clipboard as a picture in PICT data format. By default you copy the entire report window. You can copy part of any report window as described below.

If you want to copy part of a report window, select the scissors tool from the **Tools** menu and drag it diagonally to delineate the area you want to copy. The tip of the upper blade marks the starting and ending points of the selection. As you drag, a rectangle appears and selects the area to copy as shown in **Figure 12.13**.

Figure 12.13 Selecting with the Scissors Tool

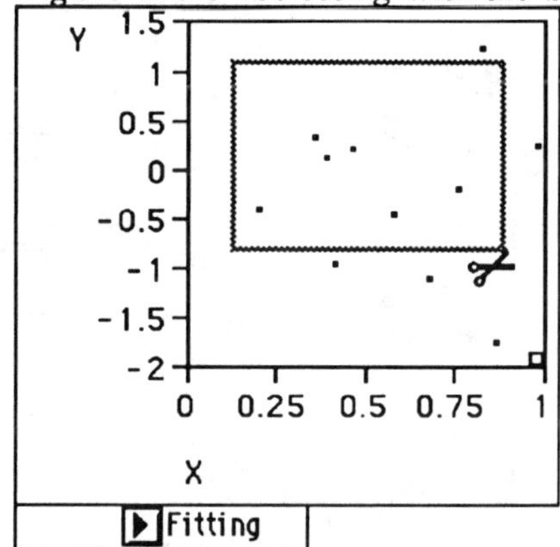

You can change the selection if necessary. Drag the scissors over a new area or extend the current selection by shift-dragging the scissors across an additional area. Extensions do not have to be contiguous, and you can add as many extensions as you want. To deselect all areas, click in the report window (without dragging).

If you want the scissors tool to select specific components of a report, hold down the option key and click or drag. Each component of the report highlights as a separate unit (see **Figure 12.14**). Hold down both the option and shift keys to extend a selection of discrete components.

Figure 12.14 Selecting with the Option Key

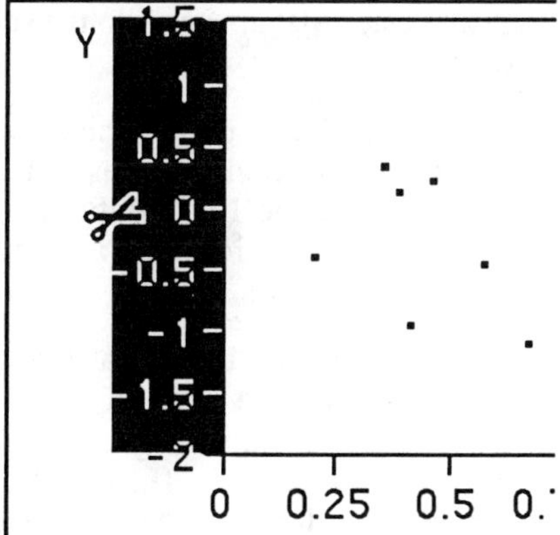

Copying Text Only

To copy text only from a report window, choose **Copy Report** from the **Edit** menu as shown in **Figure 12.15**. This command copies all text from the active report window, ignoring any current selection. Each field within a text report is separated by tabs so that you can edit the text in a word processing application.

Figure 12.15 The Copy Report Command

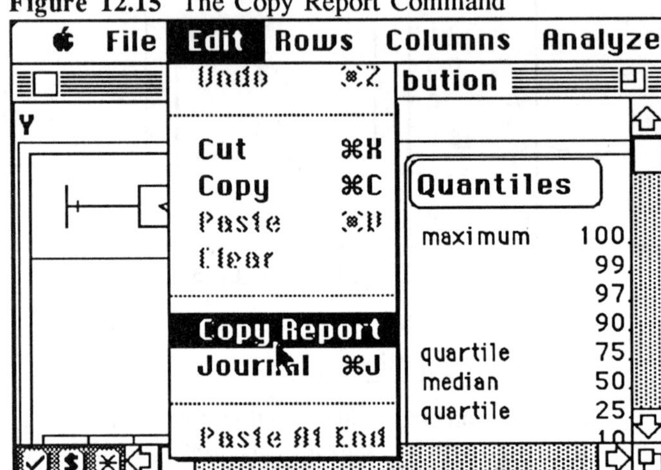

Pasting Between Applications

If you are using Multifinder and have plenty of memory, you can run JMP and other applications simultaneously to copy and paste between them. Otherwise, you can paste JMP images into the scrapbook desk accessory for intermediate storage.

JMP data is copied in PICT data format with hierarchical grouping information (used by applications such as MacDraw). However, there are some differences between the PICT data and the on-screen image. For example, points in JMP plots are saved as very small squares in order to improve their printed appearance.

Not all applications interpret PICT data in the same way. For example, some drawing programs ignore clipping information (information about invisible parts of pictures). In such programs, some lines appear longer than they appeared in JMP. For example, fitted lines and confidence curves may extend beyond axes boundaries when copied from JMP and pasted into another application.

If you use a color monitor, color images you copy from JMP map to the original eight colors supported by the Macintosh. Because some applications do not support color, copied pictures may sometimes convert to black and white or patterned images.

See the section called **Printing Reports** for information about copying JMP PICT data into other applications and printing.

Journaling Reports

Journaling is a special feature of JMP designed so you can save analysis results in a text file you create. When you select the **Journal** command from the **File** menu, all text and pictures from the active JMP analysis window are written to the journal file. If an area of an analysis window is selected with the scissors tool, then the **Journal** command saves the selected area rather than the entire window.

Journal Files

The first time you choose the **Journal** command from the **File** menu, you are asked to name the file and choose its file format. Journal files can be either in MacWrite file format or text file format. The journaling dialog is shown in **Figure 12.16**.

Figure 12.16 The Open Journal Dialog

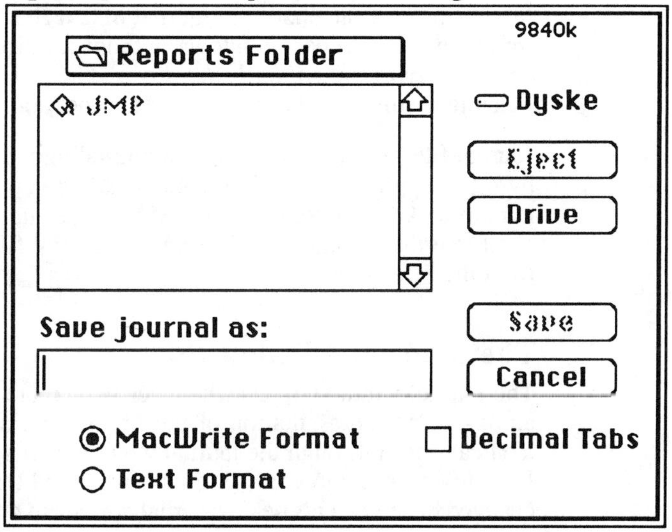

MacWrite file format stores both pictures and text and can be read by most word processing and page layout programs. The applications listed below have successfully opened JMP journal files in tests. However, because these applications are maintained by other software developers, their compatibility cannot be guaranteed

- MacWrite and MacWrite II
- Microsoft Word from Microsoft
- FullWrite from Ashton-Tate
- WriteNow (v.1)

Before journaling, you may want to choose font and size settings from the Preferences dialog in the File menu. These settings determine the font settings for MacWrite file format journals.

Because MacWrite does not support flush right tabs, the journal command saves numeric data with flush left tabs. The numbers are padded with non-breaking spaces to make numbers in monospaced fonts appear right-aligned. Most word processors allow you to convert from flush-left to flush-right tabs. If you are using MacWrite, you can right align numbers by adding more rulers to the journal file instead.

If you prefer to align numbers with decimal tabs, you can change the default tab setting when you open the journal file. This decimal tab aligns numbers by their decimal points and can approximate right alignment for numbers without decimal points. However, whole numbers aligned to a decimal tab appear left-aligned in the journal window.

If you plan to open journal file using applications not compatible with MacWrite, you can choose to journal in text file format instead. Text format stores text from report windows without font information *and without the accompanying graphical displays.* These files can be read by most Macintosh editors and can be ported to other computers.

Journal files are locked during the journaling process and cannot be opened by other applications in Multifinder. After you close a journal file, it can remain locked temporarily until Multifinder updates the file's status. You can force this update by selecting the journal file and choosing **Get Info** from the **File** menu.

The Journal Window

The journal window opens whenever you create or open a journal file. It is an on-screen representation of the journal file as shown in **Figure 12.17.** You can scroll through the journal window and append text at the end of the file. However, the journal window is not an editor. Appended text cannot be changed once you choose the journal command again.

You can hide the journal window by selecting **Hide** from the **Window** menu. If you close the journal window (using the close box or close command), you also close the journal file.

Figure 12.17 The Journal Window

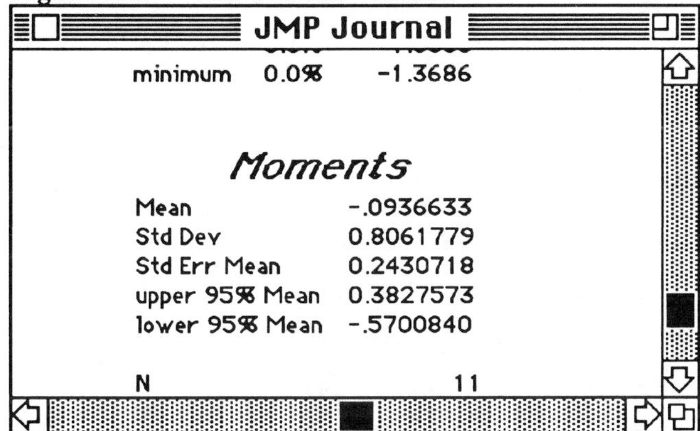

Printing Reports

Select **Print** from the **File** menu to print revealed text reports and plots from the active report window. Reports print best on Postscript® output devices such as the LaserWriter®, but they also print on Quickdraw laser printers and ImageWriters®.

When you print a JMP report, the arrangement of text reports and graphical displays will automatically rearrange on each printed page for best fit. Some controls including reveal buttons will not appear on the printouts.

The PICT files that store graphical displays generated by JMP contain special PostScript (Adobe Systems Corp.) information to make them print better. With this special information, lines and curves are thin and dashed lines have uniformly sized segments. When you copy a JMP picture into most word processing applications, this special information is retained. However, some drawing programs discard extra PostScript commands.

Statistical Commands and Platforms

There are only six analyze commands in JMP. That is something to cheer about because it means that it's easy to find the one you need.

Platform

Each command from the **Analyze** menu *launches a platform.* A platform is an interactive window that you can use to analyze data, work with points on plots, and save results. The platform persists until you close it, or close the data table to which it is linked. You can have any number of data tables open at the same time, and any number of platforms displayed for each data table. If you have several data tables then the platform you launch analyzes the *current* data table. This table is either the active data table window or the table that corresponds to an active analysis report window.

Design

The navigational instructions for the analysis component of JMP are simple, but before you begin note some points of design philosophy.

First, JMP implements general methods consistent with its key concepts (see the section called **Conceptual Keys** below). All the platforms produce a variety of results handling many situations. They are consistent throughout in how they treat data and statistical concepts. But don't expect JMP to provide all statistical methods. It is not designed to be an encyclopedic product.

Second, JMP methods are generic. They adapt themselves to the context at hand. The principal concept is that of *measurement level,* which gears all analyses. For example, JMP automatically analyzes a variable that has nominal values differently than it analyzes a variable with interval values.

Third, JMP doesn't wait for options and commands. A platform tends to give you everything you might at once. If this is more information than you want, you can conceal the tables or graphs you don't need. The advantage is that you won't often have to *mouse around* to find commands for extra statistics.

Navigating the Analyze Platforms

To begin an analysis, follow three simple steps to assign measurement levels, choose variable roles, and select an analysis platform.

1. Measurement Level

The *measurement level* assigned to a variable can be interval, nominal, or ordinal. To assign a level you can use the pop-up menus at the top of each column on the data table window as shown in **Figure 13.1**, or use the **Column Info** dialog. This assignment becomes an attribute of the column until you change it.

Figure 13.1 Assign a Measurement Level

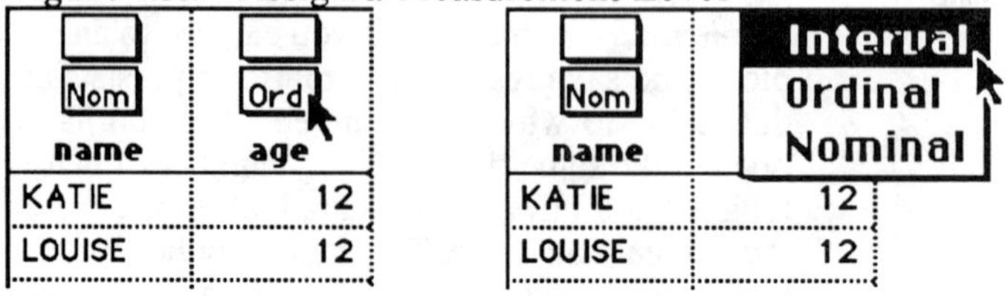

Nom name	Ord age
KATIE	12
LOUISE	12

Nom name	Interval Ordinal Nominal
KATIE	12
LOUISE	12

The measurement levels are not just descriptive tags. They tell each analysis platform how to analyze and graph the data. They are prescriptions for treating the values in a model. Furthermore, there is a difference in measurement level treatment according to whether the variable is a response (Y) variable or a factor (X) variable. Do not hesitate to change the measurement level in order to study the same variable in different ways.

The levels of measurement used by JMP are defined the following way:

Interval
: For interval columns, the numeric value is used directly. Often the variable is measured on a continuous scale. Interval variables are amenable to concepts like *expected value* or *mean*. In particular, interval responses are described as having a random component that is typically thought of as a normally distributed error term.

Ordinal
: An ordinal value is not used directly, but only as a name. The values can be either numeric or character. However, the values are interpreted by JMP as having an order. Numeric values are ordered by their increasing numeric values, and character values are ordered alphabetically or in collating sequence.

Nominal
: A nominal value is just a name, and the values are interpreted as having no particular order in the analysis conducted. The values can be either character or numeric. Numeric values are treated as characters values.

2 Choose Variable Roles

You choose X or Y roles for the variables depending on how you want JMP to treat them. A column assigned the role of Y is a response or dependent variable. An X column is a factor variable that fits, predicts, or groups a response. You can mark columns as X or Y in the data table before you launch the platform; or, if no variables are assigned X and Y roles on the spreadsheet, each analysis platform prompts you with a role selection dialog.

To choose variable roles in the data table, you can use the pop-up menu at the top of the column as shown in **Figure 13.2**, or use the **Assign Roles** command in the **Columns** menu.

Figure 13.2 Assign Variable Roles

name (Nom)	age (Ord)
KATIE	12
LOUISE	12
JANE	12
JACLYN	12

name (Nom)	
KATIE	
LOUISE	
JANE	
JACLYN	12

None
X
Y
Weight
Freq
Label

Other roles that are available include **Weight** and **Freq**, when you want to assign a weight or frequency count to each row, and **Label**, when you want column values to be used for labeling points in plots.

3
Launch the Platform

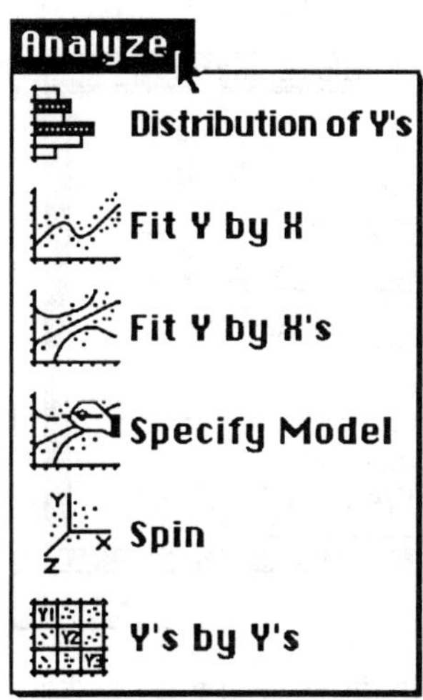

You launch an analysis platform from the **Analyze** menu according to the way you want X and Y variables treated together.

Distribution of Y's describes the distribution of each Y column using histograms and other graphical and textual reports. If a column has interval values, the graphical display consists of a histogram and quantile box plot. If a column has nominal or ordinal values, the graphical display consists of a histogram and a divided (mosaic) bar chart showing a bar for each level of the ordinal or nominal variable.

Fit Y by X describes each pair of X and Y columns. The displays and reports vary depending on the levels of measurement you specify for the X and Y columns.

Fit Y by X's fits one Y column by all the X columns selected. It is used for multiple regression analysis, main effects analysis of variance, analysis of covariance, and analysis of categorical linear models.

Specify Model displays a dialog that defines complex models automatically and lets you choose specific model effects and error terms.

Spin produces a three-dimensional view of data and an approximation to higher dimensions through principal components. The plot is a spinable display of the values of any three numeric columns you choose.

Y's by Y's describes relationships among Y's with correlations, a scatter plot matrix, and a multivariate outlier plot.

Chapters 14-19 discuss each of the analysis platforms in detail.

JMP Recipes for Textbook Methods

Some statistical software is organized like cookbooks, with a recipe for each situation. JMP is organized more as a scientific instrument. You supply general information for the solution rather than trying to match your case to a specific command in the software. As discussed in the previous section, the only information JMP needs are the roles you assign to columns, the level of measurement you assign to columns, and the analysis platform you select.

Figure 13.3 The 3-Step General Approach to Analysis in JMP

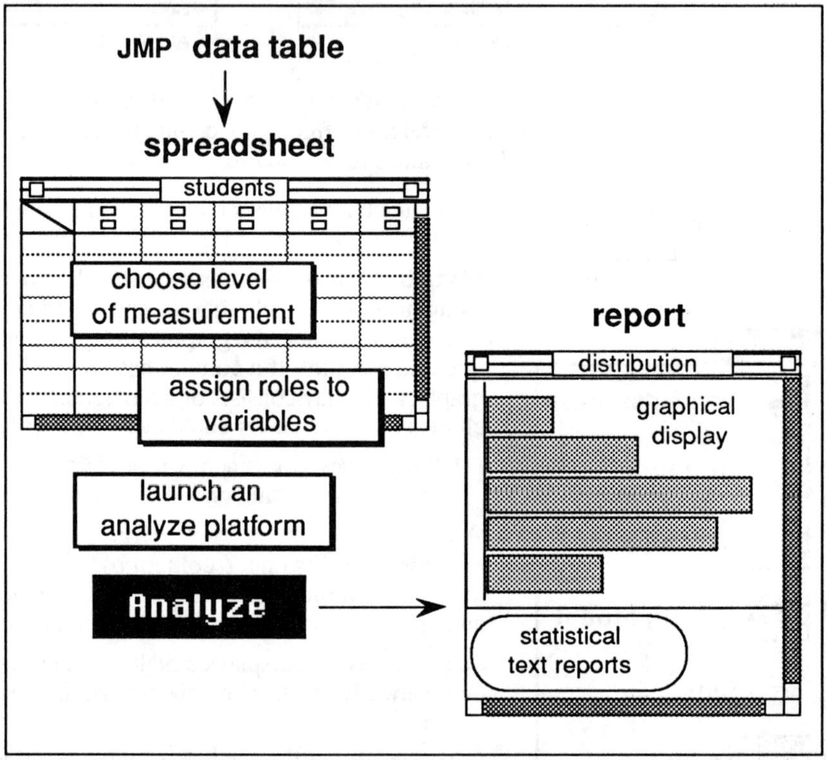

JMP data table

spreadsheet

students

choose level of measurement

assign roles to variables

launch an analyze platform

Analyze

report

distribution

graphical display

statistical text reports

Table 13.1 lists many typical situations and tells how to approach them with JMP. All the analyses listed are just specific cases of variable roles, measurement levels, and an analysis platform.

To initiate multiple analyses, you can specify multiple response variables. However, JMP does not perform multivariate analyses on multiple response specifications at this time.

Table 13.1 General JMP Methods for Selected Types of Analyses

Type of Analysis	Variable Roles	Level of Measurement	Analysis Platform
Descriptive: means, quantiles	Y	interval	**Distribution of Y**
Descriptive: frequencies	Y	nominal/ordinal	**Distribution of Y**
Cross-tabulation for each X, Y 2-way Frequency Counts	Y X	nominal nominal	**Fit Y by X**
Correlations	multiple Y's or Y X	interval interval interval	**Y's by Y's** **Fit Y by X** with Density Ellipsoids
t test for a specific mean *where Y is a column that subtracts a prescribed value from a variable*	Y	interval	**Distribution of Y** with tests report
Paired t test *where Y is a column that is the difference between two columns*	Y	interval	**Distribution of Y** with tests report
2-group t test	Y X	interval nominal	**Fit Y by X** with tests report
k-group means comparison, one-way analysis of variance	Y X	interval nominal	**Fit Y by X** with ANOVA report
Simple regression	Y X	interval interval	**Fit Y by X** with *Fit Line* option
Polynomial regression	Y X	interval interval	**Fit Y by X** with *Fit Polynomial* option
Multiple regression	Y multiple X's	interval interval	**Fit Y by X's**
2-way analysis of variance without interaction	Y multiple X's	interval nominal	**Fit Y by X's**
2-way analysis of variance with interaction	Y multiple X's	interval nominal	**Specify Model** (complete dialog)
Analysis of covariance	Y factor X covariate X	interval nominal interval	**Fit Y by X's**
Analysis of covariance with separate slopes	Y factor X covariate X	interval nominal interval	**Specify Model** (complete dialog)
2-way contingency table with Chi-square test	Y X	nominal nominal	**Fit Y by X** loglikelihood report
Multi-way contingency table	Y multiple X's	nominal/ordinal nominal/ordinal	**Specify Model** (complete dialog)
Logistic regression	Y X	nominal/ordinal interval	**Fit Y by X**

The Responses Models

The models and methods available in JMP are practical, widely used, and suit the need for a general approach in a statistical software instrument. As with all statistical software, you are responsible for learning the assumptions of the models you choose to use, and the consequences if the assumptions are not met. For more information see the section **The Usual Assumptions** later in this chapter.

Interval Responses

Response Model

When the response column is interval, JMP fits the value of the response directly. The basic model is that for each observation,

Y = (some function of the X's and parameters) + error .

To carry out statistical tests the assumption is made that the error term in the model is normally distributed.

Fitting Principle

The fitting principle is called *least squares*. The least squares method estimates the parameters in the model to minimize the sum of squared errors. The errors in the fitted model, called *residuals*, are the differences between the actual and predicted values of each observation.

The least squares method is equivalent to the maximum likelihood method of estimation if the errors have a normal distribution. This means that the analysis estimates the model that gives the most likely residuals. The negative log-likelihood is a scale multiple of the sum of squared errors for the normal distribution.

Base Model

The simplest model for interval measurement fits just one value to predict all the response values. This value is the estimate of the mean. (The mean is just the arithmetic average of the response values.) All other models are compared to this base model.

Nominal Responses

Response Model

When the response has a nominal level of measurement, JMP fits the probabilities that the response is one of r different response levels given by the data values. The basic model is that for each observation

Probability($Y = j$) = (some function of the X's and parameters) .

The probability estimates must all be positive. For a given configuration of X's, the probability estimates must sum to 1 over the response levels. The function that JMP uses to predict probabilities is a composition of a linear model and a multi-response logistic function. This is sometimes called a *log-linear model* because the logs of ratios of probabilities are linear models.

The log-linear model can be expressed as

$$\log\left(\frac{P(y=1)}{P(y=2)}\right) = \text{(function of X's and parameters)}$$

Fitting Principle

The fitting principle is called *maximum likelihood.* It estimates the parameters such that the joint probability for all the responses given by the data is the greatest obtainable by the model. Rather than reporting the joint probability (likelihood) directly, it is more manageable to report the total of the negative logs of the likelihood.

The uncertainty (–log likelihood) is the sum of the negative logs of the probabilities attributed by the model to the responses that actually occurred in the sample data. It is often denoted as H and written

$$H = \sum_{i=1}^{n} -\log P(y=y_i)$$

If you attribute a probability of 1 to each event that did occur, then the sum of the logs approaches 1 if the model fits poorly, or zero for a perfect fit. Unlike interval response data where a high R square is fairly common, in practice it is rare to get an uncertainty close to zero.

The nominal model can take a lot of time and memory to fit, especially if there are many response levels. JMP tracks the progress of its calculations with an *iteration history*, which shows the –log likelihood values becoming smaller as they converge to the estimates.

Base Model

The simplest model for a nominal response is a set of constant response probabilities fitted as the occurrence rates for each sample level. In other words, the probability that Y is response level *j* is estimated by dividing the total sample count *n* into the total of each response level n_j, and is written

$$p_j = \frac{n_j}{n}$$

All other models are compared to this base model. This base model serves the same role for a nominal response as the sample mean does for interval models.

Ordinal Responses

Response Model

With an ordinal response, as with nominal responses, JMP fits probabilities that the response is one of *r* different response levels given by the data.

Ordinal data are ordered like interval data, but the spacing between the ordered levels is not used. If you have a numeric response but want your model to ignore the spacing of the values, then you can assign the ordinal level to the response column. If you have a classification variable and the levels are in some natural order such as low, medium, and high, then you can use the ordinal level of measurement. For details about the function that is fitted, see Chapter 15, *Analyze: Fit Y by X*.

One way of thinking about an ordinal model is to imagine the response mechanism as a two-phase process. The X's (the linear model) are used to first determine some variable Z with interval values. However, Z is not observed. Instead, Z determines the ordinal response level for Y as it crosses various threshold values.

Schematically this is

Z = (some function of X's and parameters) + (error)

Y = (some level determined by the value of Z)

where the error has a logistic distribution.

Specifically, the value of Y ascends the ordinal levels as Z crosses certain threshold values. If the levels of Y are 1 through r, then

$$Y = \begin{matrix} r & \text{if } \alpha_{r-1} \leq Z \\ j & \text{if } \alpha_{j-1} \leq Z < \alpha_j \\ 1 & \text{if } Z \leq \alpha_1 \end{matrix}$$

A different, but mathematically equivalent way to envision an ordinal model is to think of a nominal model, where instead of modeling the odds, you model the cumulative probability. Instead of fitting functions for all but the last level, you fit only one function and adjust it to fit each cumulative response probability.

Fitting Principle The maximum likelihood fitting principle for an ordinal response model is the same as for the nominal response model. It estimates the parameters such that the joint probability for all the responses actually occurring is the greatest obtainable by the model. It uses an iterative method that is faster and uses less memory than nominal fitting.

Base Model The simplest model for an ordinal response, like a nominal response, is a set of response probabilities fitted as the occurrence rates of the sample levels.

The Factor Models

The way the X variables (called factors) are modeled to predict an expected value or probability is the subject of the factor side of the model.

The factors enter the prediction equation as a a linear combination of X values and the parameters to be estimated. For an interval-response model, the assumed true model for a typical observation, y_i, is written

$$y_i = \beta_0+\beta_1 x_{1i}+ \ldots+\beta_k x_{ki}+\varepsilon_i$$

where

y_i is the response,

x_{ji} are functions of the data,

ε_i is an unobservable realization of the random error,

β_j are unknown parameters.

The way the X's in the linear model are formed from the factor terms is different for each of the measurement levels. The linear model X's can also be complex effects such as interactions or nested effects, discussed in detail later.

Interval Factors

When an X variable is assigned the interval level of measurement, its values are used directly in the prediction formula. The X variables are called *regressors*, or *covariates*.

Nominal Factors

When an X variable is assigned the nominal measurement level, the response can have separate prediction coefficients fit to each level of the nominal variable. The prediction coefficients are modeled to represent differences from the average response of all the nominal values.

In the context of a linear model, there is an indicator variable for each value except the last, and the indicator is set to 1 when that level occurs. The last level causes all the indicators to be -1. This is done so that the parameters across all the levels sum to zero.

For example, suppose the nominal factor A has three levels 1, 2, and 3, and you want to fit it as a linear model. The model has the intercept and two parameters that show a value to add to the intercept if the level of A is 1 or 2. For the last level of A, you subtract both A parameters from the intercept to get the predicted value,

$$Y = \beta_0+\beta_1 X_1+\beta_2 X_2+\text{error}$$

where

β_0 is the intercept parameter

β_1 and β_2 are parameters for the A nominal effect

X_1 is 1 if A = 1, 0 if A = 2, −1 if A = 3

X_2 is 0 if A = 1, 1 if A = 2, −1 if A = 3 .

Thus, for a Parameter Estimates report showing

Term	Estimate
Intercept	7.9179166
A[1-3]	-1.205416
A[2-3]	.07020833

the actual prediction formula for the fitted Y becomes

$$7.9179+\left(\begin{cases} 1.1352, & \text{if } A=3 \\ 0.0702, & \text{if } A=2 \\ -1.2054, & \text{if } A=1 \\ \bullet, & \text{otherwise} \end{cases}\right).$$

Note that the value for A = 3 of 1.1352 is the negative sum of the other two values, .0702 and -1.2054. This is just an alternative way to implement the model that fits three means to the three groups.

Level	Least Sq Mean
1	6.712500000
2	7.988125000
3	9.053125000

For a nominal response, nominal factors have the effect of estimating different response rate profiles for each of the groups instead of different means.

Ordinal Factors

When an X variable is assigned the ordinal level of measurement, its lowest value is considered a baseline value, and each level above that baseline contributes some additional quantity used to determine the response.

In the context of a linear model, there is an indicator variable for each factor level except the first. All the indicators are set to 1 at and below the indicator for the current factor level.

There is not much difference between nominal and ordinal factors for simple models, but there are major differences when interactions are used.

The Usual Assumptions

Before you put your faith in statistics, you need to reassure yourself that you know both the value and the limitations of the techniques you use.

Statistical methods are just tools...they cannot guard you from incorrect science (invalid statistical assumptions) or bad data.

Assumed Model

Most statistics are based on the assumption that the model is correct. You assume that the equation really does determine the response up to a normally distributed random error. To the extent that your model is not correct, you must attenuate your credibility in the statistical reports that result from the model.

Relative Significance

Many statistical tests do not evaluate the model in an absolute sense. Significant test statistics might only be saying that the model fits better than some reduced model, such as the mean. The model appears to fit the data but might not describe the underlying physical model well at all.

Multiple Inferences

Often the value of the statistical results is not that you believe in the statistical results directly, but rather that they provide a key to some discovery. To confirm the discovery, you may need to conduct further studies. Otherwise you may just be 'mining the data'. For Instance, if you conduct enough analyses you can find 5% significant effects in five percent of your studies by chance alone. Similarly, to the extent that you use your data to shape your model (instead of testing the correct model for the data), you are corrupting the significance levels in your report. The random error then influences your model selection.

Validity Assessment

The following list describes various techniques to use and patterns to look for in assessing the validity of the model.

- Model validity can be checked against a *saturated* version of the factors with Lack of Fit tests. These tests are automatically done in the **Fit Y by X's** platform if you have replicated X data in a non-saturated model.
- You can check the distribution assumptions for an interval response by using the **Distribution of Y** platform on the studentized residuals and looking at the histogram with a normal curve and the normal quantile plot. The residuals are not quite independent but you can informally identify severely non-normal distributions. Residuals and studentized residuals are available from the save menu of the **Fit Y by X's** platform.
- The best all-around diagnostic tool for interval responses is the leverage plot because it shows the influence of each point on each hypothesis test. If you suspect that there is a mistaken value in your data, then this plot helps to determine if a statistical test is heavily influenced by a single point. Leverage plots are described in Chapter 16, *Analyze: Fit Y by X*.
- It is a good idea to scan your data for outlying values and examine them to see if they are valid observations. You can spot univariate outliers in **Distribution of Y** reports and plots. Bivariate outliers appear in **Fit Y by X** scatterplots and in the **Y's by Y's** scatterplot matrix. You can see trivariate outliers in a 3-dimensional spinning plot produced by the **Spin**

command. Higher dimensional outliers can be searched for with Principal Components in the **Spin** platform and with Mahalanobis distances computed and plotted in the **Y's by Y's** platform.

Alternative Methods

The statistical literature describes special nonparametric and robust methods, but JMP does not offer them at this time. These methods require fewer distributional assumptions (are nonparametric), and are more resistant to contamination (robust). However, they are less conducive to a general methodological approach, and the small-sample significance probabilities on the test statistics can be time consuming to compute.

If you are interested in linear rank tests and need only normal large sample significance approximations, then you can analyze the ranks of your data to perform the equivalent of a Wilcoxon rank-sum or Kruskal-Wallis one-way test.

If you are uncertain that an interval response adequately meets normal assumptions, you can change the level of measurement from interval to ordinal and then analyze safely, even though this approach sacrifices some richness in the presentations and some statistical power as well.

Conceptual Keys

There are two conceptual keys that unify classical statistics and encapsulate statistical properties and fitting principles into forms you can visualize.

- a unifying concept of uncertainty
- the two basic fitting machines.

These two ideas help unlock the understanding of statistics with intuitive concepts that are based on the foundation laid by mathematical statistics.

Statistics is to science what accounting is to business. It is the craft of weighing and balancing observational evidence. Statistical tests are like credibility audits. But statistical tools can do more than that. They are instruments of discovery that can show unexpected things about data and lead to interesting new ideas. Before using these powerful tools, you need to understand a bit about how they work.

Uncertainty, a Unifying Concept

When you do accounting you total money amounts to get summaries. When you look at scientific observations in the presence of uncertainty or noise, you need some statistical measurement to add up and summarize. Just as money is additive, uncertainty is additive if you choose the right measure for it.

The best measure is not the probability directly, because to get a joint probability you have to assume the observations are independent and then multiply probabilities rather than add them. However, if you take the log of each probability, then you can sum them and the total is the log of the joint probability.

However, the log of a probability is negative because it is the log of a number between 0 and 1. So to keep the numbers positive, JMP uses the negative log of the probability. As the probability becomes smaller, the negative log of it becomes larger. Thus this measure is called *uncertainty*, and it is measured in reverse fashion from probability.

In business, you want to maximize revenues and minimize costs. In science you want to minimize uncertainty so think of uncertainty as a cost. All statistical methods are organized to fit models such that uncertainty is minimized.

It is not difficult to visualize uncertainty. Just think of flipping a series of coins where each toss is independent. The probability of tossing a head is .5, and –log(.5) is 2, for base-2 logarithms. The probability of tossing h heads in a row is simply,

$$p = (1/2)^h \ .$$

Solving for h, produces

$$h = -\log_2(p) \ .$$

You can think of the uncertainty of some event as the number of coins you have to flip and turn up heads to get an equally rare event.

Almost everything we do statistically has uncertainty, $-\log(p)$, at the core. Statistical literature refers to uncertainty as *negative log-likelihood.*

The Two Basic Fitting Machines

An amazing fact about statistical fitting is that most of the classical methods reduce to using two simple machines, the spring and the pressure cylinder.

Springs

First, the springs useful for explaining fits to an interval response model. Suppose that you have n points, and you want to know the *expected value* (mean) of the points. Envision what happens if you lay the points out on a scale, connect one end of a spring to each point, and join the other ends of the springs together at a common junction and let go (see **Figure 13.4**). The springs wiggle the junction point around and then bring it to rest at the mean! This is what must happen according to physics!

Figure 13.4 Connect Springs to Data Points

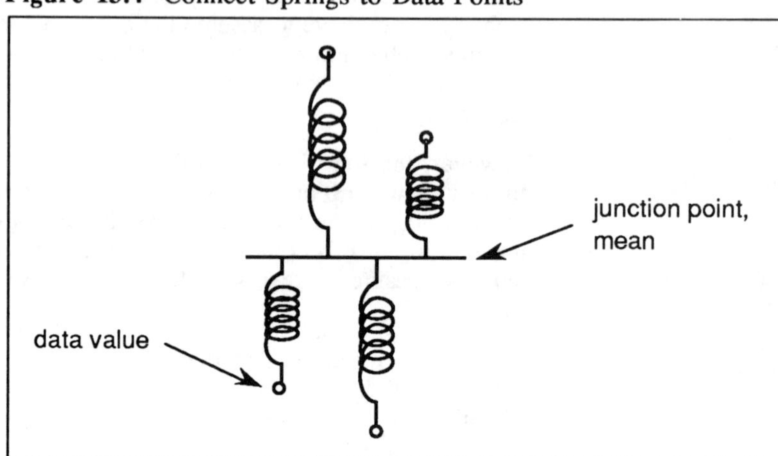

If the data are normally distributed with a mean at the junction point where springs are tied, then the physical energy in each spring as it is stretched from each observation point to the mean is proportional to the uncertainty of the data point. All you have to do to calculate the energy in the springs, (the uncertainty) is to compute the sum of squared distances of each point to the mean.

To chose an estimate that attributes the least uncertainty to the observed data, the spring-settling point is chosen as the estimate of the mean. That is the point that requires the least energy to stretch the springs and is equivalent to the least squares fit.

That is how you fit one mean or fit several means. That is how you fit a line, or a plane, or a hyperplane. That is how you fit almost any model to continuous data. You measure the energy or uncertainty by the sum of squares of the distances you must stretch the springs.

Statisticians put faith in the normal distribution because it is the one that requires the least faith. It is, in a sense, the most random. It has the most non-informative shape for a distribution. It is the one distribution that has the most expected uncertainty of all for a given variance. It is the distribution whose uncertainty is measured in squared distance. In many cases it is the limiting distribution when you have a mixture of distributions or a sum of independent quantities. It is the distribution that leads to test statistics that can be measured fairly easily.

You test hypotheses by measuring this same spring energy as the fit is constrained. Suppose that you have responses from four different treatments in an experiment, and you want to test if the means are significantly different. First, envision your data in groups as shown in **Figure 13.5**, with springs connected to a separate mean for each treatment. Then exert pressure against the spring force to move the individual means to the common mean. Presto! The energy needed to constrain the means to be the same is the test statistic you need. That energy is the main ingredient in the F test for that hypothesis that tests whether the means are the same.

Figure 13.5 Comparing Group Means

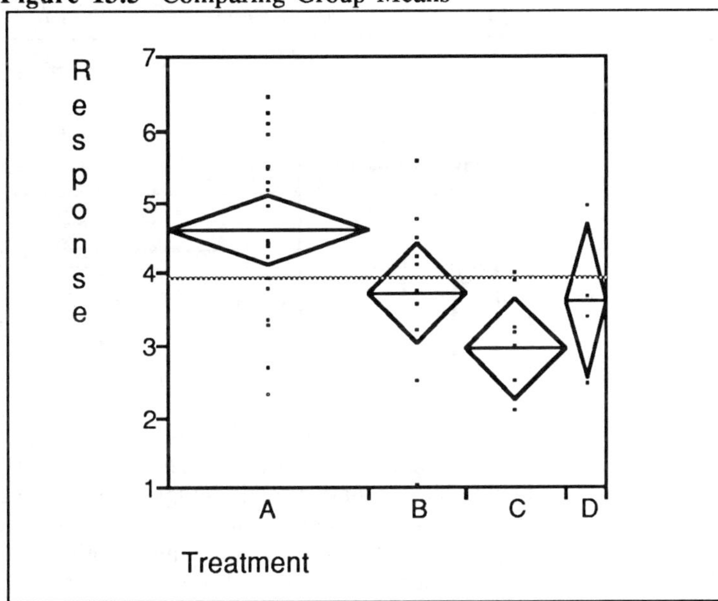

Pressure Cylinders.

What if your response is categorical instead of interval? For example, suppose that the response is the country of origin for a sample of cars. For your sample, there are probabilities for the three response levels, American, European, and Asian. You can set these probabilities for country of origin to some estimate and then evaluate the uncertainty in your data. This uncertainty is found by summing the negative logs of the probabilities of the responses given by the data and is written

$$H = \Sigma\, h_{y(i)} = \Sigma -\log p_{y(i)} \ .$$

The idea of springs illustrates how a mean is fit to interval data. When the response is categorical, statistical methods estimate the response probabilities directly and choose the estimates that minimize the total uncertainty of the data. The probability estimates must be nonnegative and sum to 1. You can picture the response probabilities as the composition along a scale whose total length is 1. For each response observation, load into its response area a gas pressure cylinder (a tire pump). Let the partitions between the response levels vary until an equilibrium of lowest potential energy is reached. The sizes of the partitions that result then estimate the response probabilities.

Figure 13.6 shows what the situation looks like when there are thirteen responses in three categories. The first response level has 6 responses, the next has 2, and the last has 5 responses. The response probabilities become 6/13, 2/13, and 5/13, respectively, as the pressure against the response partitions balances out to minimize the total energy.

Figure 13.6 Effect of Pressure Cylinders in Partitions

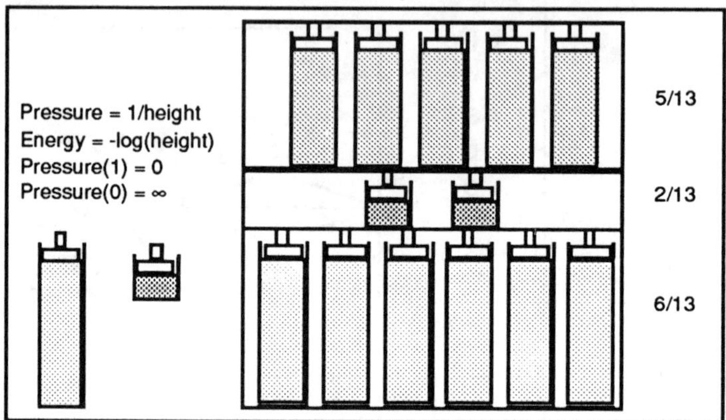

As with springs for interval data, you can divide your sample by some factor and fit separate partitionings. Then test that the response rates are the same across the groups by measuring how much additional energy you need to push the partitions to be equal. Imagine the pressure cylinders for car origin probabilities grouped by the size of the car. The energy required to force the partitions in each group shown in the mosaic plot below to align horizontally tests whether the variables have the same probabilities.

Figure 13.7 A Mosaic plot for Categorical Data

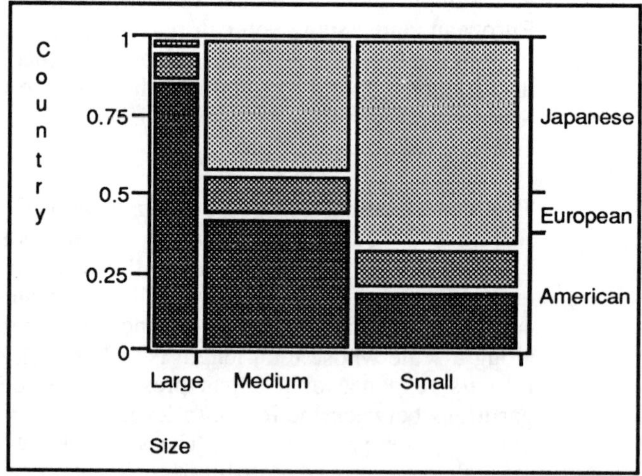

Chapter 14.
Analyze: Distribution of Y's

The **Distribution of Y's** analysis describes a distribution of column values using histograms and other graphical and textual reports.

- If a column has interval values, the graphical display consists of a histogram and quantile box plot.
- If a column has nominal or ordinal values, the graphical display consists of a histogram and divided (mosaic) bar chart showing a bar for each level of the ordinal or nominal variable.

The distribution plots are supported with text reports. For interval variables, the reports show selected quantiles and moments. For ordinal and nominal variables, the reports show counts, proportions, and total log likelihood.

There are options which let you save ranks, probability scores, and normal quantile values as new columns in the data table.

Chapter Contents

The Display

The **Distribution of Y's** command produces a graphical description and supporting descriptive statistics for each column assigned as a Y variable. An example of the display is shown in **Figure 14.1.** The graphical display consists of a histogram for each variable accompanied by either a mosaic bar for ordinal or nominal columns or a quantile plot for interval variables. There are text report reveal buttons for all variables, and these reports display supporting tables for interval variables by default.

The graphs for each response variable can be resized using the plot frame stretch box located in the lower right corner of the plot. The heights of both a histogram and its accompanying mosaic bar or quantile plot always adjust as one unit even though the graphs have separate stretch boxes. However, the widths of all plots can be adjusted independently.

You can also access special pop-up menu options which are located by the icons to the left of the horizontal scroll bar.

Figure 14.1 The Distribution of Y Platform

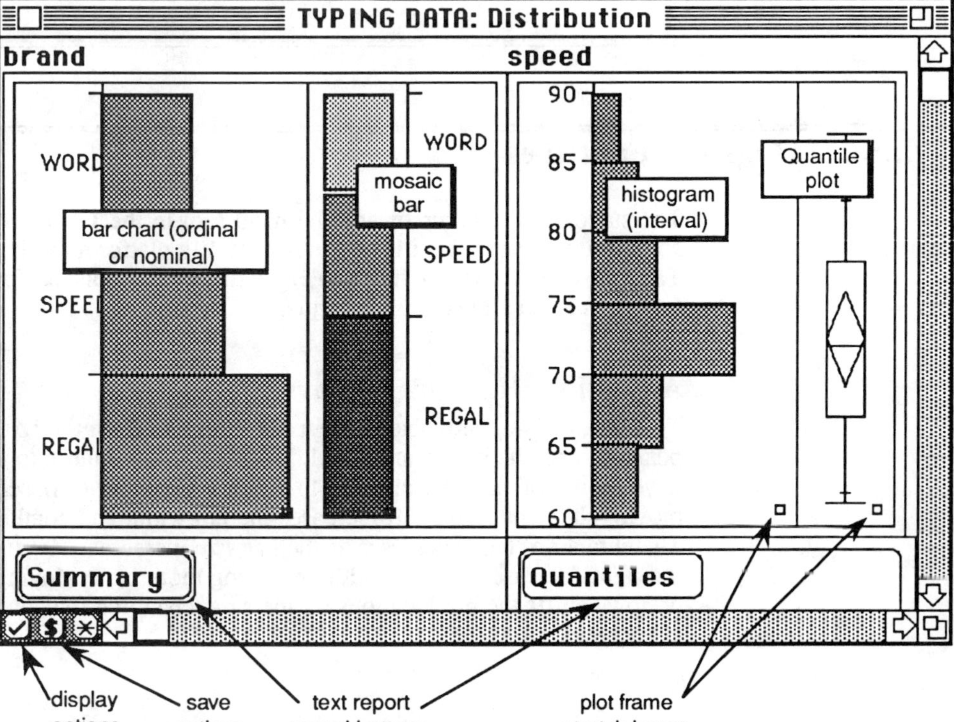

You can select one or more columns for analysis by assigning each the role of Y using the role assignment pop-up menus on the spreadsheet. If you do not preassign roles on the spreadsheet, you are prompted with the selection dialog shown in **Figure 14.2.** You can choose any columns in the data table as response variables. Use standard mouse selection actions to select from the dialog.

Figure 14.2 The Select Columns For Distribution Dialog

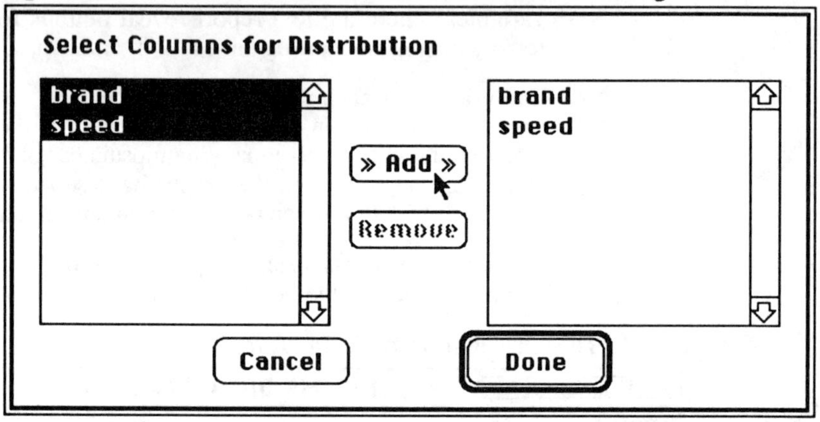

Dynamic Display Features

The standard arrow cursor and the hand tool in the **Tools** Menu perform special functions on the **Distribution of Y's** platform as discussed below. The brush tool has no function, and the other tools perform the same functions as on all display platforms.

Adjusting Histogram Bars

To adjust a histogram's bars, select the hand tool from the **Tools** menu and position it on the histogram. Hold down the mouse button to *grab* the plot. If you think of each bar as a bin that holds a number of observations, then moving the hand to the left increases the bin width and combines intervals. The number of bars decreases as the bar size increases. Moving the hand to the right decreases the bin width producing more bars. Moving the hand up or down shifts the bin locations on the axis, changing the contents and size of each bin.

Figure 14.3 Adjusting the Number of Bars

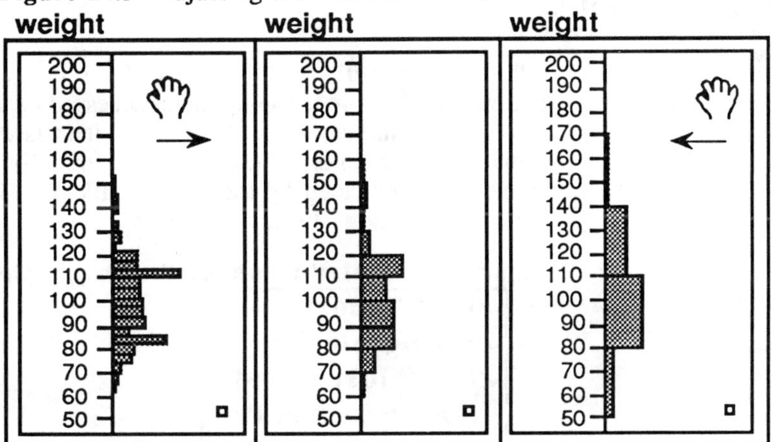

Highlighting Bars and Selecting Rows

Clicking on any histogram bar with the arrow cursor highlights the bar and selects the corresponding rows in the data table. The appropriate portions of all other graphical displays also highlight to show the selection. **Figure 14.4** shows the results of highlighting a bar in the height histogram. The corresponding rows are selected in the data table, and the same values highlight on the sex histogram to show the conditional distribution.

Use command-click to deselect any highlighted histogram bar.

Figure 14.4 Highlighted Bars and Rows

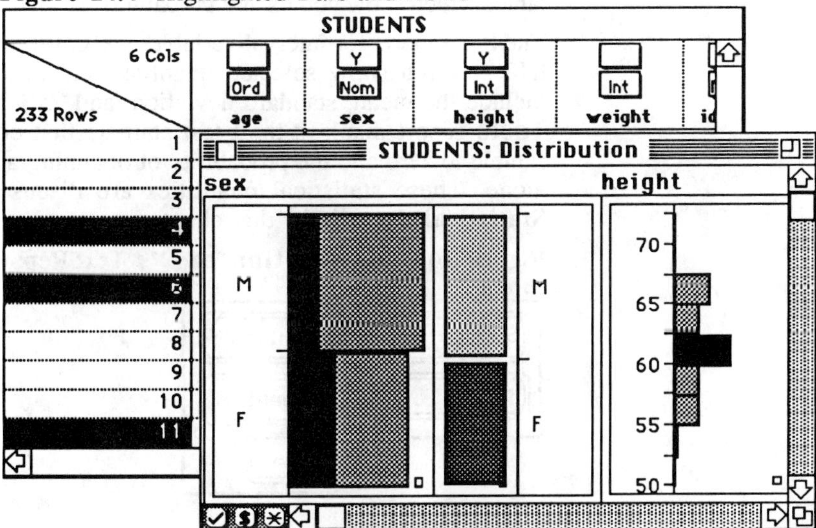

Display Options

All parts of the **Distribution of Y's** report are optional. The display options are listed in a pop-up menu accessed by the check mark to the left of the horizontal scroll bar. Each menu item is a toggle, which alternately turns the option on or off when selected. Initially, the first three options are in effect as indicated by the check marks in **Figure 14.5**. The interval variable text reports display by default.

Figure 14.5 **Distribution of Y's** Display Options Menu

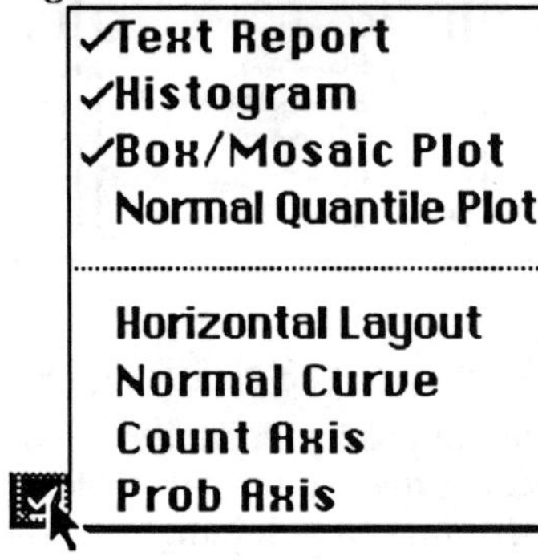

Text Report displays the reveal/conceal buttons shown in **Figure 14.6**. These are used to display statistical summary tables for the response variables.

The text reports for nominal and ordinal values are Summary and Profile tables which show counts and probabilities.

The text reports for interval variables are Quantiles, Moments, and Tests tables which show selected quantiles and descriptive moments. These include the mean, standard deviation, and 95% confidence intervals. The Tests report displays the t-test statistic that compares the mean of the sample to zero and the probability of obtaining a greater t-value by chance alone. These statistical quantities are discused in the section called **Statistical Details** in this chapter.

Figure 14.6 **Distribution of Y's** Text Report Buttons

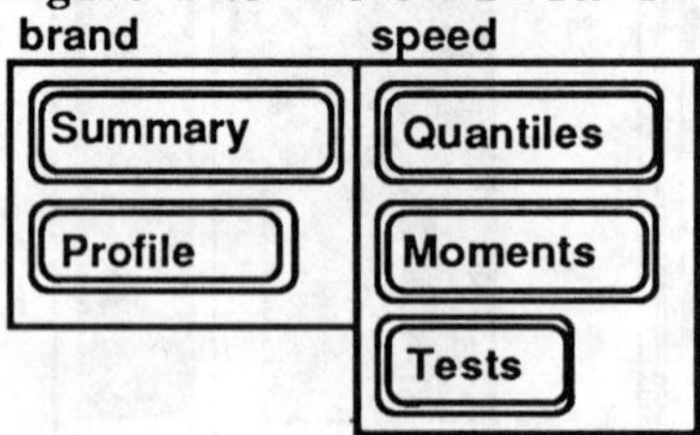

Histogram displays histograms for all response variables. Histograms of nominal and ordinal variables are bar charts with a bar for each value (level) of the variable. Histograms of interval variables show bars for a selected number of intervals. The number of bars or intervals can be changed using the hand tool from the **Tools** menu.

Figure 14.7 Histograms of Response Columns

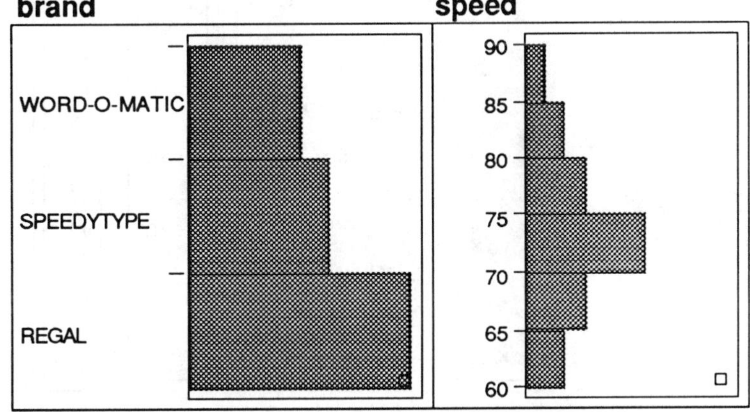

Box/Mosaic Plot displays either a box plot or a mosaic bar plot for each response variable. The display in **Figure 14.8** shows the **Box/Mosaic Plot** option selected alone. The variable brand is nominal and speed is interval.

- The plot for ordinal or nominal variables is a mosaic stacked bar chart where each segment is proportional to its group's sample size.
- The quantile box plot shows selected quantiles on the response axis. The box shows the median as a line across the middle and the quartiles (25th and 75th percentiles) as its ends. The means diamond identifies the mean of the sample and the 95% confidence interval about the mean.

For more information about quantiles and box plots, see the section called **Statistical Details** in this chapter.

Figure 14.8 Mosaic and Box Plots of Response Columns

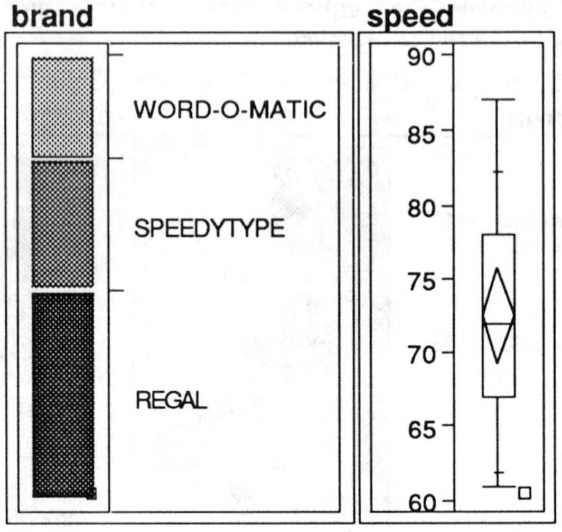

Normal Quantile Plot produces an additional graph for each interval variable which is useful for visualizing the extent to which the variable is normally distributed. If a variable is normal, the normal quantile plot is a diagonal straight line. This plot is also referred to as a theoretical quantile-quantile plot or Q-Q plot. More examples of normal quantile plots are discussed in the section called **Statistical Details** in this chapter.

Figure 14.9 Normal Quantile Plot

3.5
3
2.5
2
1.5
1
0.5
0
-0.5
-1
-1.5
-2
-2.5

-3 -2 -1 0 1 2 3

Normal Quantile

Horizontal Layout arranges text reports to the right of their corresponding graphical displays. In **Figure 14.10**, the options for **Horizontal Layout**, **Text Report**, and **Histogram** are checked.

Figure 14.10 Horizontal Layout of Graphical Display

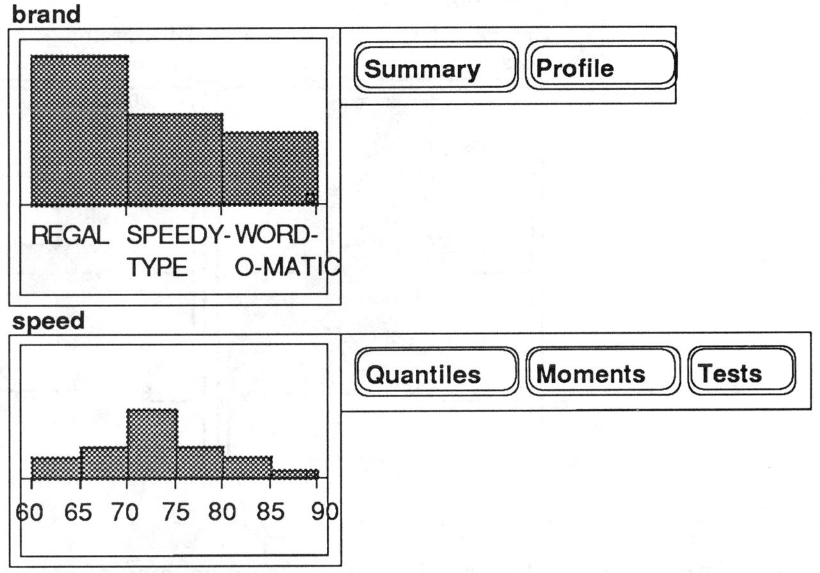

Normal Curve superimposes a normal curve on the histogram of each interval variable. The curve is computed from the mean and standard deviation of the column. **Figure 14.11** displays a histogram with the options for **Normal Curve** and **Horizontal Layout** checked.

Figure 14.11 Normal Curve on Interval Response Histogram

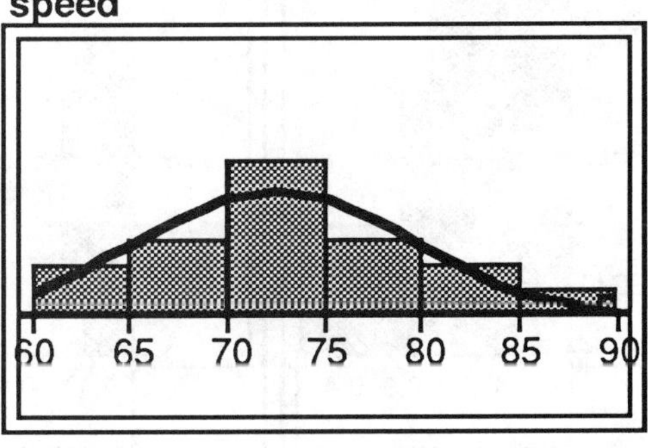

Count Axis adds an axis showing the frequency of column values represented by the histogram bars. **Figure 14.12** illustrates the use of the ruler tool from the **Tools Menu** and the **Count Axis** option to estimate the frequency of rows for speed values from 70 to 75. Fractional values on the axis are due to scaling.

Figure 14.12 Count Axis on Histograms

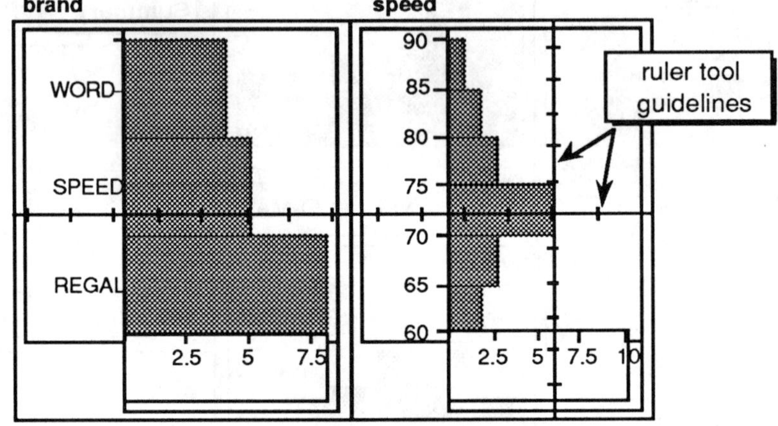

Prob Axis adds an axis showing the proportion of column values represented by histogram bars. The **Prob Axis** option can be used alone or along with the **Count Axis** option.

Figure 14.13 Probability Axis on Histograms

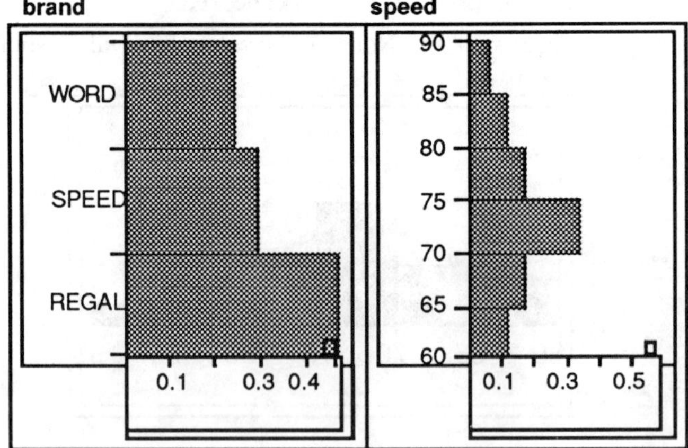

Save Commands

You can save information computed from response variables using the save ($) pop-up menu at the left of the horizontal scroll bar. Each command generates a new column in the current data table. However, all commands except **Save Level Numbers** apply only to interval data.

Note➜ The save commands are not toggles and can be selected repeatedly, allowing you to apply the *exclude* status to rows before saving each new set of analysis information. However, it is important to realize that for each response variable, every new column you generate using a particular save command has the same name. If you use a save command multiple times, it is best to rename each new column in the data table to reflect each specific analysis.

Figure 14.14 Statistical Save Commands

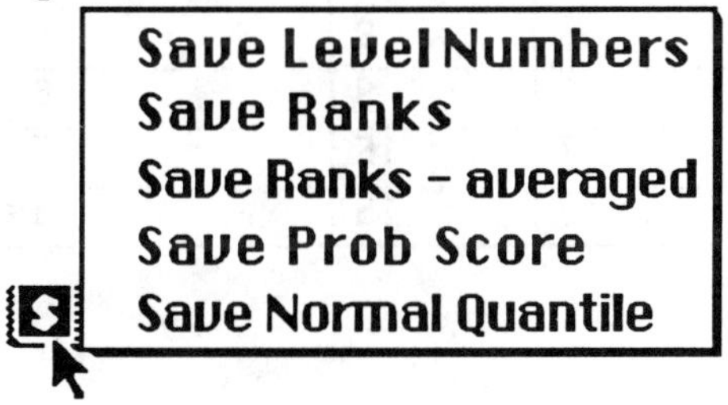

Save Level Numbers
Save Ranks
Save Ranks – averaged
Save Prob Score
Save Normal Quantile

Save Level Numbers creates a new column, called Level *colname*, for each response variable in the current data table. The level number of each observation corresponds to the histogram bar that contains the observation. The histogram bars are numbered from low to high beginning with 1.

Figure 15.15 Level Number Column

6 Cols / 8 Rows	Y Int nums	Int Level nums
1	0	1
2	4	1
3	4	1
4	4	1
5	9	2
6	9	2
7	15	3
8	15	3

Save Ranks creates a new column for each interval response variable in the current data table. Each new column, called Ranked *colname*, contains a ranking for each of the corresponding column's values. Duplicate response values are assigned consecutive ranks in order of their occurrence in the spreadsheet.

Figure 14.16 Ranks Column

6 Cols / 8 Rows	Y Int nums	Int Ranked nums
1	0	1
2	4	2
3	4	3
4	4	4
5	9	5
6	9	6
7	15	7
8	15	8

Save Ranks -averaged creates a new column, called RankAvgd *colname*, for each interval response variable in the current data table. If a value is unique, the averaged rank is the same as the rank. If a value occurs *k* times, the average rank is computed as the sum of the value's ranks divided by *k*.

Figure 14.17 Ranks - averaged Column

6 Cols / 8 Rows	Y Int nums	Int RankAvgd nums
1	0	1
2	4	3
3	4	3
4	4	3
5	9	5.5
6	9	5.5
7	15	7.5
8	15	7.5

Save Prob Score creates a new column, called Prob *colname*, for each response variable in the current data table. For N non-missing scores, the probability score of a value is computed as the averaged rank of that value divided by (N + 1). This column is similar to the empirical cumulative distribution function.

Figure 14.18 Probability Score Column

6 Cols / 8 Rows	Y Int nums	Int RankAvgd nums	Int Prob nums
1	0	1	.111111111
2	4	3	.333333333
3	4	3	.333333333
4	4	3	.333333333
5	9	5.5	.611111111
6	9	5.5	.611111111
7	15	7.5	.833333333
8	15	7.5	.833333333

Save Normal Quantile creates a new column, called N-Quantile *colname*, for each response variable in the current data table. The normal quantile values are computed as

$$\Phi^{-1}\left(\frac{r_i}{N+1}\right)$$

where Φ^{-1} is the cumulative probability distribution function for the normal distribution, r_i is the rank of the ith observation, and N is the number of non-missing observations. These normal scores are Van Der Waerden approximations to the expected order statistics for the normal distribution.

Figure 14.19 Normal Quantile Score Column

6 Cols / 8 Rows	Y Int nums	Int N-Quantile nums
1	0	-1.2206403
2	4	-.43072729
3	4	-.43072729
4	4	-.43072729
5	9	.282216147
6	9	.282216147
7	15	.967421566
8	15	.967421566

Statistical Details

The Box Plot and Quantiles Report

If a column has interval values, the **Distribution of Y's** analysis produces a histogram and quantile box plot. These distribution plots are supported with text reports which show selected quantiles and moments of interval values.

The Quantiles report lists the values of selected quantiles, sometimes called percentiles. These quantiles are also shown on the box plot accompanying the histogram. Note that JMP box plots are skeletal box plots and are not the same as Tukey schematic box plots which are used for outlier identification.

Quantiles are values that divide a distribution into two groups where the Pth quantile is larger than P% of the values. For example, half the data are below and half the data are above or equal to the 50th quantile, also called the median.

Figure 14.20 The Box Plot and Quantiles Report

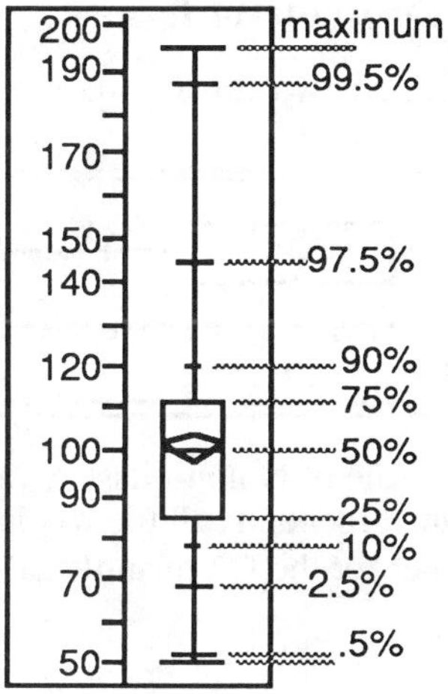

Quantiles

maximum	100.0%	196.00
	99.5%	188.35
	97.5%	145.30
	90.0%	121.00
quartile	75.0%	112.00
median	50.0%	100.00
quartile	25.0%	85.00
	10.0%	78.40
	2.5%	69.00
	0.5%	53.21
minimum	0.0%	51.00

If a distribution is normal, the quantiles shown in a JMP box plot are approximately equidistant from each other, particularly for 90% of the central values. This lets you see at a glance if the distribution is normal. For example, if the quantile marks are grouped closely at one end but have greater spacing at the other end, the distribution is *skewed* toward the end with more spacing (as in **Figure 14.20**).

Warning→ The quantile estimates for the extreme ends of the distribution tails are highly variable and, thus, you cannot expect equal spacing unless the sample is large. **Table 14.1** shows the spacing of quantiles for a normal (0,1) distribution. The two-tailed outside area is computed as

$$2\,(1 - \text{probability})$$

which approximately corresponds to equal normal quantile spacing.

Table 14.1 Spacing of Quantiles in The Normal Distribution

Probability	Normal Quantile	* Normal Quantile ÷ .674	Outside Area (Both Tails)
.5	0	0	1
.75	*.674	1	.5
.90	1.282	1.9	.2
.975	1.960	2.9	.05
.995	2.576	3.8	.01

* Approximate equal spacing between selected quantiles in normal data

The quantile plot shown in **Figure 14.21** is the result of a simulation of 2000 normal (0,1) points. When the sample is large, the quantiles are approximately equidistant from each other and spaced according to the normal quantile values in **Table 14.1**.

Figure 14.21 Quantiles Shown for Simulation of Normal Distribution

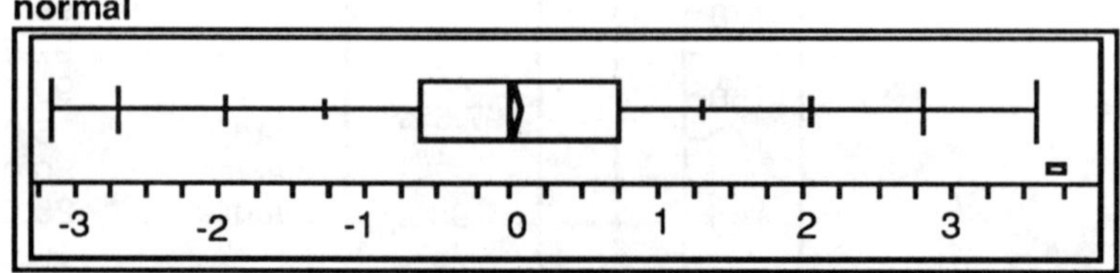

Computation of quantiles

To compute the Pth quantile of N non-missing values in a column, arrange the N values in ascending order and call these column values $y_1, y_2, \ldots, y_N$. Compute the rank number for the Pth quantile as

$$\frac{P(N+1)}{100} \quad .$$

- If the result is an integer, the Pth quantile is that rank's corresponding value.
- If the result is not an integer, the Pth quantile is found by interpolation. Denote the integer portion of the computed rank number as I and the fractional portion as f. The Pth quantile, denoted q_P, is computed

$$q_P = (1-f)y_I + (f)y_{I+1} \quad .$$

If I = N, then y_N is taken as the quantile.

For example, suppose a data table has 15 rows, and you want to find the 75th and 90th quantile values of an interval column. After the column is arranged in ascending order, the ranks that contain these quantiles are computed as

$$.75(16)=12$$

$$.90(16)=14.4 \quad .$$

The value, y_{12}, is the 75th quantile.

The 90th quantile is interpolated by computing a weighted average of the 14th and 15th ranked values as follows:

$$q_{90} = .6y_{14} + .4y_{15}$$

The Normal Quantile Plot

The normal quantile plot shows the column value on the Y axis and the expected normal scores for each value on the X axis. The examples in **Figure 14.22** show normal quantile plots for simulations of 400 points from two distributions. The first example is the normal quantile plot for a normal distribution and appears as a diagonal linear pattern. The second example is for a uniform distribution. It is *heavy tailed* and produces an S shaped quantile plot. A very peaked or *light-tailed* distribution produces an inverted S shaped quantile plot (not shown).

Figure 14.22 Normal Quantile Plot Comparison

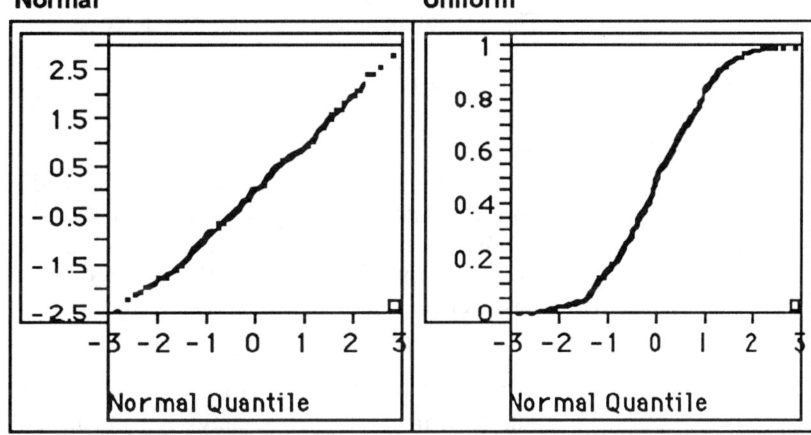

Squaring a normal distribution yields a new distribution that is skewed to the right. This produces the concave normal quantile plot which is labeled **Normal Squared** in **Figure 14.23**. A distribution that is skewed to the left produces the convex pattern similar to the one shown in the example labeled **–Normal Squared.**

Figure 14.23 Normal Quantile Plot Examples of Skewed Distributions

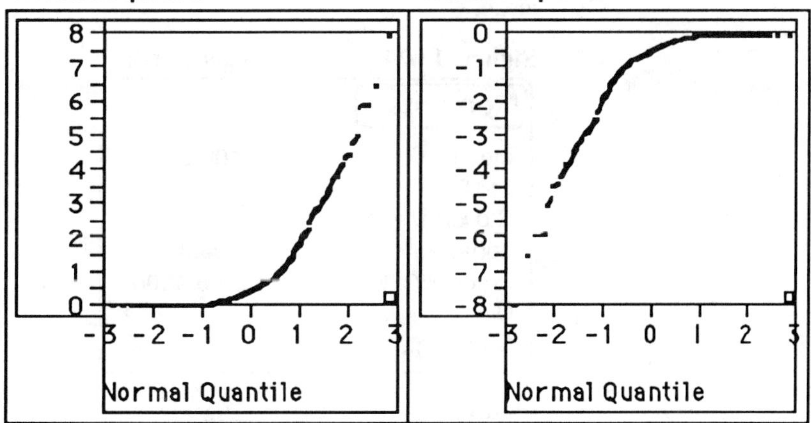

One other pattern not shown here is a 'staircase' pattern which is indicative of data that have been rounded or have discrete values.

These example normal quantile plots were generated from simulated distributions to provide you with a guide for interpretation of normal quantile plots. As with all statistical patterns they can occur purely by chance, particularly when the sample is small.

The normal quantile values are computed as

$$\Phi^{-1}\left(\frac{r_i}{N+1}\right)$$

where Φ^{-1} is the cumulative probability distribution function for the normal distribution, r_i is the rank of the ith observation, and N is the number of non-missing observations. These normal scores are Van Der Waerden approximations to the expected order statistics for the normal distribution.

The Moments Report

The Moments report displays the mean, standard deviation, and other summary statistics. If the sample is normally distributed, the values are arranged symmetrically around the mean (arithmetic average). The mean and standard deviation completely define a normal distribution, giving an easy way to summarize data as the sample becomes large.

- 68% of the values are within one standard deviation of the mean.
- 95% of the values are within two standard deviations of the mean.
- More than 99% of the values are within three standard deviations of the mean.

Figure 14.24 The Moments Report

Moments	
Mean	100.9142
Std Dev	19.1847
Std Err Mean	1.2568
upper 95% Mean	103.3776
lower 95% Mean	98.4508
N	233
Sum Wgts	233

Mean

estimates the expected value of the underlying distribution for the response variable. It is the arithmetic average of the column's values and is computed as the sum of the non-missing values divided by the number of non-missing values.

If the data table has a column assigned the role of weight, the mean is computed by multiplying each column value by its corresponding weight, adding these values, and dividing by the sum of the weights. The formula for the mean of a column Y with N non-missing values and a weight column W is written

$$\bar{y} = \frac{\sum_{i=1}^{N} w_i y_i}{\sum_{i=1}^{N} w_i} .$$

Alternatively, if no weight column is specified, then w_i is 1 for all i and $\Sigma w_i = N$.

Std Dev

measures the spread of a distribution around the mean. It is frequently denoted as s and is the square root of the sample variance, denoted s^2. For a normal distribution, about 68% of the distribution is within one standard deviation of the mean. The sample standard deviation is calculated as

$$s = \sqrt{s^2}, \text{ where } s^2 = \frac{\sum_{i=1}^{N} w_i (y - \bar{y})^2}{N - 1} .$$

Std Err Mean

estimates the standard deviation of a distribution of means, each calculated from a sample size of N. Std Err Mean, the standard error of the mean, is computed by dividing the sample standard deviation, s, by the square root of N. If a column is assigned the role of weight, then the denominator is the square root of Σw_i. This statistic is used in t-tests of means and in constructing confidence intervals for means. It is computed as

$$s_{\bar{y}} = \frac{s}{\sqrt{\Sigma w_i}}, \text{ where } s^2 = \frac{\sum_{i=1}^{N} w_i (y_i - \bar{y})^2}{N - 1} .$$

If no weight column is specified, then w_i is 1 for all i and $\Sigma w_i = N$.

upper 95% Mean
and lower 95% Mean

are 95% confidence limits about the mean. They define an interval which is very likely to contain the true population mean. If many random samples are drawn from the same population and each 95% confidence interval is determined, you expect 95% of the confidence intervals so computed to contain the true population mean. The upper and lower limits are computed as

$$\overline{y} + t\, s_{\overline{y}}$$

$$\overline{y} - t\, s_{\overline{y}}$$

where t is the Student t value for $\alpha = .05/2$ and $N - 1$ degrees of freedom.

N

is the total number of non-missing values and is used to compute statistics in the Moments Table when no column is assigned the role of weight.

Sum Wts

is the sum of a column assigned the role of weight and is used in the denominator for computations of the mean instead of N.

The Tests Report

The Tests report presents a t-test statistic that applies only when you want to compare the mean of a column to zero. It is used most often as a paired t-test where the column represents the difference of two paired values for a series of observations. The data shown in **Figure 14.25** represent the yields from popping eight batches of popcorn. Each batch was divided in half with one half specially treated (gourmet) and the other left untreated (plain). The difference in yield for each batch is computed and selected as the response (Y) variable.

The Moments report summarizes the distribution for the difference column, and the mean and standard error of the mean are used to produce the Tests report shown in **Figure 14.25**.

Figure 14.25 The Moments and Tests Reports

POPCORN

3 Cols / 8 Rows	plain yield (Int)	gourmet yield (Int)	difference (Int)
1	8.2	8.6	0.4
2	10.4	9.2	-1.2
3	9.9	12.1	2.2
		18.0	7.4
		8.2	-0.6
		9.8	1
		15.9	5.8
		16.0	8.6

Moments

Mean	2.950000
Std Dev	3.789836
Std Err Mean	1.339909
upper 95% Mean	5.576222
lower 95% Mean	0.323778
N	8
Sum Wgts	8

Tests

t Ratio for mean=0	2.20164
Prob > |t|	.063576
Prob > t	.031788
Prob < t	.968211

t Ratio for mean=0

is the test statistic that compares the sample mean to zero using statistics from the Moments report. It is computed as

$$\frac{\text{Mean}}{\text{Std Err Mean}} \quad .$$

Prob > |t|

is the probability of obtaining a greater absolute t-value by chance alone when the sample mean is not different from zero. This is the P-value for observed significance of the two-tailed t test.

Prob > t

is the probability of obtaining a t value *greater than* the computed sample t ratio by chance alone when the sample mean is not zero. This is the P-value for observed significance of a one-tailed t test.

Prob < t

is the probability of obtaining a t value *less than* the computed sample t ratio by chance alone when the sample mean is not zero. This is the P-value for observed significance of a one-tailed t test.

The Summary Report

The Summary report describes the overall characteristics of an ordinal or nominal response variable.

Figure 14.26 The Summary Report

Summary	
Number of Levels	4
Total Frequency	32
Total LogLikelihood	40.365690

Number of Levels
: is the number of different values found for a nominal or ordinal response variable.

Total Frequency
: is the total non-missing values, N, found for a nominal or ordinal response variable.

Total LogLikelihood
: is a measure of spread, called uncertainty, for a nominal or ordinal sample distribution. The total log likelihood is computed as

$$\sum_{i=1}^{k} n_i(-\log_2 p_i)$$

where k is the number of levels, n_i is the size of the ith level, and

$$p_i = \frac{n_i}{N} .$$

The total log likelihood is analogous to a corrected sum of squares for an interval response.

The Profile Report

The Profile report lists specific information for each level of a nominal or ordinal response variable.

Figure 14.28 The Profile Report

Profile

Level	Count	Probability	Cum Prob
A	4	0.12500	0.12500
B	6	0.18750	0.31250
C	7	0.21875	0.53125
D	15	0.46875	1.00000

Level
lists each value found for a response variable.

Count
lists the number of rows found for each level of a response variable.

Probability
lists the probability of occurrence for each level of a response variable. The probability is computed as the count divided by the total frequency of the variable (as displayed in the summary table for the variable).

Cum Prob
contains the cumulative sum of the column of probabilities.

Chapter 15. Analyze: Fit Y by X

The **Fit Y by X** command analyzes pairs of variables, examining the distribution of a response variable Y as conditioned by the values of a factor X. There are plots with accompanying text reports presented for each combination of Y and X. The displays and reports vary depending on the levels of measurement you specify for the X and Y columns, where the level of measurement for each column can be *interval*, *ordinal*, or *nominal*.

The two variable roles (X and Y) and the three levels of measurement, (interval, ordinal and nominal) produce five different analysis combinations, each with a graphical display and corresponding statistical text reports as summarized in **Table 15.1**.

This chapter begins with an overview of the **Fit Y by X** platform and follows with sections that address each kind of graphical display and analysis.

Table 15.1 Five Kinds of Display and Analysis

Variable Role and Measurement Level		Report Platform
Y	X	PLOT AND ANALYSIS
1. interval	interval	scatterplot and regression, curve fitting
2. interval	nominal or ordinal	means plot and analysis of variance
3. nominal or ordinal	nominal or ordinal	contingency table mosaic plot and categorical data analysis
4. nominal	interval	logistic probability plot and logistic regression
5. ordinal	Interval	cumulative probability plot and logistic regression

Chapter Contents

The Display

The **Fit Y by X** command launches a report platform for each pair of columns given X and Y roles in the active data table. If there are several variables assigned X and Y roles, each X and Y combination produces a plot and accompanying text reports.

The type of platform varies according to the measurement levels of the Y and X variables. **Figure 15.1** shows the platforms as they correspond to the five measurement level combinations listed in **Table 15.1.** The family of logistic regression curves applies to either a nominal or ordinal response given an interval factor.

Figure 15.1 Plots for Different Response, Factor Combinations

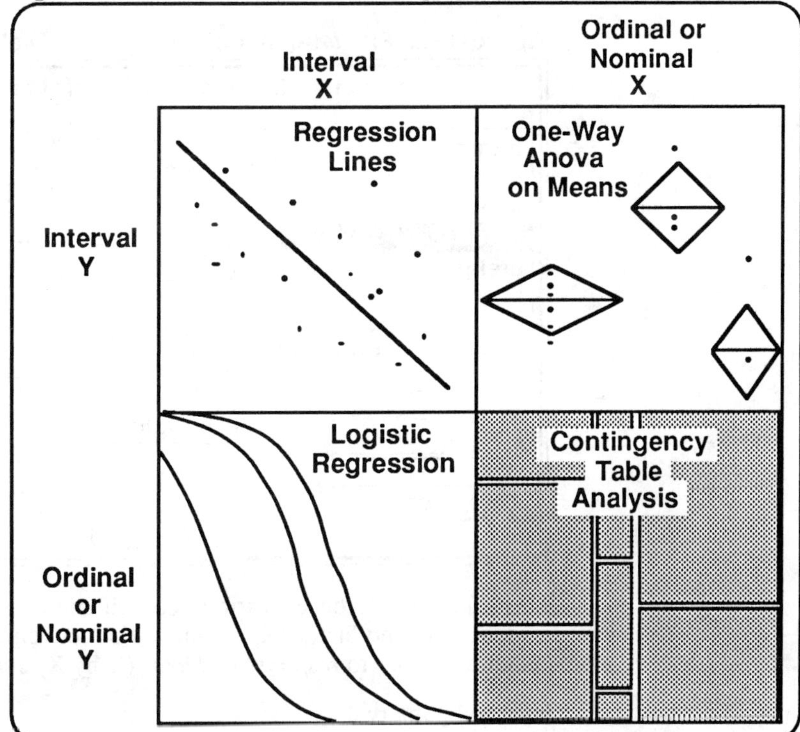

If you have not already assigned at least one X and one Y column, JMP prompts you to assign roles with the dialog shown in **Figure 15.2.** The selector list on the left side of the dialog shows all columns in the current data table. The **X**, **Y**, **Weight**, **Freq**, and **Label** buttons are dimmed until you select one or more variables from this variable selector list. Click on the appropriate role button after selecting variables for that role. Be sure to assign at least one X and one Y column. Each role is is listed below.

- Y identifies one or more response variables.
- X identifies one or more factors.
- **Weight** is an optional role that identifies one column whose values supply weights for the response in each Y by X combination.
- **Freq** is an optional role that identifies one column whose values assign a frequency to each row for the analysis.
- **Label** is an optional role that identifies a column whose values specify a labeling convention for points in plots.

To remove unwanted variables from assigned roles, select them in their role box and click on the **Remove** button.

After assigning roles, click on **Done** to see the analysis for each Y by X combination.

Figure 15.2 **Fit Y by X** Role Assignment Dialog

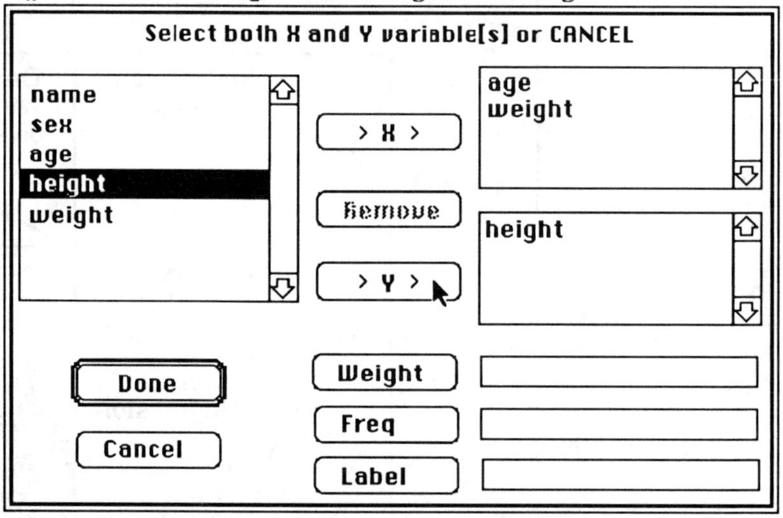

This chapter references various combinations of measurement level for response (Y) and factor (X) columns. For convenience, the following short-cut phrases refer to *response by factor* (Y by X) combinations.

- interval by interval
- interval by nominal/ordinal
- nominal/ordinal by nominal/ordinal
- nominal by interval
- ordinal by interval

The remainder of this chapter covers each kind of Y by X analysis combination. The data for the examples are in the JMP file called BIG CLASS found in the SAMPLE data folder.

Regression Analysis and Curve Fitting

Regression analysis and curve fitting apply to interval by interval Y by X combinations. This platform's initial stage is a scatterplot of points for each X, Y pair. You can superimpose various fits on the points, which approximate or predict the Y variable as a function of the X variable.

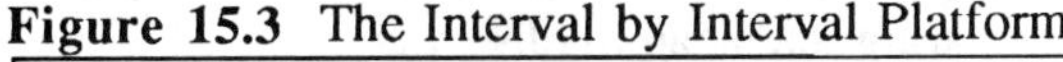

Figure 15.3 The Interval by Interval Platform

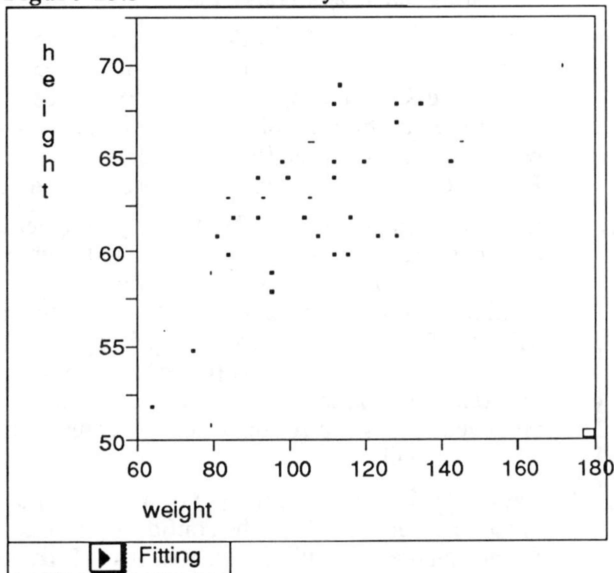

Most JMP displays include features which allow you to resize graphs, highlight points with the cursor or brush tool, and label points. See Chapter 12, *The Report Window* for a description of these features.

Fitting Options

The Fitting pop-up menu, shown in **Figure 15.4**, lists a **Show Points** toggle and five fit commands.

Note➔ The fit commands are not toggles and can be selected repeatedly. This allows you to try fits with different points excluded to compare them on the same plot.

Figure 15.4 Interval by Interval Fitting Pop-up Menu

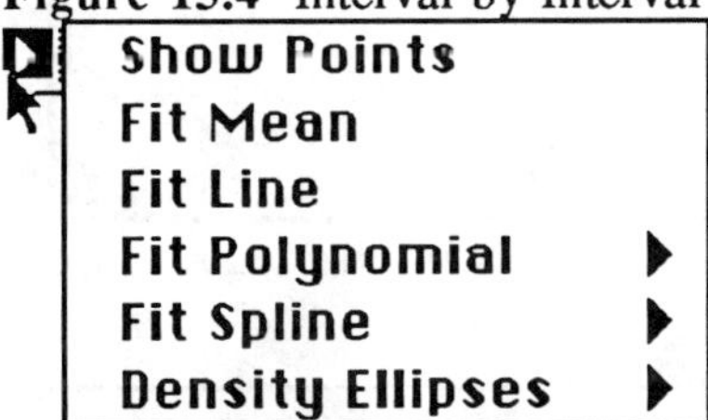

The following list describes each of the fitting options.

Show Points alternately hides or displays the points in the plot.

Fit Mean adds a horizontal line to the plot at the mean of the response variable (Y). You can add confidence lines to the mean with the **Confidence Limits** display option described in the next section. It is useful to fit the mean first as a reference line for other fits. See **Figure 15.5** for an example comparing the mean to other fits.

Fit Line adds a straight-line fit to the plot using least squares regression. Curves representing the 95% confidence limits for the line of fit also appear. You can suppress them using the **Confidence Limits** display option described in the next section. Additional options let you save predicted values and residuals for the linear fit as new columns in the current data table. See **Figure 15.5** for an example of a linear fit compared to other fits.

Fit Polynomial fits a polynomial curve of the degree you select from the **Fit Polynomial** submenu. After you select the polynomial degree, the curve is fit to the data points using least squares regression. Curves representing the 95% confidence limits for the line of fit also appear. You can suppress them using the **Confidence Limits** display option described in the next section. You can select the **Fit Polynomial** option multiple times and choose different polynomial degrees for comparison. Also, options let you save predicted values and residuals as new columns in the current data table for each polynomial fit.

Fit Polynomial Submenu

- 2-quadratic
- 3-cubic
- 4-quartic
- 5
- 6

Figure 15.5 shows an example of a polynomial fit of degree 2, compared to the mean and line fits. The confidence curves for the line and polynomial fits are suppressed with the **Confidence Curves** display option described in the next section.

Figure 15.5 Example of Selected Fits

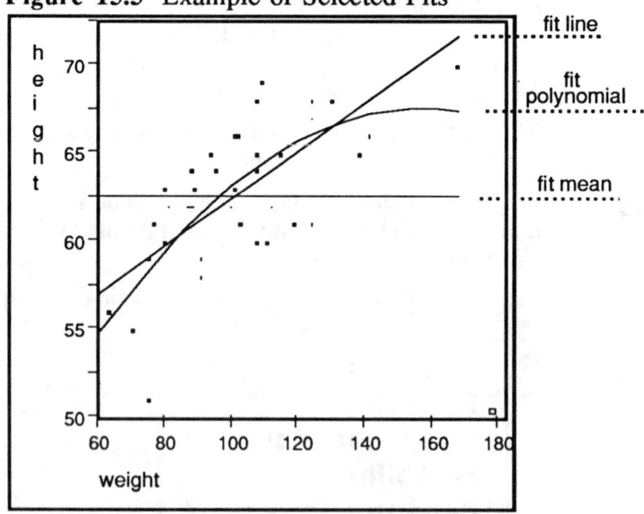

Fit Spline

Fit Spline Submenu

1000000-stiff
100000
10000
1000
100
10
1
.1
.01-flexible
other...

fits a smoothing-spline using a smoothness parameter, called lambda, which you choose from the **Fit Spline** submenu. If you want to use a lambda value not listed on the submenu, choose the **other** selection and enter the value when prompted. As the value of lambda decreases, the error term of the fit has more weight and the fit becomes more curved. Higher values of lambda make the fit stiff (less curved), approaching a straight line.

The spline fit is a set of third degree polynomial segments that are spliced together such that the resulting curve is continuous and smooth at the splices (knot points).

Figure 15.6 Example of Selected Spline Fits

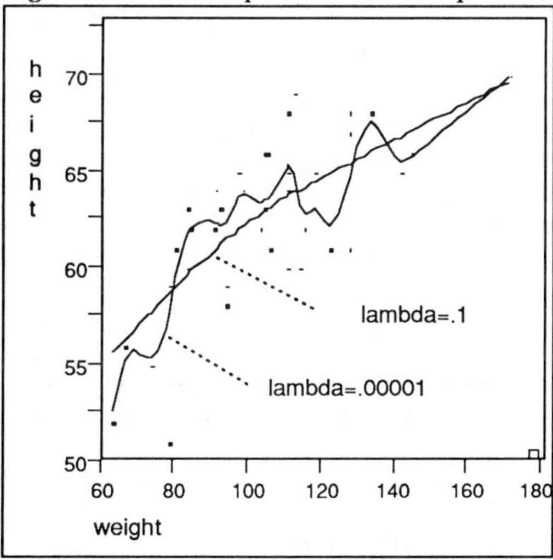

Density Ellipses

Density Ellipses Submenu

.99
.95
.90
.50
other...

draws an ellipse which contains the specified mass of points as determined by the probability you choose from the **Density Ellipses** submenu. The density ellipsoid is computed from the bivariate normal distribution fit to the X and Y variables. The bivariate normal density is a function of the means and standard deviations of the X and Y variables and the correlation between them. The **other** selection lets you specify any probability greater than zero and less than or equal to one.

Figure 15.7 Example of Selected Density Ellipses

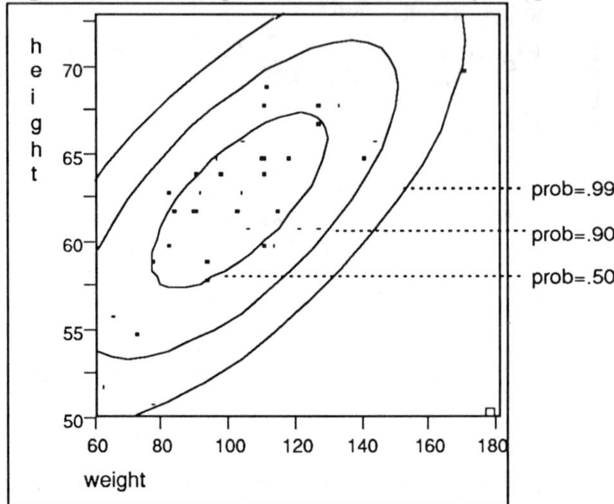

Display Options and Save Commands

Each fit is drawn on the plot with a solid line. The **Fit Line** and **Fit Polynomial** commands automatically produce confidence curves with dashed lines. Confidence curves can be added to the **Fit Mean** command by request. Each fit you select adds a pop-up menu with additional display features and a text report reveal button beneath the scatterplot. For example, the options shown in **Figure 15.8** result when you select **Fit Mean**, then **Fit line**, and finally **Fit Polynomial** with degree 2.

Figure 15.8 Example of Fit Information and Text Reveal Buttons

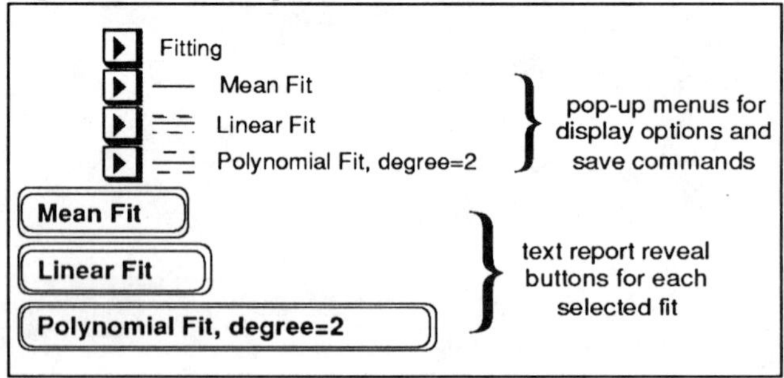

The pop-up menu for each requested fit offers the following display options and save commands when they are appropriate to the fit.

Figure 15.9 Display Options and Save Commands for Each Fit

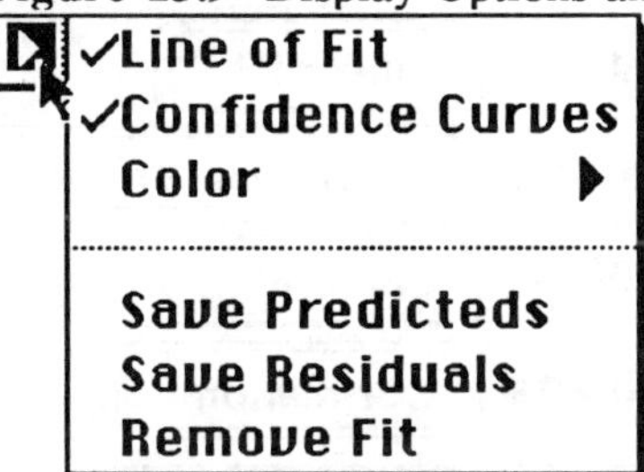

Line of Fit displays or hides the line of fit.

Confidence Curves displays or hides confidence curves for mean, line, and polynomial fits. The **Confidence Curves** option does not apply to the spline and density ellipse fits and is dimmed on those menus.

Color > changes the color of the line of fit on color monitors. There are 15 colors from which to choose.

Save Predicteds creates a new column in the current data table called Predicted *colname* that contains the predicted values of the Y variable called *colname*. **Save Predicteds** does not apply to the mean and density ellipse fits and is dimmed on those menus. For linear and polynomial fits, this option saves the prediction formula with the predicted column.

Save Residuals creates a new column in the current data table called Residuals *colname* that contains the residual values of the Y variable called *colname*. **Save Residuals** does not apply to the mean fit or density ellipses and is dimmed on those menus.

Remove Fit removes the fit from the graph and removes its text reports.

Note➜ You can use the **Save Predicteds** and **Save Residuals** commands for each fit. If you use these commands multiple times, it is best to rename the columns in the data table to reflect each fit.

Statistical Details for Regression Analysis

Linear regression analysis is the process of fitting equations, called linear models, to data values. These equations fit the data such that the sum of squared distances to the resulting regression line or curve is minimized. JMP offers a variety of regression equations.

Each time you select a fit, a text report reveal button is added to the platform. The type of report varies depending on the kind of fit you request.

Figure 15.10 Text Report Reveal Buttons for Each Kind of Fit

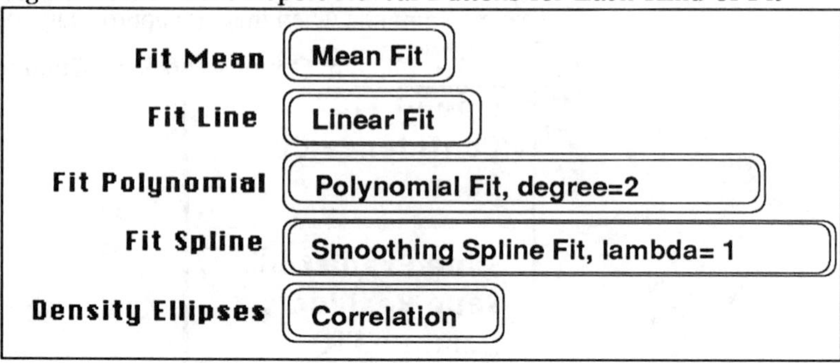

Fit Mean fits a horizontal straight line at the response mean. This is equivalent to a polynomial fit of degree 0.

Fit Line finds the parameters β_0 and β_1 for the straight line that fits the points and minimizes the residual sum of squares. The line is drawn with a set of confidence curves showing its 95% confidence intervals. If this confidence area includes the response mean, then the slope of the line of fit is not significantly different from zero at the .05 significance level. The predicted value of the *i*th row is computed as

$$y_i = \beta_0 + \beta_1 x_i + \varepsilon_i$$

The **Fit Line** command is equivalent to a polynomial fit of degree 1.

Fit Polynomial fits a polynomial curve of a specified degree (k)to the data where $k > 1$. A polynomial of degree 2 is a parabola, and degree 3 is a cubic curve. For degree k, the model for the *i*th observation is written as

$$y_i = \sum_{j=0}^{k} \beta_j x_i^j + \varepsilon_i$$

Fit Spline fits is a smoothing spline which varies in smoothness (or flexibility) according to lambda (λ), a tuning parameter in the spline formula that you specify. Lower values of lambda produce a flexible line with more curves, and higher values result in a stiff spline, closer to a straight line. You may find yourself trying several lambda values before you settle on the spline curve that seems best for your data.

The cubic spline method uses a set of third degree polynomials spliced together such that the resulting curve is continuous and smooth at the splices (knot points). The estimation is done by minimizing an objective function that is a combination of the sum of squares error and a penalty for curvature integrated over the curve extent. This method is described in the text by Reinsch (1967).

The smoothing spline is an excellent way to get an idea of the shape of the expected value of the distribution of Y across X. Each piece of the fitted curve is influenced most by the points closest to it, especially if the tuning parameter is set to produce a highly flexible curve (low value of lambda).

Density Ellipses represent the data as if X and Y were distributed according to a bivariate normal distribution. These ellipsoids are both density contours and confidence curves. As confidence curves, they show where a certain percentage of the data are expected to lie, assuming the bivariate normal distribution. The bivariate normal density is a function of the means and standard deviations of each variable and the correlation between them.

The density ellipsoid is a good graphical indicator of the correlation of two variables because the ellipsoid collapses diagonally as the correlation between the two variables approaches either 1 or –1. The ellipsoid is more circular (less diagonally oriented) if the two variables are uncorrelated. To see a matrix of ellipsoids and correlations for many pairs of variables, use the **Y's by Y's** platform in the **Analyze** Menu.

The Mean Fit Report

The Mean Fit report summarizes information about the total response. This is the simple mean model used as a comparison with other models.

Figure 15.11 The Mean Fit Text Report

Mean Fit	
Mean	62.55
Std Dev [RMSE]	4.242338
Std Error	.6707726
SSE	701.9

Mean
: is the sample mean (arithmetic average) of the response variable. It is the predicted response when no model effects are specified.

Std Dev [RMSE]
: is the standard deviation of the response variable. It is the square root of the mean square of the response variable, also referred to as the root mean square error (RSME).

Std Error
: is the standard deviation of the response mean. It is found by dividing the RMSE by the square root of the number of non-missing values.

SSE
: is the error sum of squares for the simple mean model. It is the sum of the squared distances from the mean and appears as the C Total sum of squares in the analysis of variance tables for all other regression fits. C Total is the total sum of square corrected for the response mean, shown in the Analysis of Variance Table (see **Figure 15.15**). Each

new fit that is an improvement over the simple mean model shows a reduction in the SSE in its text report.

Regression Reports for Fit Line and Fit Polynomial

There are three additional text report reveal buttons for linear (**Fit Line**) and polynomial (**Fit Polynomial**) regression fits.

Figure 15.12 Linear and Polynomial Text Reports

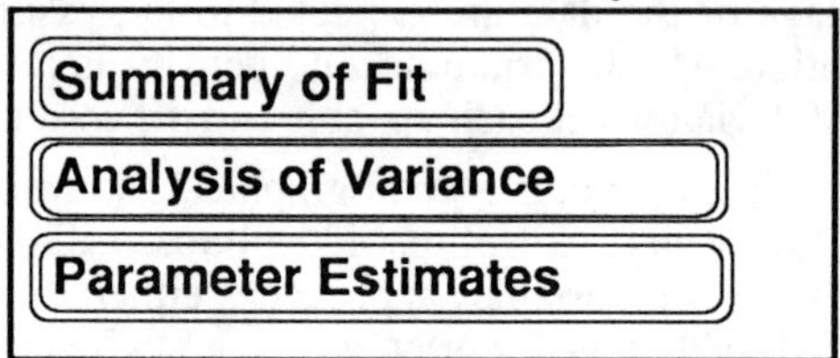

The Summary of Fit Table

The Summary of Fit tables in **Figure 15.13-14** show the numeric summaries of the response for the linear fit and polynomial fit of degree 2 for the same data. Comparing Summary of Fit tables lets you see the improvement of one model over another as indicated by a larger Rsquare value and smaller Root Mean Square Error.

Figure 15.13 The Summary of Fit Table for a Linear Fit

Summary of Fit	
Rsquare	.5029171
Root Mean Square Error	3.030122
Mean of Response	62.55
Observations (or Sum Wgts)	40

Figure 15.14 The Summary of Fit Table for a Polynomial Fit

Summary of Fit	
Rsquare	.5585102
Root Mean Square Error	2.893990
Mean of Response	62.55
Observations (or Sum Wgts)	40

Rsquare

measures the proportion of the variation around the mean accounted for by the linear or polynomial model. The remaining variation is attributed to random error. Rsquare is 1 if the model fits perfectly. An Rsquare of 0 means that the fit is no better than the mean. Using quantities from the corresponding analysis of variance table, the Rsquare for any interval response fit is always calculated as

$$\frac{\text{Sum of Squares for Model}}{\text{Sum of Squares for C Total}}$$

The Rsquare values seen in **Figures 15.13-14**, indicate that the polynomial fit of degree 2 is a small improvement over the linear fit.

Root Mean Square Error
: estimates the standard deviation of the random error. It is the square root of the mean square for Error found in the Analysis of Variance tables shown in **Figures 15.15-16**.

Mean of Response
: is the sample mean (arithmetic average) of the response variable. This is the predicted response when no model effects are specified.

Observations
: is the number of observations used in estimating the fit. If weights are used, this is the sum of the weights.

The Analysis of Variance Table

The Analysis of Variance table for a linear regression *partitions* the total variation of a sample into components. The components are used to compute an F ratio that evaluates the effectiveness of the model. If the probability associated with the F ratio is small, then the model is considered a better statistical fit for the data than the response mean alone. The Analysis of Variance tables in **Figures 15.15-16** show a comparison of the linear fit (**fit Line**) and a second degree polynomial fit. Both fits are statistically different from a horizontal line at the mean.

Figure 15.15 The Analysis of Variance Table for Linear Fit

Analysis of Variance

Source	DF	Sum of Squares	Mean Square	F Ratio
Model	1	352.99756	352.998	38.4460
Error	38	348.90244	9.182	**Prob > F**
C Total	39	701.90000		0.0000

Figure 15.16 The Analysis of Variance Table for Polynomial Fit

Analysis of Variance

Source	DF	Sum of Squares	Mean Square	F Ratio
Model	2	392.01835	196.009	23.4036
Error	37	309.88165	8.375	**Prob > F**
C Total	39	701.90000		0.0000

The following list describes the quantities in the Analysis of Variance table.

Source
: lists the three sources of variation called C Total, Model, and Error.

DF
: records the associated degrees of freedom for each source of variation.

- A degree of freedom is subtracted from the total number of non-missing values (N) for each parameter estimate used in the computation. The computation of the total sample variation uses an estimate of the mean, so one degree of freedom is subtracted from the total, leaving 39. The total degrees of freedom are partitioned into the Model and Error terms.

- One degree of freedom from the total (shown on the Model line) is used to estimate a single regression parameter (the slope) for the linear fit. Two degrees of freedom are used to estimate the parameters (β_1 and β_2) for a polynomial fit of degree 2.
- The Error degrees of freedom are the difference between the total DF and Model DF.

Sum of Squares

records an associated sum of squares (SS for short) for each source of variation.

- The total (C Total) sum of squared distances of each response from the sample mean is 701.9 as shown in **Figure 15.15-16**. That is the sum of squares for the base model (or simple mean model) used for comparison with all other models.
- For the linear regression the sum of squared distances from each point to the line of fit reduces from 701.90 to 348.90. This is the unexplained Error (*residual*) SS after fitting the model. The residual SS for a second degree polynomial fit is 309.88, accounting for slightly more variation than the linear fit.
- The total SS less the error SS gives the sum of squares attributed to the model.

Mean Square

is a sum of squares divided by its associated degrees of freedom. The F ratio for a statistical test is computed as the ratio of mean squares.

- The Model mean square is 352.998 and estimates the variance, but only under the hypothesis that the model parameters are zero.
- The Error mean square is 9.182 and estimates the variance of the error term.

F Ratio

is the model mean square divided by the error mean square. If the hypothesis is true that all the regression parameters (except the intercept) are zero, then the mean square for error and the mean square for model both estimate the error variance, and their ratio has an F distribution. If there is a significant effect in the model, the F Ratio is higher than expected by chance alone.

Prob > F

is the observed significance probability (P-value) of obtaining a greater F value by chance alone if the specified model fits no better than the overall response mean. Observed significance probabilities of .05 or less are often considered evidence of a regression effect.

The Parameter Estimates Table

The terms in the Parameter Estimates table for a linear fit (**Figure 15.17**) are the intercept and the single X variable. **Figure 15.18** shows the parameter estimates for a polynomial model of degree 2. For a polynomial fit of order *k*, there is an estimate for the model intercept and a parameter estimate for each of the *k* powers of the X variable.

Figure 15.17 The Parameter Estimates Table for Linear Fit

Parameter Estimates

Term	Estimate	Std Error	t Ratio	Prob>|t|
Intercept	48.321691	2.34419	20.61	0.0000
weight	.13550769	.021854	6.20	0.0000

Figure 15.18 The Parameter Estimates Table for Polynomial Fit

Parameter Estimates

Term	Estimate	Std Error	t Ratio	Prob>|t|
Intercept	32.336011	7.73695	4.18	0.0002
weight	.43492353	.140276	3.10	0.0037
weight^2	-.0013430	.000622	-2.16	0.0374

Term
lists the name of each parameter in the requested model. The intercept is a constant term in all models.

Estimate
lists the parameter estimates of the linear model. The prediction formula is the linear combination of these estimates with the values of their corresponding variables.

Std Error
lists the estimates of the standard errors of the parameter estimates. They are used in constructing tests and confidence intervals.

t Ratio
lists the test statistics for the hypotheses that each parameter is zero. It is formed by the ratio of the parameter estimate to its standard error. If the hypothesis is true, then this statistic has a Student's t distribution. Looking for a t ratio greater than 2 in absolute value is a common rule-of-thumb for judging significance because it approximates the .05 significance level.

Prob>|t|
lists the observed significance probability calculated from each t ratio. It is the probability of getting by chance alone a t ratio greater (in absolute value) than the computed value, given that the hypothesis is true. Often, a value below .05 (or sometimes .01) is interpreted as evidence that the parameter is significantly different from zero.

The Spline Report

The spline text report contains only the R-Square for the spline fit and the Sum of Squares Error. These values are useful for comparing the spline fit to other fits, or for comparing different spline fits to each other.

Figure 15.19 The Spline Fit Text Report

Smoothing Spline Fit, lambda=	0.1
R-Square	.7021995
Sum of Squares Error	209.0261

R-Square
: measures the proportion of variation accounted for by the spline model. In the example shown in **Figure 15.19**, 70.22% of the total variation is explained by fitting a spline with lambda =.1 .

Sum of Squares Error
: is the sum of squared distances from each point to the spline fit, reducing from the C Total of 701.9 to 209.02 in this example (see the Analysis of Variance tables in **Figures 15.15-16**). This is the unexplained error (*residual*) after fitting the spline model.

The Density Ellipse Report

The text report which accompanies each **Density Ellipse** shows the correlation coefficient for the X and Y variables and the probability that the correlation between the variables is significant. To see a matrix of ellipsoids and correlations for many pairs of variables, use the **Y's by Y's** command.

Figure 15.20 The Density Ellipse Correlation Report

Correlation	
Correlation	**Signif. Prob**
.7091665	0.0000

Correlation
: is the Pearson correlation coefficient, denoted r , and is computed as

$$r_{xy} = \frac{s_{xy}^2}{\sqrt{s_x^2 s_y^2}} \quad \text{where} \quad s_{xy}^2 = \Sigma w_i(x_i - \bar{x}_i)(y_i - \bar{y}_i)$$

and w_i is either the weight of the ith observation if a weight column is specified or 1 if no weight column is assigned. If there is an exact linear relationship between two variables then the correlation is 1 or –1 depending on whether the variables are positively or negatively related. If there is no relationship the correlation tends toward zero.

Signif. Prob
: is the probability of obtaining a correlation greater absolute value than the computed value by chance alone if no linear relationship exists between the X and Y variables.

One-Way Analysis of Variance

When the measurement level of the factor (X) is either nominal or ordinal and the response is interval, the **Fit Y by X** command fits the mean response for each level and then performs a one-way analysis of variance. This tests whether the factor group means are significantly different. In the example shown in **Figure 15.21** the response variable height is interval and the factor age is assigned the ordinal measurement level.

The graphical display for an interval by nominal/ordinal analysis shows the vertical distribution of response points for each factor value (X). The distinct values of X are sometimes called levels, and groups of identical values are called categories or samples of the X variable. The basic fitting approach is to compute the mean response within each group. If the variance of the response is assumed to be the same across the groups, then a one-way analysis of variance tests whether the means of the groups are equal.

Figure 15.21 The Interval by Nominal/Ordinal Platform

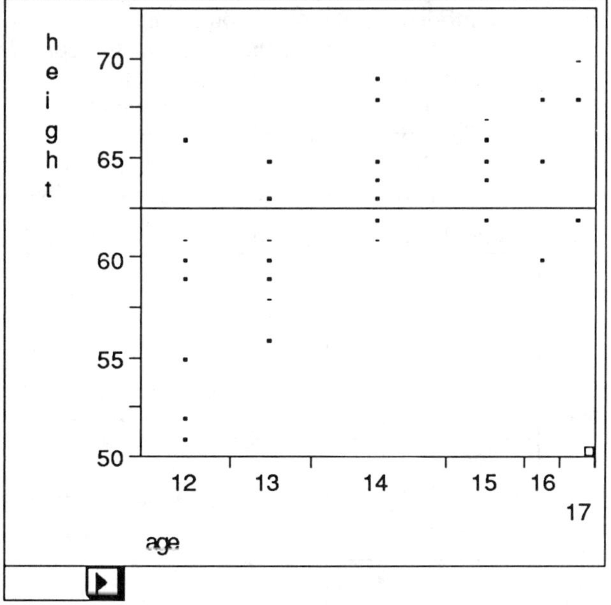

Fitting Options

The options triangle below the scatterplot accesses a pop-up menu that lists the options described below. Each option is a toggle that alternately displays or hides the fitting feature and its text reports. There are no save commands with the interval by ordinal/nominal fit.

Figure 15.22 Options for Interval by Nominal/Ordinal Fit

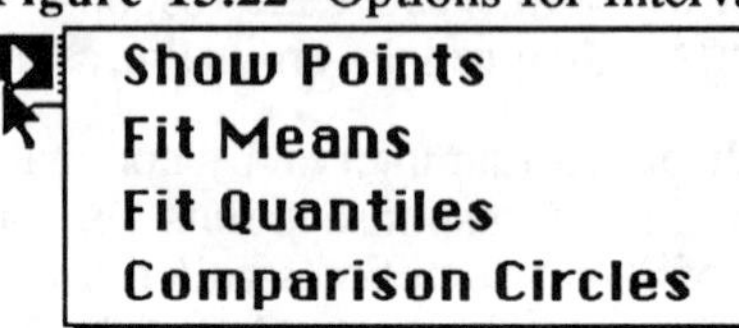

Show Points toggles the display of the data points.

Fit Means displays a *means diamond* schematic of the mean for each group as shown in **Figure 15.23.** The line across each diamond represents the group mean. The height of each diamond represents the 95% confidence interval for each group, and the diamond width represents the group sample size.

The confidence interval computation assumes that variances are equal across observations. Therefore the height of the confidence interval (diamond) is proportional to the reciprocal of the square root of the number of observations in the group. The horizontal extent of each group along the X axis (the horizontal size of the diamond) is made proportional to the sample size of each level. Thus the narrower diamonds are also the taller ones, since fewer data points yield a less precise estimation of the means.

Figure 15.23 Diagram of Means Diamonds

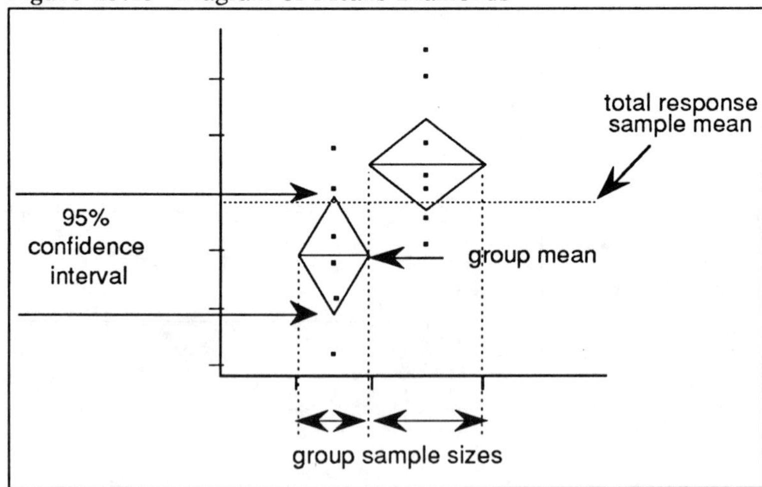

Fit Quantiles overlays quantile box plots that summarize the distribution of points at each factor level. For a detailed description of box plots, see the discussion in Chapter 14, *Analyze: Distribution of Y's.*

Figure 15.24 Diagram of Quantile Box Plots

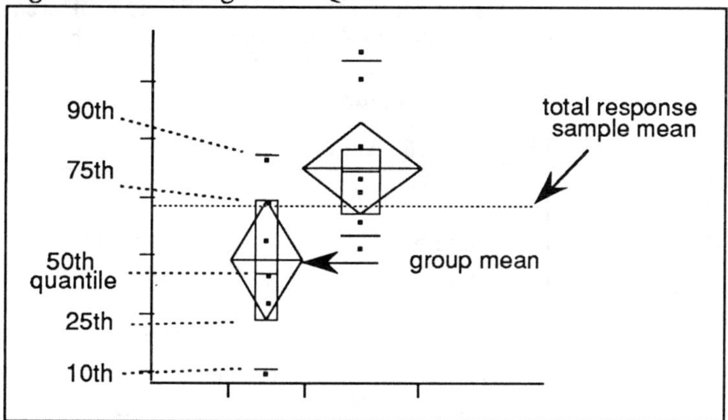

Comparison Circles draws 95% confidence circles useful for comparing group means. For each level of the response variable, a circle is drawn with its center at the mean and its diameter corresponding to the 95% confidence interval for the mean.

Figure 15.25 Diagram of Confidence Circles

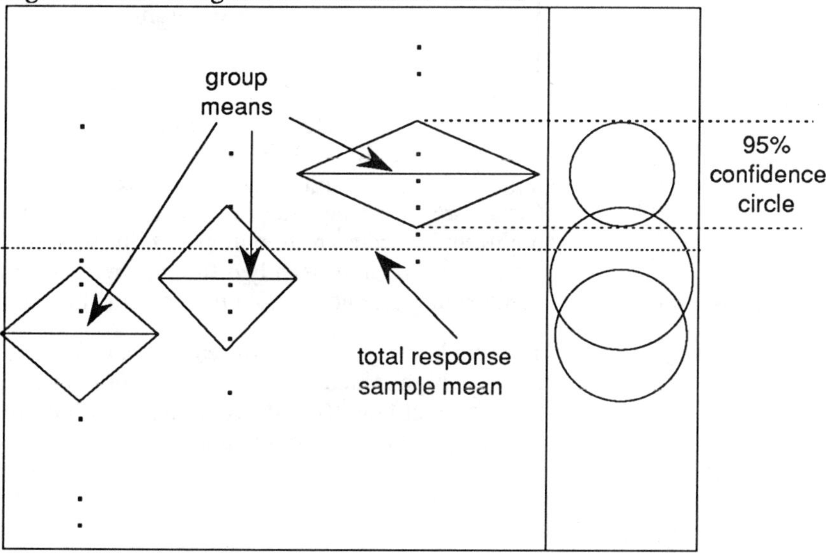

Visual Comparison of Group Means

You can compare each pair of group means visually by examining how the comparison circles intersect. The outside angle of intersection tells you whether or not group means are significantly different (see **Figure 15.26**). Circles for means that are significantly different either do not intersect or intersect slightly so that the outside angle of intersection is less than 90 degrees. If the circles intersect by an angle of more than 90 degrees or if they are nested, the means are not significantly different.

Figure 15.26 Angle of Intersection and Significance

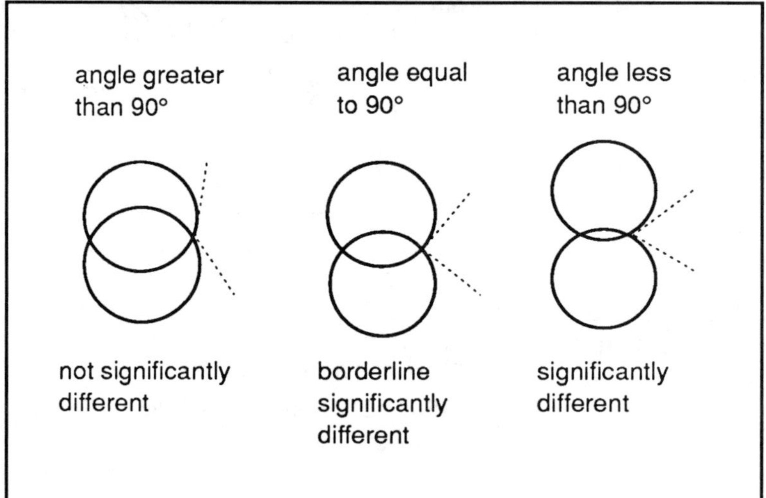

If the intersection angle is close to 90 degrees, you can verify whether the means are significantly different by clicking on the confidence circle to highlight it. The highlighted circle is drawn with a thick solid red line. Circles representing means that are not significantly different from the highlighted circle are drawn with thin line red lines. Circles representing means that are significantly different are drawn with a thick gray pattern. To deselect circles, click in the graph outside the circles.

The labels on the X axis for each level also show the comparison when a circle is highlighted. The selected level is in bold red type, and levels that are not statistically different are in plain red type. Levels that are statistically different are in italics.

Figure 15.27 Highlighting Confidence Circles for Comparison

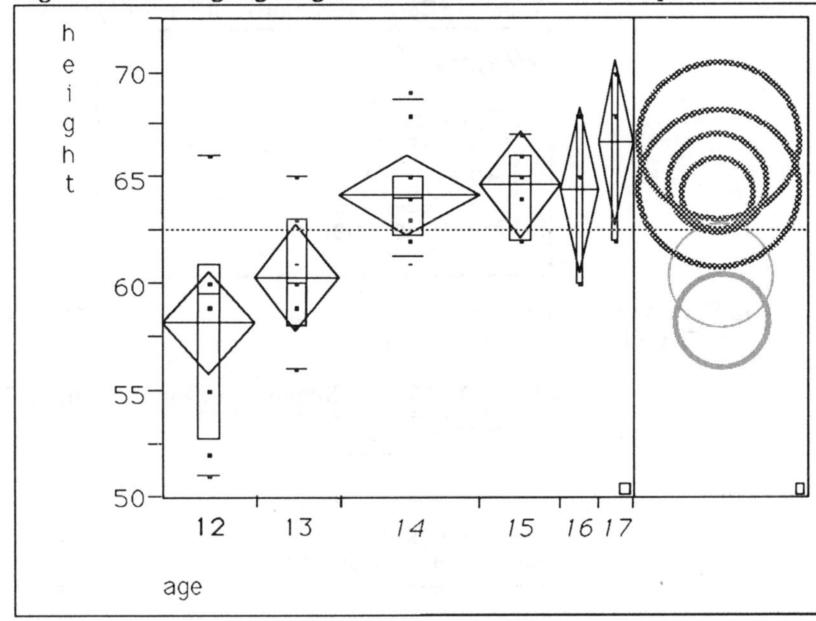

For more information about comparison circles, see the discussion in the section called **Geometric Comparison of Means** in the statistical details that follow.

Statistical Details for One-Way Analysis of Variance

A one-way analysis of variance is the attribution and test that some of the variability in a response is due to the difference in mean responses among the factor groups. The analysis of variance technique computes the mean within each group and the sum of squared distances from each point to the group mean. These sums of squares are totaled (pooled) to form an overall estimate of the variation unaccounted for by grouping the response. The total response variation less this residual determines the amount of variation explained by the analysis of variance model.

The Fit Means Reports

The **Fit Means** command produces the analysis of variance reports. The Means reveal button displays three further text report reveal buttons with the first report, the Summary of Fit table, displayed automatically.

Figure 15.28 Text Reports for One-way Analysis of Variance

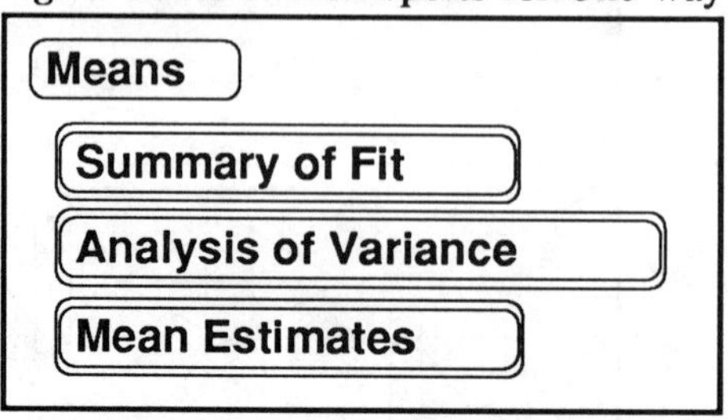

The Summary of Fit table

The Summary of Fit table in **Figure 15.29** shows a numeric summary for a one-way analysis of variance.

Figure 15.29 The Summary of Fit Table for a One-Way ANOVA

Summary of Fit

Rsquare	.4457645
Root Mean Square Error	3.382558
Mean of Response	62.55
Observations (or Sum Wgts)	40

Rsquare

measures the proportion of the variation accounted for by fitting means to each factor level. The remaining variation is attributed to random error. The Rsquare value is 1 if fitting the group means accounts for all the variation with no error. An Rsquare of 0 indicates that the fit serves no better as a prediction model than the overall response mean. Using quantities from the Analysis of Variance table for the model, the Rsquare for any interval response fit is always calculated

$$\frac{\text{Sum of Squares for Model}}{\text{Sum of Squares for C Total}}$$

Root Mean Square Error

estimates the standard deviation of the random error. It is the square root of the mean square for Error found in the Analysis of Variance table shown in **Figure 15.30.**

Mean of Response

is the overall mean (arithmetic average) of the response variable. This is the predicted response for the base model.

Observations

is the number of observations used in estimating the fit. If weights are used, this is the sum of the weights.

The Analysis of Variance table

The Analysis of Variance table *partitions* of the total variation of a sample into components. The ratio of the Mean Square components forms an F Ratio that evaluates the effectiveness of the model fit. If the probability associated with the F ratio is small, then the model is a statistically better fit than the overall response mean.

Figure 15.30 Example One-way Analysis of Variance Table

Analysis of Variance

Source	DF	Sum of Squares	Mean Square	F Ratio
Model	5	312.88214	62.5764	5.4692
Error	34	389.01786	11.4417	**Prob > F**
C Total	39	701.90000		0.0008

Source

lists the three sources of variation, called C Total, Model, and Error.

DF

records an associated degrees of freedom (DF) for each source of variation.

- A degree of freedom is subtracted from the total number of non-missing values N for each parameter estimate used in a model computation. For example, an estimate of the mean parameter is used in the computation of the total sample variation. Thus, the DF shown on the C Total line above is 39 after subtracting 1 from the total number of observations (see **Figure 15.29**).

  The total degrees of freedom are partitioned into the Model and Error terms.

- If the factor has k levels, then $k-1$ degrees of freedom are used to compute the variation attributed to the analysis of variance model. In the example shown in **Figure 15.30** there are 6 nominal age levels, which results in 5 degrees of freedom for the Model.

- The Error degrees of freedom is the difference between the total DF and the Model DF.

Sum of Squares

records an associated sum of squared distances (SS for short) for each source of variation.

- The total (C Total) SS of each response from the overall response mean is 701.90 in this example. The C Total SS is the base model used for comparison with all other models.

- The sum of squared distances from each point to its respective group mean is 312.88 in this example. This is the remaining unexplained Error (*residual*) SS after fitting the analysis of variance model.

- The total SS less the error SS gives the sum of squares attributed to the model.

Mean Square
is a sum of squares divided by its associated degrees of freedom.

- The Model mean square for this example is 62.58. It estimates the variance of the error, but only under the hypothesis that the group means are equal.
- The Error mean square for this example is 11.44. It estimates the variance of the error term independently of the model mean square and is unconditioned by any model hypothesis.

F Ratio
is the model mean square divided by the error mean square. Assuming the hypothesis that all the group means are equal, the mean square for error and the mean square for model both estimate the error variance, and their ratio has an F distribution. If the analysis of variance model results in a significant reduction of variation from the total, the F Ratio is higher than expected.

Prob > F
is the probability of obtaining by chance alone an F value greater than the one calculated if, in reality, fitting group means accounts for no more variation than using the overall response mean. Observed significance probabilities of .05 or less are often considered evidence of a regression effect.

The Mean Estimates table

The Mean Estimates table summarizes response information for each level of the nominal or ordinal factor.

Figure 15.31 The Mean Estimates Text Report

Mean Estimates

Level	number	Estimate	Std Error
12	8	58.1250	1.1959
13	7	60.2857	1.2785
14	12	64.1667	0.9765
15	7	64.5714	1.2785
16	3	64.3333	1.9529
17	3	66.6667	1.9529

Level
lists the values of the ordinal or nominal variable that name or describe the categories for fitting each group mean.

number
lists the number of observations in each level used to calculate the group mean estimate.

Estimate
lists the estimates of the mean response (sample mean) of each group.

Std Error

lists the estimates of the standard deviations of the group means. This Std Error is estimated assuming that the variance of the response is the same in each level. It is formed by dividing the root mean square error by the square root of the number of values used to compute the mean.

Geometric Comparison of Group Means

One approach to comparing two means is to see if their actual difference is greater than their *least significant difference* (LSD). This least significant difference is a Student's t statistic multiplied by the standard error of the difference of the two means and is written

$$LSD = t_{\alpha/2}\, std(\hat{\mu}_1 - \hat{\mu}_2) \, .$$

The standard error of the difference of two independent means is calculated from the relationship

$$\left[std(\hat{\mu}_1 - \hat{\mu}_2)\right]^2 = \left[std(\hat{\mu}_1)\right]^2 + \left[std(\hat{\mu}_2)\right]^2 \, .$$

When the means are uncorrelated these quantities have the relationship

$$LSD^2 = \left[t_{\frac{\alpha}{2}} std(\hat{\mu}_1 - \hat{\mu}_2)\right]^2 = \left[t_{\frac{\alpha}{2}} std(\hat{\mu}_1)\right]^2 + \left[t_{\frac{\alpha}{2}} std(\hat{\mu}_2)\right]^2 \, .$$

These squared values form a Pythagorean relationship and are shown graphically with a right triangle in **Figure 15.32.**

Figure 15.32 Relationship of the Difference Between Two Means

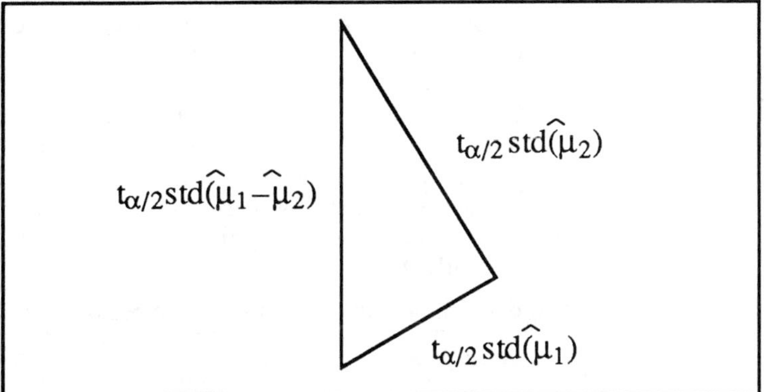

The hypotenuse of this triangle is a measuring stick for comparing means. The means are significantly different if and only if the actual difference is greater than the hypotenuse (LSD).

Now suppose that you have two means that are exactly on the borderline, where the actual difference is the same as the least significant difference. Draw the triangle with vertices at the means measured on a vertical scale.

Also, draw circles around each mean such that the diameter of each is equal to the confidence interval for that mean.

Figure 15.33 Geometric Relationship of t test Statistics

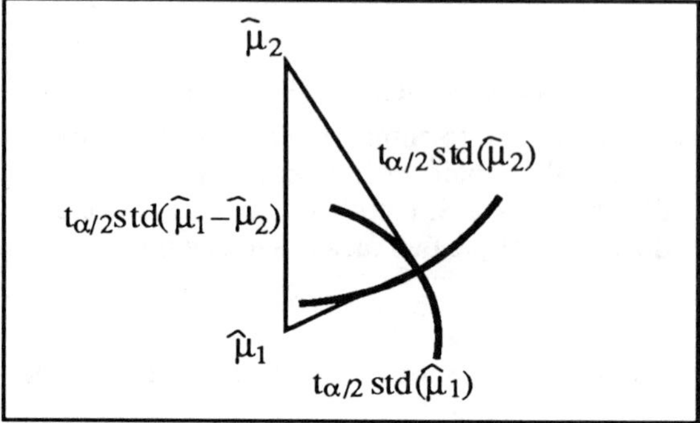

The radius of each circle is the length of the corresponding leg of the triangle, which is

$$t_{\alpha/2}\text{std}(\hat{\mu}_i)$$

Note that the circles must intersect at the same right angle as the triangle legs, giving the following relationship:

> If the means differ exactly by their least significant difference, then the confidence interval circles around each mean intersect at a right angle. That is, the angle of the tangents is a right angle.

Now consider the way these circles must intersect if the means are different by greater than or less than the least significant difference. If the circles intersect such that the outside angle is greater than a right angle, then the means *are not* significantly different. If the circles intersect such that the outside angle is less than a right angle then the means *are* significantly different. An outside angle of less than 90 degrees indicates that the means are farther apart than the least significant difference. If the circles do not intersect, then they are significantly different. If they nest, they are not significantly different.

Figure 15.34 Angle of Intersection and Significance

angle greater than 90°
not significantly different

angle equal to 90°
borderline significantly different

angle less than 90°
significantly different

Contingency Table Analysis

When both the response and the factor are nominal or ordinal, then the data are summarized by frequency counts. These counts are used to estimate the response rates of the Y levels for each level of the X factor. They are graphically portrayed in a mosaic plot. A mosaic plot is any plot divided into small rectangles such that the area of each rectangle is proportional to some frequency count of interest. JMP divides the Y counts within each X level by the total of the level to compute each X, Y proportion. This estimates the response probability for each X, Y level.

In the mosaic plot shown in **Figure 15.35**, the proportions shown on the X axis represent the relative sizes of the female and male totals. Likewise, the proportions shown on the right Y axis represent the relative sizes of each age group for the combined levels. The mosaic plot is equivalent to side-by-side divided-bar charts. For each sex, the proportion of each age divides the bars vertically.

The scale of the left Y axis shows response probability, with the whole axis being probability 1 (the total sample). To see how the response probabilities vary for the different levels of X, compare the heights of Y levels across the X levels. If the response divisions line up horizontally across the X levels, then the response rates are the same. A statistical test for equal response rates is described in the section, **Statistical Details for Contingency Tables.**

Figure 15.35 The Nominal/Ordinal by Nominal/Ordinal Platform

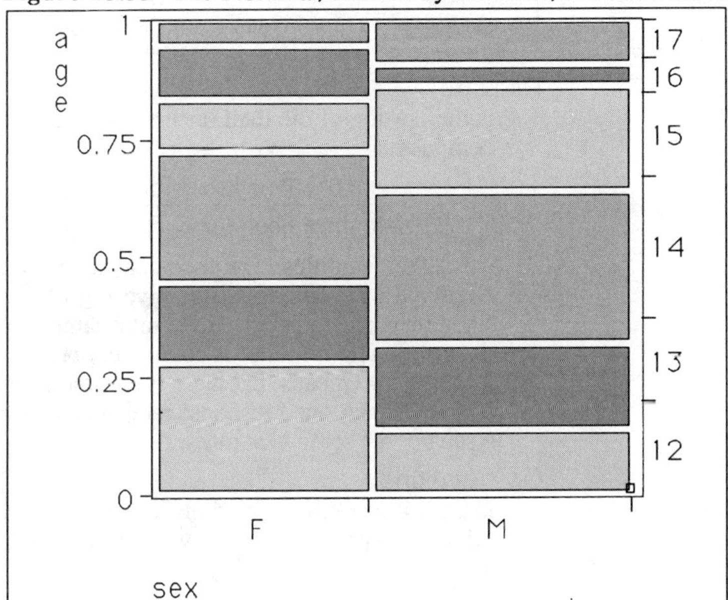

Statistical Details for Contingency Tables

This platform includes four text reports that compare the distribution of responses across factor levels and summarize the response counts and proportions. There are no display options or save commands with the nominal/ordinal by nominal/ordinal platform.

Note➔ JMP categorical data analysis uses the negative log likelihood, sometimes referred to as *uncertainty*, as the unit of variation.

Figure 15.36 Text Reports for Nominal/Ordinal by Nominal/Ordinal

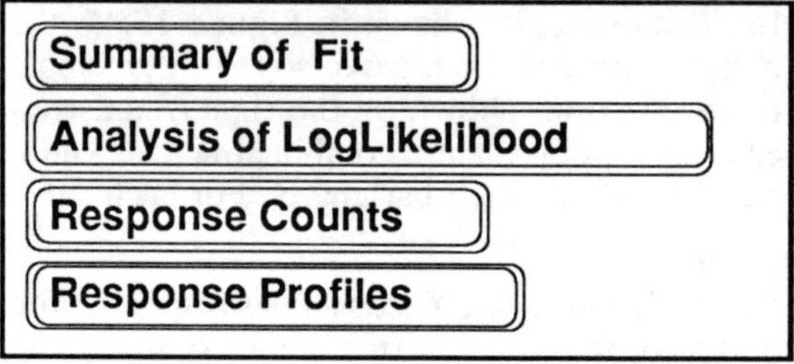

The Summary of Fit table

The Summary of Fit table shows the Rsquare and the total number of observations.

Figure 15.37 The Summary of Fit Table

Summary of Fit

Rsquare (U)	.0192577
Observations (or Sum Wgts)	40

Rsquare

is the portion of the total uncertainty attributed to the model fit. It is computed as

$$\frac{-\text{loglikelihood for Model}}{-\text{loglikelihood for C Total}}$$

using the quantities from the Analysis of Loglikelihood table in **Figure 15.38**. The total negative log likelihood is found by fitting fixed response rates across the total sample.An Rsquare of 1 means that the categorical response is completely predicted by the factors. An Rsquare of 0 means that there is no gain from using the model instead of fixed background response rates. In categorical data analysis, high Rsquares are somewhat rare.

Observations

is the number of observations in the current data table with non-missing response values. If weights are used, this is the sum of the weights.

The Analysis of LogLikelihood table

The Analysis of LogLikelihood table is analogous to the Analysis of Variance table for interval data. The negative log likelihood for categorical data plays the same role as sums of squares in interval data. chi-square tests evaluate how well the categorical model fits the data.

Figure 15.38 The Analysis of LogLiklihood Table

Analysis of LogLikelihood

Source	DF	-LogLikelihood
Model	5	1.295399
Error	30	65.970952
C Total	35	67.266350

Test	ChiSquare	Prob>ChiSq
Likelihood Ratio	2.591	0.7628
Pearson	2.554	0.7683

Source

lists the names of the three sources of uncertainty, called C Total, Model, and Error.

DF

records the degrees of freedom (DF) associated with each source of uncertainty.

- The total observations shown in **Figure 15.37** is 40. The degrees of freedom (DF) associated with the C Total in **Figure 15.38** are 35, or N – (r – 1) where r is the number of response levels. The total degrees of freedom are also partitioned into Regression and Error terms.
- The degrees of freedom shown on the Model line are used to compute the response rates for each factor level and are equal to (s – 1)(r – 1), where r is the number of response levels and s is the number of factor levels.
- The Error degrees of freedom are the difference between the total DF and Regression DF.

– LogLikelihood

measures fit and uncertainty in the same manner that sums-of-squares does in interval response situations.

- The C Total –LogLikelihood is the negative log likelihood (or uncertainty) when the probabilities are estimated by fixed rates for each response level. This corresponds to the total variation in an Analysis of Variance table. It is computed as

$$-\sum_{ij} n_{ij}\log(p_j) \quad \text{where} \quad p_j = \frac{N_j}{N} \ .$$

Here n_{ij} is the count for the jth response level of the ith factor level (or sample), N_j is the total of the jth response level, and $N = \sum_{ij} n_{ij}$.

Note that j indexes Y response levels and i indexes X factor levels.

- The Error –LogLikelihood is the uncertainty calculated after fitting a model. This term is found by using separate observed response rates for each sample. This value is minimized by the log likelihood fitting process and is computed as

$$-\sum_{ij} n_{ij}\log(p_{ij}) \text{ where } p_{ij} = \frac{n_{ij}}{N_j} \ .$$

- The Regression –LogLikelihood is the difference between the C Total and Error –LogLikelihoods, and it measures the reduction in uncertainty due to fitting the model. It is computed as

$$\sum_{ij} -n_{ij} \log p_j - \sum_{ij} -n_{ij} \log p_{ij} \ .$$

This value is used to construct the Likelihood Ratio chi-square test statistic.

Test

lists two chi-square statistical tests of the hypothesis that the response rates are the same in each sample category.

- The Likelihood Ratio chi-square test is computed as twice the negative log likelihood for Model in the Analysis of Loglikelihood table. Some books use the notation G^2 for this statistic. In terms of the difference of two –LogLikelihoods, one with whole-population response probabilities and one with each-population response rates, this is written as

$$G^2 = 2\left[\sum_{ij} -n_{ij} \log p_j - \sum_{ij} -n_{ij} \log p_{ij}\right]$$

$$\text{where } p_{ij} = \frac{n_{ij}}{N_i} \text{ and } p_j = \frac{N_j}{N} \ .$$

Often textbooks word this formula more compactly as

$$G^2 = 2\sum_{ij} n_{ij} \log\left(\frac{n_{ij}}{\frac{N_j N_i}{N}}\right) .$$

- Pearson is another chi-square test of the hypothesis that the response rates are the same in each sample category. This is calculated by summing the squares of the differences between observed and expected cell counts. The Pearson chi-square distribution approximates normality in large samples.

The chi-square statistic is

$$X^2 = \sum_{ij} \frac{(n_{ij}-m_{ij})^2}{m_{ij}} \quad \text{where } m_{ij} = \frac{n_{ij}}{N_i N_j}$$

or in a more familiar form

$$X^2 = \sum_{\text{cells}} \frac{(\text{observed}-\text{expected})^2}{\text{expected}} .$$

No continuity correction is done here, as is sometimes done in 2-by-2 tables.

Prob>ChiSq
lists the probability of obtaining (by chance alone) a chi-square value greater than the one computed if no relationship exists between the response and factor.

The Response Counts table

The Response Counts table is a two-way frequency table. There is a row for each response level and a column for each factor (sample) level. The borders of the table display the row, column and grand totals.

Figure 15.39 The Response Counts Table

Response Counts

age	F	M	Total
12	5	3	8
13	3	4	7
14	5	7	12
15	2	5	7
16	2	1	3
17	1	2	3
	18	22	40

The Response Profiles table

The Response Profiles table shows the response rates (proportions) for each factor level such that each column sums to 1. The total frequency for each column is shown at the bottom.

Figure 15.40 The Response Profiles Table

Response Profiles

age	F	M
12	0.2778	0.1364
13	0.1667	0.1818
14	0.2778	0.3182
15	0.1111	0.2273
16	0.1111	0.0455
17	0.0556	0.0909
	18	22

Nominal Logistic Regression

When the response variable is nominal and the factor variable is interval, the **Fit Y by X** command performs a nominal logistic regression.

To illustrate a logistic regression, suppose you want to guess the value of age for a given weight value. In **Figure 15.41**, the age column is assigned the nominal measurement level, and weight is interval. The left Y axis represents probability, and the proportions shown on the right Y axis represent the relative sizes of each age group in the total sample.

The logistic platform displays S-shaped multiple-logistic curves that partition the total response probability (1) at each X level. The **Statistical Details** section that follows describes the interpretation of logistic probability curves .

Figure 15.41 The Nominal by Interval Platform

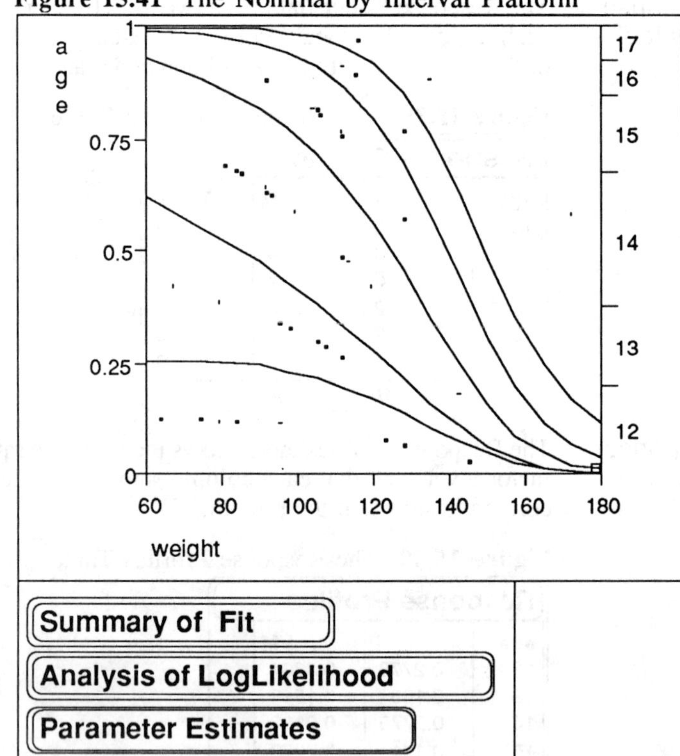

Statistical Details for Nominal Logistic Regression

Nominal logistic regression estimates the probability of choosing one of the response levels as a smooth function of the factor. The probabilities must be between 0 and 1, and they must sum to 1 across the response levels for a given factor value.

In a logistic probability plot, the Y axis represents probability. For k response levels, $k - 1$ smooth curves partition the total probability (1) among the response levels. The fitting principle for a nominal logistic regression minimizes the total negative log likelihood of the response events that occur.

Summary and analysis information accessed by the text report reveal buttons shown beneath the graph. There are no display options or save commands with the nominal by interval report.

Interpretation of a Logistic Probability Plot

If the X variable has no effect on the response, then the fitted lines are horizontal, and the probabilities are constant for each response across the interval factor range. **Figure 15.42** shows a logistic probability plot for selected ages that each have the same weight distribution. The weight factor has no predictive value.

Figure 15.42 Logistic Plot Showing No Y by X Relationship

Nom age	Int weight
12	59
12	63
12	59
12	63
13	59
13	63
14	59
14	63

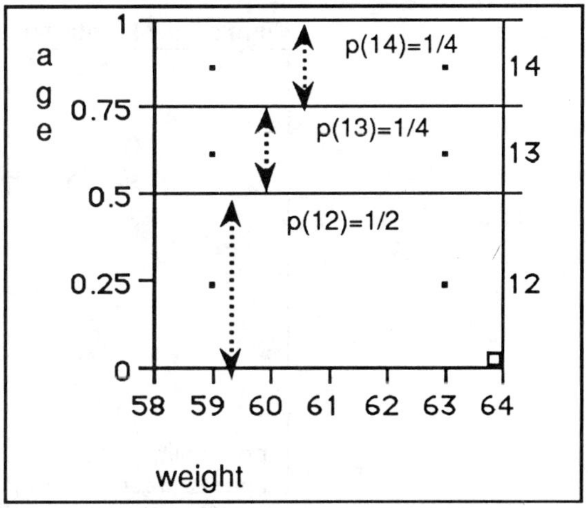

If the response is completely predicted by the value of the factor, then the logistic curves are effectively vertical. In this case, the parameter estimates become very large, and they are marked *unstable*. The prediction of a

response is certain (the probability is 1) at each of the factor levels, as illustrated in **Figure 15.43**.

Figure 15.43 Logistic Plot Showing a Given Y by X Relationship

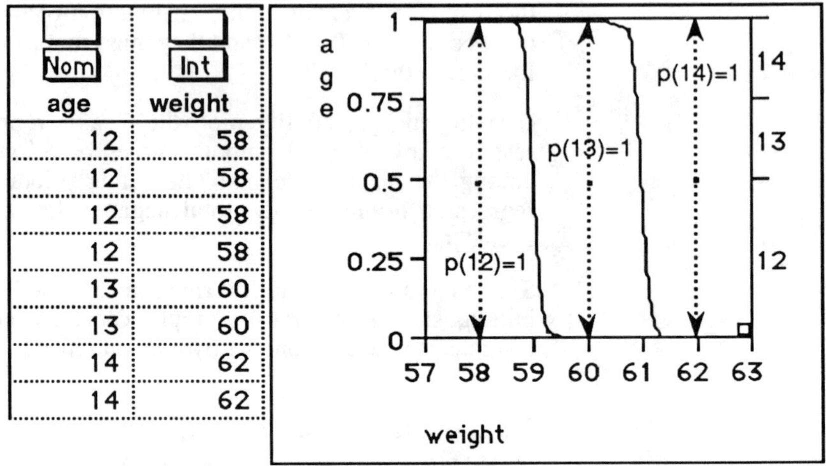

A set of logistic probability curves partitions the probability axis on the left at each point on the factor (X) axis. At any point on the factor axis, the probability associated with each response is the vertical distance on the probability axis corresponding to that response.

Figure 15.44 Interpretation of a Logistic Probability Plot

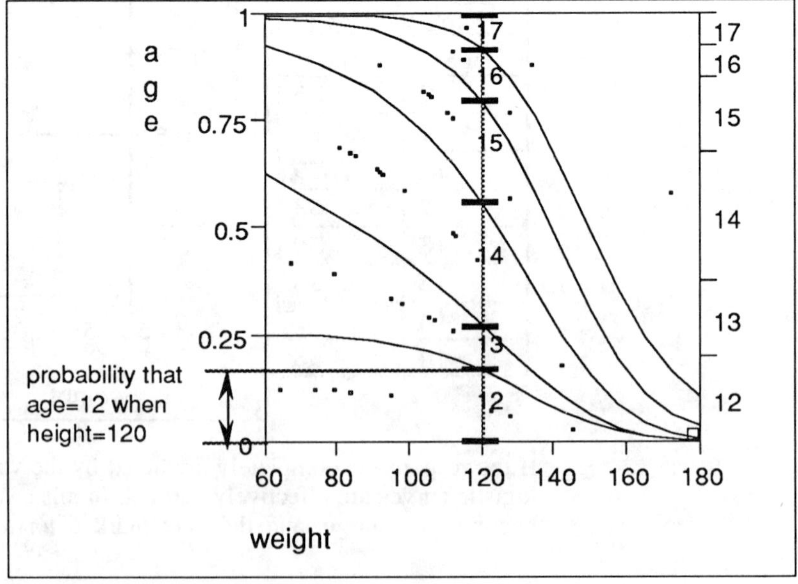

Nominal by Interval Text Reports

The Nominal by Interval platform includes three text reports similar to the reports described previously for categorical models.

Figure 15.45 The Nominal by Interval Text Reports

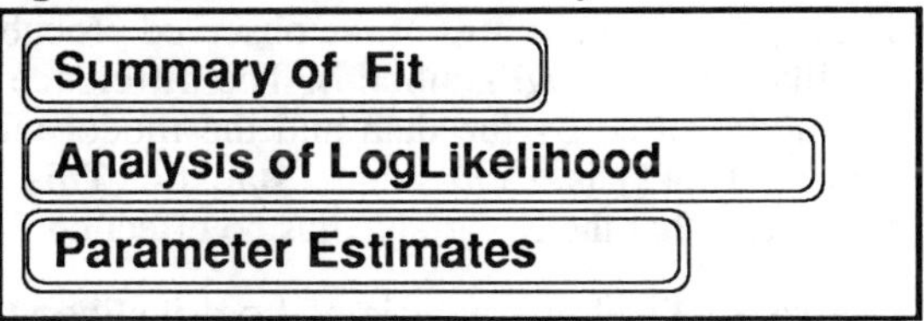

The Summary of Fit table

The Summary of Fit table lists only the model Rsquare and the number of nonmissing response values. The Rsquare values range from 0 to 1, where 1 is a perfect fit and 0 means there is no gain by using the model over using fixed background response rates. In **Figure 15.43**, all probabilities of response events are 1 and Rsquare is 1. In **Figure 15.42**, all fitted lines are horizontal and Rsquare is 0.

High Rsquares are rare in categorical models. An Rsquare of 1 means that the categorical response is completely predicted by the factor. An Rsquare of 0 means that there is no gain by using the model over using fixed background response rates.

Figure 15.46 The Summary of Fit Table

Summary of Fit	
Rsquare (U)	.0916992
Observations (or Sum Wgts)	40

Rsquare
: is the proportion of the total uncertainty that is attributed to the model fit. The difference between the log likelihood from the fitted model and the log likelihood that uses horizontal lines is a test statistic to examine the hypothesis that the factor variable has no effect on the response. The ratio of this test statistic to the background log likelihood is subtracted from 1 to calculate Rsquare. More simply, it is computed as

$$\frac{-\text{loglikclihood for Model}}{-\text{loglikelihood for C Total}}$$

using quantities from the Analysis of Loglikelihood table below.

Observations
: is the number of observations. If weights are used, this is the sum of the weights.

The Analysis of LogLikelihood table

The Analysis of LogLikelihood table corresponds to the Analysis of Variance table for interval data. The negative log likelihood for categorical data plays the same role as the sum of squares does for interval data. A chi-square test evaluates how well the categorical model fits the data.

The negative sum of logs of the observed probabilities is called the negative log likelihood (–loglikelihood). The difference in the log likelihood from the model fitted by the data and the model with equal probabilities (see **Figure 15.42**) is a chi-square statistic. This test statistic examines the hypothesis that the X variable has no effect on the response.

Figure 15.47 The Analysis of LogLikelihood Table

Analysis of LogLikelihood

Source	DF	-LogLikelihood	ChiSquare
Model	5	6.168277	12.33655
Error	30	61.098073	**Prob>ChiSq**
C Total	35	67.266350	0.030456

The **Source** and **DF** columns of the Analysis of LogLikelihood table are discussed previously in the section called **Statistical Details for Contingency Tables.**

–LogLikelihood
measures the variation, sometimes called uncertainty, in the nominal by interval sample.

- C Total is the negative loglikelihood (or uncertainty) for the case when the probabilities are estimated by fixed background rates. This is the background uncertainty when the model has no effects.
- Error is the negative loglikelihood (or uncertainty) calculated after fitting the model. The fitting process involves predicting response rates with a linear model and a logistic response function. This value is minimized by the fitting process.
- The difference of these two –LogLikelihoods is the reduction due to fitting the model. This value is the Likelihood Ratio chi-square test statistic.

ChiSquare
is the Likelihood Ratio chi-square test of the hypothesis that the model fits no better than fixed response rates across the whole sample. This is computed as twice the –Loglikelihood for Model as shown in **Figure 15.47**. It is the difference of two negative log likelihoods, one with whole-population response probabilities and one with each-population

response rates. This statistic is sometimes denoted G2 and is written

$$G^2 = \sum_i -\log \text{p(background)} - \sum_i -\log \text{p(model)}$$

where i indexes each observation rather than each cell.

Prob>ChiSq
is the observed significance probability (often referred to as the P-value) for the chi-square test. It is the probability of getting (by chance alone) a chi-square value greater than the one computed. Models are often judged significant when this probability is below .05.

The Parameter Estimates table

The nominal logistic model fits a parameter for the intercept and slope for each of $k - 1$ logistic comparisons, where k is the number of response levels. Each parameter estimate can be examined and tested individually, though this is seldom of much interest.

Figure 15.48 The Nominal by Interval Parameter Estimates Table

Parameter Estimates

Term	Estimate	Std Error	ChiSquare	Prob>ChiSq
Intercept	13.4041297	5.8347741	5.28	0.0216
height	-.10543709	.04681750	5.07	0.0243
Intercept	14.6094734	5.9440578	6.04	0.0140
height	-.11926243	.04852061	6.04	0.0140
Intercept	13.2340182	5.6838346	5.42	0.0199
height	-.09967641	.04489319	4.93	0.0264
Intercept	10.3190474	5.6576148	3.33	0.0682
height	-.07694640	.04430837	3.02	0.0825
Intercept	6.26425316	5.7788496	1.18	0.2784
height	-.04861281	.04456795	1.19	0.2754

Term
lists each parameter in the logistic model. There is an intercept and a slope term for the factor at each level of the response variable.

Std Error
lists the standard error of each parameter estimate. They are used to compute the statistical tests that compare each term to zero.

ChiSquare
lists the Wald tests which test the hypotheses that each of the parameters are zero.

Prob>ChiSq
lists the observed significance probabilities for the chi-square tests.

Ordinal Logistic Regression

When Y is ordinal, then a modified version of logistic regression is used for fitting. The cumulative probability of being at or below each response level is modeled by a curve. The curves are the same for each level except they are shifted to the right or left.

The ordinal logistic model fits a different intercept but the same slope for each of $r - 1$ cumulative logistic comparisons, where r is the number of response levels. Each parameter estimate can be examined and tested individually, though this is seldom of much interest.

To illustrate a nominal logistic regression, suppose you want to guess the value of age for a given weight value. In **Figure 15.49**, age is assigned the ordinal measurement level and weight is interval. This platform is interpreted the same way as the nominal by interval platform. See the section called **Interpretation of a Logistic Probability Plot** in this chapter for more details about nominal logistic regression.

Figure 15.49 The Ordinal by Interval Platform

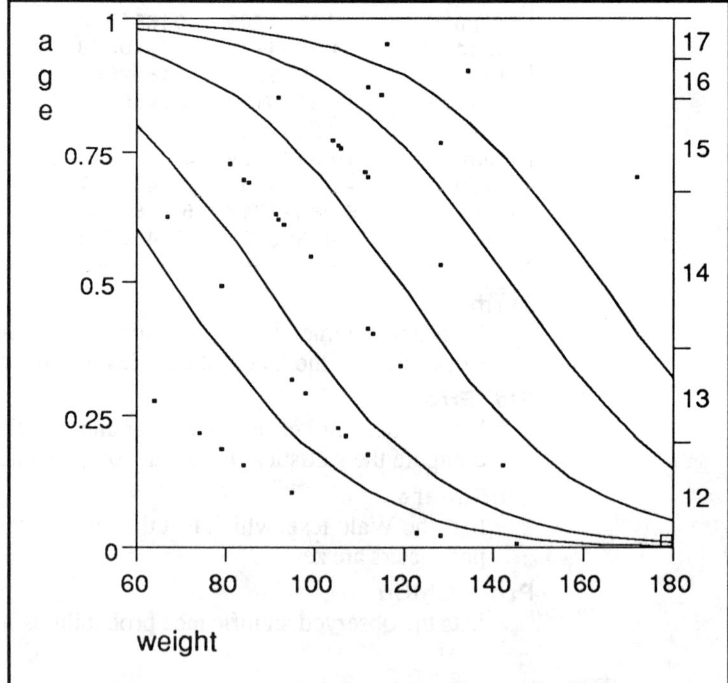

The ordinal by interval text reports are the same as those for a nominal by interval analysis. See the section in this chapter, called **Statistical Details for Logistic Regression,** for a discussion of the Summary of Fit and Analysis of Loglikelihood tables.

Figure 15.50 The Ordinal by Interval Text Reports

Summary of Fit

Rsquare (U)	.0748933
Observations (or Sum Wgts)	40

Analysis of LogLikelihood

Source	DF	-LogLikelihood	ChiSquare
Regression	1	5.037801	10.07560
Error	34	62.228549	**Prob>ChiSq**
C Total	35	67.266350	0.001502

Parameter Estimates

The Parameter Estimates table

In the Parameter Estimates table, an intercept parameter is estimated for every response level except the first, but there is only one slope parameter. The intercept parameters show the spacing of the response levels. They always increase monotonically.

Figure 15.51 The Parameter Estimates Table

Parameter Estimates

Term	Estimate	Std Error	ChiSquare	Prob>ChiSq
Intercept	3.31044008	1.4996113	4.87	0.0273
Intercept	4.28554430	1.5348405	7.80	0.0052
Intercept	5.74764136	1.6404794	12.28	0.0005
Intercept	6.95162592	1.7497634	15.78	0.0001
Intercept	7.92223618	1.8595254	18.15	0.0000
height	-.04822943	.01482909	10.58	0.0011

Term
lists each parameter in the ordinal logistic model.

Std Error
lists the standard error of each parameter estimates. The standard errors are used to compute the statistical tests that compare each term to zero.

ChiSquare
lists Wald tests which examine the hypothesis that each of the parameters is zero.

Prob>ChiSq
lists the observed significance probabilities for the chi-square tests.

Chapter 16.
Analyze: Fit Y by X's

This chapter covers the kinds of fits appropriate for the **Fit Y by X's** command. These include simple linear regression, multiple regression, main effects analysis of variance and analysis of covariance models. To fit more complex models see Chapter 17, *Analyze: Specify Model.*

The display options and save commands available for the general fitting platform and a discussion the leverage plots are also included in this chapter.

There is one fitting system and a single report platform that is launched from two commands:

- **Fit Y by X's**, discussed in this chapter, performs a fit of one Y variable by all the X variables selected. It is used for for multiple regression and main effects analysis of variance models, and for analysis of covariance.
- **Specify Model,** discussed in the next chapter, displays a general dialog that includes more complex models and lets you choose specific model effects and error terms.

The primary purpose of the general fitting platform is to fit a model to an interval response. If Y has interval values a least squares fit is done with accompanying analysis of variance reports and leverage plots.

If a nominal or ordinal response is specified, logistic regression is done by maximum likelihood and summarized with text reports, but no graphical display is available. To see a logistic probability plot for a nominal or ordinal response and a single effect, use the **Fit Y by X** command described in Chapter 15, *Analyze: Fit Y by X.*

Chapter Contents

The Display

Selecting Variables to Fit

There are three ways to specify the variables for the model you want to fit.

- Assign X and Y roles to the variables in the data table, then launch the **Fit Y by X's** command from the **Analyze** menu.
- Launch **Fit Y by X's** without assigning X and Y roles and respond to the role assignment dialog.
- Use the **Specify Model** platform discussed in Chapter 17. This platform lets you to specify more complex effects. It is also more convenient if you want to add or remove terms in the model.

The Report Layout

The display for interval-response models is organized as shown in **Figure 16.1**. Nominal and ordinal response models are discussed later. The first column of the display is for summary and other reports that deal with the general fit. The second column shows whether the model as a whole is significant. The remaining columns show the details for each effect in the model.

Figure 16.1 The **Fit Y by X's** Display Schematic

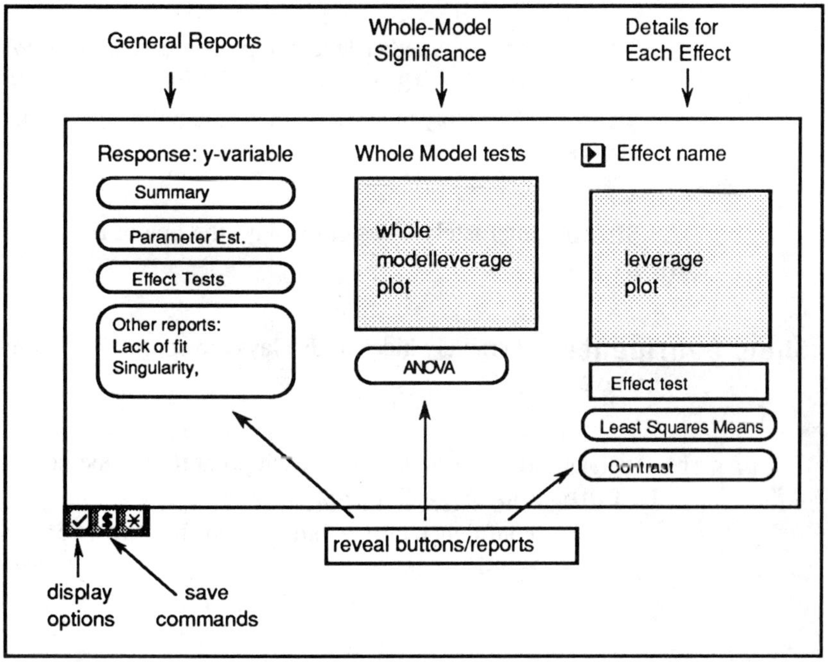

General Reports
Whole-Model Significance
Details for Each Effect
Response: y-variable
Summary
Parameter Est.
Effect Tests
Other reports: Lack of fit Singularity,
Whole Model tests
whole modelleverage plot
ANOVA
Effect name
leverage plot
Effect test
Least Squares Means
Contrast
reveal buttons/reports
display options
save commands

Display Options

All parts of the **Fit Y by X's** and **Specify Model** report are optional. When you click on the check-box icon on the lower-left of the window border, a pop-up menu displays the options shown in **Figure 16.2**. Each menu item is a toggle, turning the option on or off when selected. By default, all options except **Connect Points to Line** are in effect when the model report is presented, as indicated by the check marks on the menu.

Figure 16.2 Fit Y by X's and **Specify Model** Display Options Menu

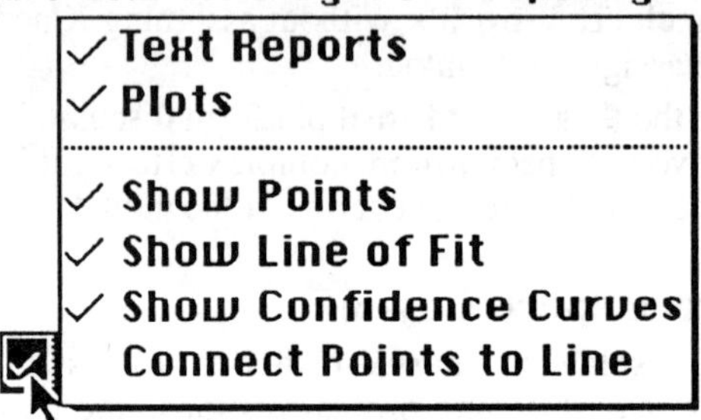

Each of the display options is described below

Text Reports is unchecked only if you want to remove most of the reports from the window in order to concentrate on the plots.

Plots toggles whether leverage plots are displayed. When **Plots** is not checked, the **Fit Y by X's** and **Specify Model** platform shows only text reports.

Show Points alternately hides or displays the points on all the leverage plots.

Show Line of Fit alternately hides or displays the line of fit on all the leverage plots.

Show Confidence Curves alternately hides or displays the confidence curves on all the leverage plots.

Connect Points to Line if checked, connects each point to the line of fit on all the leverage plots.

The example in **Figure 16.3** shows a whole model leverage plot with the confidence curves hidden and the **Connect Points to Line** option in effect.

Figure 16.3 The **Connect Points to Line** Display Option

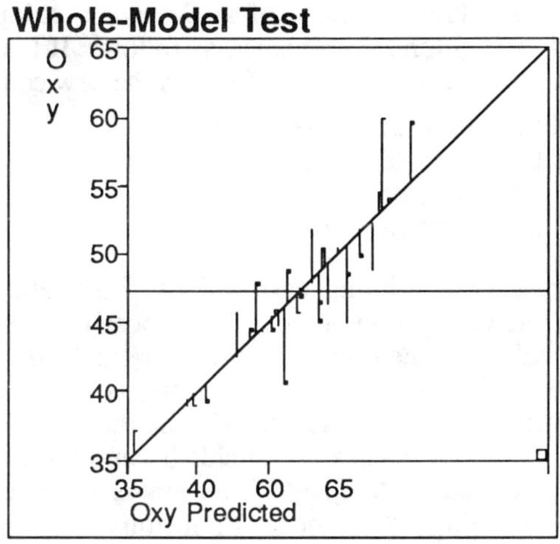

Save Commands

When you click on the dollar sign *save* icon on the lower-left of the window border, a pop-up menu offers the following choices. Each choice generates one or more new columns in the current data table titled as shown, where *colname* is the name of the response variable.

Save Predicted Values creates a new column called Predicted *colname* containing the predicted values computed by the specified model.

Save Residuals creates a new column called Residual *colname* containing the residuals, i.e. the observed response values – predicted values.

Save Prediction Confidence creates two new columns called Lower95% Pred *colname* and Upper95% Pred *colname*. The new columns contain the lower and upper 95% confidence limits for line of fit.

Save Indiv Confidence creates two new columns called Lower95% Indiv *colname* and Upper95% Indiv *colname*. The new columns contain lower and upper 95% confidence limits for individual response values.

Save Studentized Residual creates a new column called Studentized Resid *colname*. The new column values are the residuals divided by their standard error.

Save h[i] creates a new column called h *colname*. The new column values are the diagonal values of the matrix, $X(X'X)^{-1}X'$, sometimes called *hat* values.

Save Std Error of Predicted creates a new column, called StdErr Pred *colname*, containing the standard errors of the predicted values.

Save Std Error of Residual creates a new column called, StdErr Resid *colname*, containing the standard errors of the residual values.

Save Std Error of Individual creates a new column, called StdErr Indiv *colname*, containing the standard errors the individual predicted values

Save Effect Leverage Pairs creates a set of new columns that contain the values for each leverage plot. The new columns consist of an X and Y column for each effect in the model. The columns are named as follows. If the response column name is R and the effects are X1 and X2, then the new columns are named

X Leverage of X1 for R
Y Leverage of of X1 for R
X Leverage of X2 for R
Y Leverage of of X2 for R

Save Prediction Formula creates a new column, called Pred Formula *colname* containing the predicted values computed by the specified model. This column has the same values as the column created by the **Save Predicted Values** but differs in that the prediction formula is saved with the column. Use the **Column Info** command and click on the formula box to see the prediction formula. Both options are available because the prediction formula itself can require considerable space if the model is large. Use **Save Predicted Values** if you do not need the formula. For information about working with formulas see Chapter 11, *The Calculator*.

Leverage Plots

To focus attention on whether an effect is significant, you want a view of the data from the point of view of the hypothesis for that effect. You might say that you want more of an X-ray picture of the inside of the data rather than a surface view from the outside. The leverage plot is this kind of view of your data and the fit, a view calculated to give maximum insight on how the fit is carrying the data.

When an effect in a model is tested for significance, it is tested by comparing the sum of squared residuals with the sum of squared residuals that would be computed if that effect were removed from the model. If the residual errors become much larger without the effect, then the effect is confirmed as a significant contribution to the fit.

The graphical display of an effect's significance test is called a *leverage plot*, which shows for each point what the residual would be both with and without that effect in the model.

A leverage plot can be constructed as in **Figure 16.4**. The distance from a point to the line of fit shows the actual residual. The distance from the point to the horizontal line of the mean shows what the residual error would be if the effect were removed from the model, i.e. if its parameters were constrained to the hypothesized value of zero.

Figure 16.4 General Leverage Plot

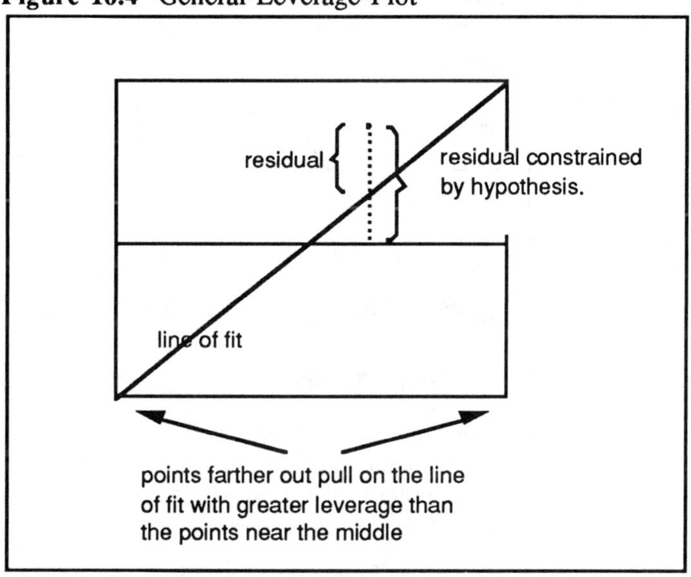

Historically, leverage plots are an extension of the display called a *partial-regression residual leverage plot* by Belsley Kuh and Welsch (1980) or an *added variable plot* by Cook and Weisberg(1982).

The term *leverage* is used because a point exerts more influence on the fit if it is farther away from the middle of the plot in the horizontal direction; at the extremes the differences of the residuals (before and after being constrained by the hypothesis) are greater, and constitute a larger part of the sums-of-squares for the hypothesis test for that effect.

The fitting platform produces a leverage plot for each effect in the model. In addition, there is one special leverage plot titled *Whole Model* that shows the actual values of the response plotted against the predicted values. This Whole Model leverage plot dramatizes the test that all the parameters (except intercepts) in the model are zero, the same test reported in the Analysis of Variance report.

The leverage plot for the linear effect in a simple regression is no different than the traditional plot of actual values of the response and the regressor. The example leverage plots shown in **Figure 16.5** are the result of a simple linear regression to predict the variable Oxy with RunPulse. The plot on the left is the Whole Model test for all regressors, and the plot on the right is the leverage plot for the effect Runpulse. Because there is only one regressor, the two plots are equivalent; the X scales of the two plots are related by a simple linear transformation.

Figure 16.5 Whole Model Leverage on Left, Effect Leverage on Right

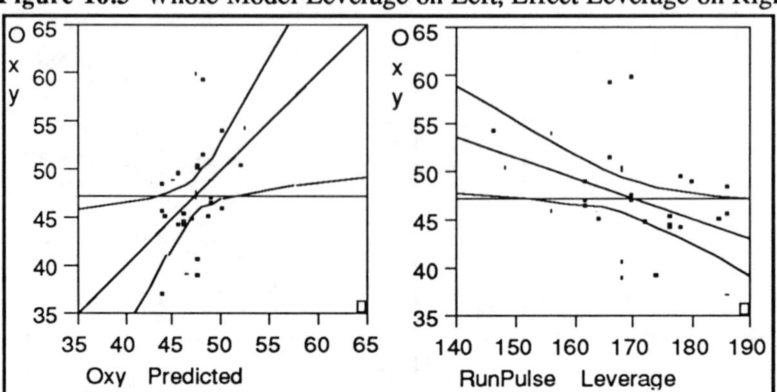

Figure 16.6 is a generalized version of **Figure 16.5**. Recall that the distance from a point to the line of fit is the actual residual, and the distance from the point to the mean is the residual error if the regressor is removed from the model.

Figure 16.6 Leverage Plots for a Simple Regression

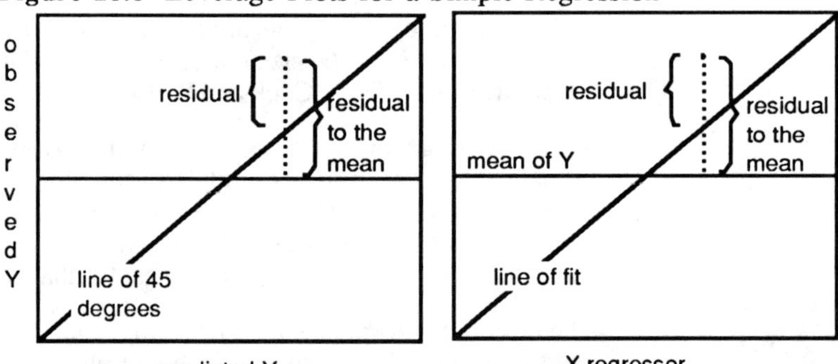

When the leverage plot is for simple regression, the points represent actual data coordinates, and the horizontal line for the constrained model is the sample mean of the response. But when the leverage plot is for one of multiple effects, the points are no longer actual data values, and the horizontal line represents a partially constrained model, rather than a model fully constrained to one mean value. Whether for simple regression or multiple regression, however, the intuitive interpretation of the plot is the same. The idea is to judge if the line of fit carries the points significantly better than the horizontal line.

Confidence Curves

The leverage plots are shown with confidence curves. These indicate whether the test is significant at the 5% level by showing a confidence region for the line of fit. If the confidence region between the curves contains the horizontal line, then the effect is not significant. If the curves cross the line, the effect is significant.

Figure 16.7 Comparison of Significance Shown in Leverage Plots

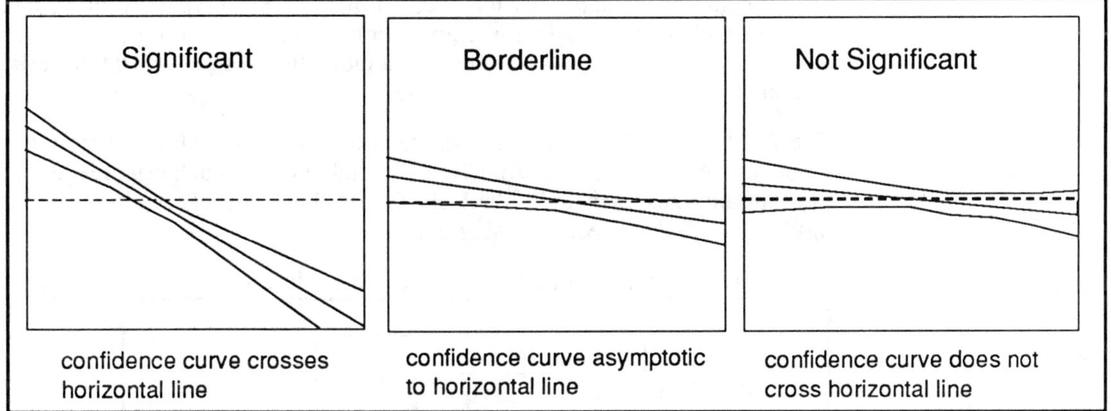

Interpretation of X Scales

If the measurement level of the regressor is interval, then the x-axis is scaled like the regressor, and the the slope of the line of fit in the leverage plot is the parameter estimate for the regressor.

If the effect is a nominal or ordinal effect or a complex effect like an interaction instead of a simple regressor, then the X axis cannot represent the values of the effect directly. In this case the X axis is scaled like the Y axis, and the line of fit is a diagonal with a slope of 1. The Whole Model leverage plot is a version of this where the X axis turns out to be the predicted response of the whole model.

With all leverage plots the influential points are the ones far out on the X axis. If an effect is closely related to another effect in the model, then the effect as a whole doesn't have much leverage. This problem is called *colinearity*. By scaling regressor axes by their original values, you can see colinearity by the shrinkage of points in the X direction.

The details of the construction of leverage plots are covered in Appendix A.

Statistical Reports

The report window contains summary table reveal buttons for the model. There are different reveal buttons available depending on the characteristics of the model. For example, if the model includes exact replicates of effects then the error term can be partitioned and a lack of fit component extracted. In this situation the table list includes a **Lack of Fit** reveal button shown in **Figure 16.8**. If one effect is a linear combination of another, then there are singularity problems and the list includes the **Singularity Details** button.

The Analysis of Variance table for the whole model is shown below the whole model leverage plot. An Effect Test table below each effect leverage plot summarizes each individual effect and the Effect Test table for the model lists all the effects in a single table.

Figure 16.8 Text Reports for **Fit Y by X's** Platform

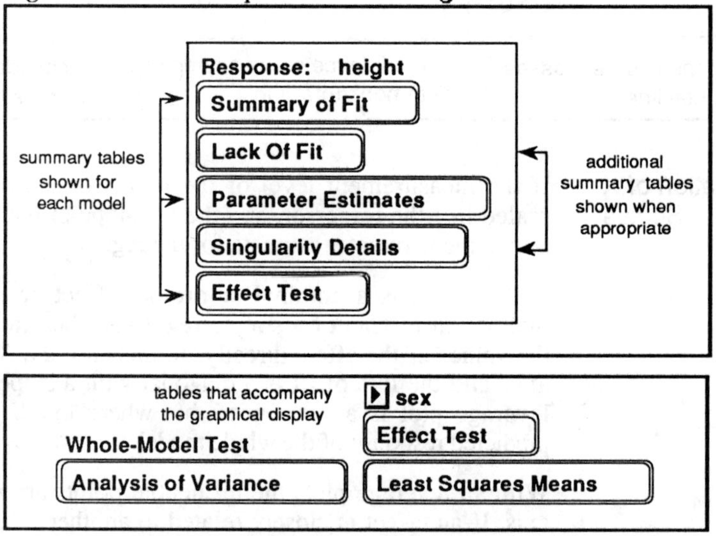

Response: height
Summary of Fit
Lack Of Fit
Parameter Estimates
Singularity Details
Effect Test
summary tables shown for each model
additional summary tables shown when appropriate

tables that accompany the graphical display
Whole-Model Test
Analysis of Variance
sex
Effect Test
Least Squares Means

The next sections describe these tables in the following order:

- Summary of Fit
- Analysis of Variance
- Lack of Fit
- Parameter Estimates
- Singularity Details
- Effect Test
- Least Squares Means

The examples in the following sections use the CLASS data table from the SAMPLE data folder. The model specifies the weight variable as predicted by age and height, where age is assigned the nominal measurement level.

The Summary of Fit Table

The Summary of Fit table in **Figure 16.9** shows numeric summaries of the response for the multiple regression model. This table is always revealed when the platform is displayed.

Figure 16.9 Leverage Plots for a Simple Regression

Summary of Fit	
Rsquare	.8741566
Root Mean Square Error	2.227565
Mean of Response	62.33684
Observations (or Sum Wgts)	19

The four statistics in the Summary of Fit table are described below.

Rsquare
: estimates the proportion of the variation in the response around the mean that can be attributed to terms in the model rather than random error. Using quantities from the corresponding Analysis of Variance table seen in **Figure 16.10**, Rsquare is calculated

$$\frac{\text{Sum of Squares for Model}}{\text{Sum of Squares for C Total}}$$

It is also the square of the correlation between the actual and predicted response. An Rsquare of 1 occurs when there is a perfect fit (the errors are all zero). An Rsquare of 0 means that the fit is no better than the overall response mean for predicting a response.

Root Mean Square Error
: estimates the standard deviation of the random error. It is the square root of the mean square for error in the corresponding Analysis of Variance table, and it is commonly denoted as *s*.

Mean of Response
: is the overall mean of the response values. It is important as a base model for prediction, since all other models are compared to it. The variance measured around this mean is the *Corrected Total* (C Total) mean square in the Analysis of Variance table.

Observations (or Sum of Weights)
: records the number of observations used in the fit. If there are no missing values and no *excluded* rows, then this is the same as the number of rows in the data table. If there is a column assigned the role of weight, then this is the sum of the weight column values.

The Analysis of Variance Table

The Analysis of Variance table shows the basic calculations of fit in a linear model. The table shows how much better the full model fits compared to a simple fit to a single mean.

Figure 16.10 The Analysis of Variance Table for the Whole Model

Analysis of Variance

Source	DF	Sum of Squares	Mean Square	F Ratio
Model	6	413.61966	68.9366	13.8928
Error	12	59.54455	4.9620	**Prob > F**
C. Total	18	473.16421		0.0001

The columns of the Analysis of Variance table are described in the following list.

Source

lists the three sources of variation, called Model, Error, and C Total.

DF

records an associated *degrees of freedom* (DF) for each source of variation.

- The C Total degrees of freedom is for the simple mean model. There is only one degree of freedom used (the estimate of the mean parameter) in the calculation of variation so the C Total DF is always one less than the number of observations.

  The total degrees of freedom are partitioned into the Model and Error terms.
- The Model degrees of freedom is the number of parameters (except for the intercept) used to fit the model.
- The Error degrees of freedom is the difference between the C Total DF and Model DF.

Sum of Squares

records an associated sum of squares (SS for short) for each source of variation. The SS column accounts for the variability measured in the response. It is the sum of squares of the differences between the fitted response and the actual response.

- The total (C Total) SS is the sum of squared distances of each response from the sample mean (473.16 in the example shown in **Figure 16.10**). That is the base model (or simple mean model) used for comparison with all other models.
- The Error SS is the sum of squared differences between the fitted values and the actual values, and is 59.54 in the example above. This sum of squares corresponds to the unexplained Error (*residual*) after fitting the regression model.

- The total SS less the error SS gives the sum of squares attributed to the Model.

One common set of notation for these is SSR, SSE, and SST for sum of squares due to regression (model) , error, and total respectively.

Mean Square
is a sum of squares divided by its associated degrees of freedom. This computation converts the sum of squares to an average (mean square).

- The Model mean square is 68.93 in **Figure 16.10** .
- The Error mean square is 4.96 and estimates the variance of the error term. It is often denoted as MSE or s^2

F Ratio
is the model mean square divided by the error mean square. It is used to test the hypothesis that all the regression parameters (except the intercept) are zero. Under this whole-model hypothesis, the two mean squares have the same expectation. If the random errors are normal, then under this hypothesis the values reported in the SS column are two independent chi-squares. The ratio of these two chi-squares divided by their respective degrees of freedom (reported in the DF column) has an F distribution. If there is a significant effect in the model, the F Ratio is higher than expected by chance alone.

Prob > F
is the probability of obtaining a greater F value by chance alone if the specified model fits no better than the overall response mean. Significance probabilities of .05 or less are often considered evidence of a there is at least one significant regression factor in the model.

Note that large values of Model SS as compared with small values of Error SS lead to large F Ratios and low P values, which is what you want if the goal is to declare that terms in the model are significantly different than zero. Most practitioners check at this F test first and make sure that it is significant before delving further into the details of the fit.

The Lack of Fit Table

The Lack of Fit table is a special diagnostic test that appears only when the data and the model provide the opportunity. The idea is that sometimes you can estimate the error variance independently of whether you have the right form of the model. This occurs when you have observations that are exact replicates of each other in terms of the X variables. The error that you can measure for these exact replicates is called *pure error*. This is the portion of the sample error that cannot be explained or predicted.

The difference between the residual error from the model and the pure error is called lack of fit error. If you have the wrong functional form of some regressor, or if you don't have enough interaction effects in an analysis of variance model, then the lack of fit error measured in fitting the model shows up significantly greater than the pure error, if you have enough data to detect it. If lack of fit is significant then you should consider adding interaction terms if appropriate, or trying to better capture the functional form of a regressor.

There are two common situations where a lack of fit test is not produced:

- there are no exactly replicated points with respect to the X data, and therefore there are no degrees of freedom for pure error.
- the model is saturated, meaning that the model itself has a degree of freedom for each different X value, therefore there are no degrees of freedom for lack of fit.

A lack of fit test may not be of much use if there are only a very few degrees of freedom for it.

Figure 16.11 The Lack of Fit Table

Lack Of Fit

Source	DF	Sum of Squares	Mean Square	F Ratio
Lack Of Fit	9	41.299551	4.58884	0.7545
Pure Error	3	18.245000	6.08167	Prob > F
Total Error	12	59.544551		0.6743

The quantities in the Lack of Fit Table are described in the following list.

Source

lists the three sources of variation called Lack of Fit, Pure Error, and Total Error.

DF

records an associated *degrees of freedom* (DF) for each source of error.

- The Total Error DF is the degrees of freedom found on the Error line of the Analysis of Variance table (see **Figure 16.10**). It is the difference between the total DF and Model DF found in that table. The error DF are partitioned into degrees of freedom for lack of fit and for pure error.
- The Pure Error DF is pooled from each group where there are multiple rows with the same values for each effect. There are 3 instances in the CLASS data table where 2 subjects have the same age and height value, giving 3(2–1) = 3 DF for Pure Error. In general, if there are g groups having multiple rows with identical values for each effect, the pooled DF, denoted DF_p, is written

$$DF_p = \sum_{i=1}^{g} (n_i - 1)$$

where n_i is the number of subjects in the ith group.

- The Lack of Fit DF is the difference between the Total Error and Pure Error degrees of freedom.

Sum of Squares

records an associated sum of squares (SS for short) for each source of error.

- The Total Error SS is the sum of squares found on the Error line of the corresponding Analysis of Variance table.
- The Pure Error SS is pooled from each group where there are multiple rows with the same values for each effect. This estimates the portion of the true random error that is not explained by model effects. In general, if there are *g* groups having multiple rows with like values for each effect, the pooled SS, denoted SS_p, is written

$$SS_p = \sum_{i=1}^{g} SS_i$$

where SS_i is the sum of squares for the *i*th group corrected for the mean of that group.

- The Lack of Fit SS is the difference between the Total Error and Pure Error sum of squares. If the lack of Fit SS is large, the model may not be appropriate for the data. The F ratio described below is used to test whether the variation due to lack of fit is small enough to be accepted as a negligible portion of the pure error.

Mean Square

is a sum of squares divided by its associated degrees of freedom. This computation converts the sum of squares to an average (mean square). F ratios for statistical tests are computed as the ratios of mean squares.

F Ratio

is the ratio of mean square for Lack of Fit to mean square for Pure Error. It is used to test the hypothesis that lack of fit error is zero.

Prob > F

is the probability of obtaining a greater F value by chance alone if the there is no lack of fit, i.e. if the lack of fit variance and the pure error variance are the same.

The Parameter Estimates Table

The Parameter Estimates table (see **Figure 16.12**) shows the estimates of the parameters in the linear model and a t test for the hypothesis that each parameter is zero. Simple interval-valued regressors have only one parameter. Models with complex classification effects have a parameter for each anticipated degree of freedom.

Figure 16.12 The Parameter Estimates Table

Parameter Estimates

Term	Estimate	Std Error	t Ratio	Prob>|t|
Intercept	46.033454	4.24996	10.83	0.0000
Age[11-16]	-2.734257	2.02595	-1.35	0.2020
Age[12-16]	-2.060846	1.06933	-1.93	0.0780
Age[13-16]	.87189087	1.34505	0.65	0.5291
Age[14-16]	2.1743788	1.08617	2.00	0.0684
Age[15-16]	.35971171	1.22583	0.29	0.7742
Height	.16384949	.040724	4.02	0.0017

The following list describes the Parameter Estimates table.

Term

names the estimated parameter. The first parameter is always the intercept. Simple regressors show up as the name of the data table column. Regressors that are *dummy* indicator variables constructed from nominal effects are labeled with the names of the levels in brackets. The dummy variables are coded as 1 except for the last level, which is coded as –1 across all the other dummy variables for that effect.

Estimate

lists the parameter estimates for each term. They the coefficients of the linear model and are determined by least squares.

Std Error

is the standard error, an estimate of the standard deviation of the distribution of the parameter estimate. It is used to construct t tests and confidence intervals for the parameter.

t Ratio

is a statistic that tests whether the true parameter is zero. It is the ratio of the estimate to its standard error and has a Student's t distribution under the hypothesis given the usual assumptions about the model.

Prob>|t|

is the probability of getting an even greater t statistic (in absolute value), given the hypothesis that the parameter is zero. This is the two-tailed test against the alternatives in each direction. Probabilities less than 0.05 are often considered as significant evidence that the parameter is not zero.

The Singularity Details Table

This table details the problems that occur when a regressor variable is a linear combination of other regressors. The columns of this report show exactly what kind of relationship holds. The column label shows the regressor in which this singularity was detected as the system brought terms into the model in sequence. When there are singularities, then the least squares estimates are not unique, and some of the parameters may be nonestimable. This means that you can still get a value for the parameter that minimizes the sum of squared errors, but the expected value of this number is not necessarily the parameter it is trying to estimate. The Parameter Estimates table flags the affected parameter as *Biased* when the number is not uniquely estimable, and *zeroed* when it is a linear function of those above it and is thus zeroed by the solution method employed.

Figure 16.13 The Singularity Details Table

Singularity Details

Term	Sex[F-M]*Age[14-15]
Intercept	-1
Sex[F-M]	1
Age[11-15]	-4
Age[12-15]	1
Age[13-15]	1
Age[14-15]	1
Sex[F-M]*Age[11-15]	4
Sex[F-M]*Age[12-15]	-1
Sex[F-M]*Age[13-15]	-1
Sex[F-M]*Age[14-15]	-1

The Effect Tests Table

The effect tests are joint tests that all the parameters that make up an individual effect are zero. If an effect has only one parameter, as is the case with simple regressors, then the tests are no different from the t tests in the Parameter Estimates table. Details concerning the parameterization and handling of singularities, which are different than the SAS GLM procedure, are discussed in Appendix A.

Figure 16.14 The Effect Test Table

Effect Tests

Source	Nparm	DF	Sum of Squares	F Ratio	Prob > F
Age	5	5	49.043397	1.9767	0.1547
Height	1	1	80.321615	16.1872	0.0017

The columns in the Effect Tests table are described in the following list.

Source

lists the names of the effects in the model.

Nparm

is the number of parameters associated with the effect.

DF

is the degrees of freedom for the effect test. Ordinarily **Nparm** and **DF** are the same. They are different if there are linear combinations found among the regressors such that an effect cannot be tested to its fullest extent. Sometimes the DF will even be zero, indicating that no part of the effect is testable. Whenever DF is less than Nparm, the note **Lost DF's** appears to the right of the line in the report.

Sum of Squares
is the sum of squares for the hypothesis that the listed effect is zero.

F Ratio
is the F statistic for testing that the effect is zero. It is formed as the ratio of the mean square for the effect divided by the mean square for Error. The mean square for the effect is the Sum of Squares for the effect divided by its degrees of freedom

In situations where a subsequent term is specified to be an *error term*, the mean square for this error term is used instead of Mean Square for Error as the denominator of the F Ratio. Effects designated as error terms are marked with the message {Error Effect} on the right. This specification is used in split-plot and repeated-measure models where there are different observational units for different layers of the model.

Prob > F
is the significance probability for the F Ratio. It is the probability that, given that the hypothesis is true, that an ever larger F statistic would occur due to random error. If the value is less than .0005, then it appears as .0000 and represents a probability that is conceptually zero.

The Least Squares Means Table

Least squares means are predicted values from the specified model across the levels of a categorical effect where the other model factors are controlled by being set to *neutral* values. Least squares means are the values to examine to see which levels produce higher or lower responses, holding the other variables in the model constant. Least squares means are also called adjusted means or population marginal means.

The neutral values that other effects are set to are the sample means (possibly weighted) for regressors with interval values, and the average coefficient over the levels for unrelated categorical effects.

Least squares means are the statistics that are compared when effects are tested. They may not reflect typical real-world values of the response if the values that the factors are set to do not reflect prevalent combinations of values in the real world. They are mainly used for the purpose of comparison in experimental situations.

A Least Squares Means table (see **Figure 16.15**) is produced for all categorical effects in the model. The standard error is also reported beside it. For main effects, the sample means are also reported. It is common for the least squares means to be closer together than the sample means, just as regressors are more often substitutes rather than complements.

For further details on least squares means, see Appendix A.

Figure 16.15 The Least Squares Means Table

Least Squares Means

Age

Level	Least Sq Mean	Std Error	Mean
11	59.68845794	2.051534812	54.4000
12	60.36186898	1.022208438	59.4400
13	63.29460607	1.366759807	61.4333
14	64.59709403	1.116324179	64.9000
15	62.78242691	1.318971233	65.6250
16	63.81183723	3.017277235	72.0000

Level
lists the names of each categorical level.

Least Sq Mean
lists the least squares mean for each level of the categorical variable.

Least Sq Mean
lists the standard error for each level of the categorical variable.

Mean
lists the response sample mean for each level of the categorical variable. This will be different from the least squares mean if the values of other effects in the model do not balance out across this effect.

The Contrast Dialog and Report

A contrast is a set of linear combinations of parameters that you want to jointly test to be zero. JMP builds contrasts in terms of the least squares means of the effect. By convention each column of the contrast is normalized to have sum zero and an absolute sum equal to two.

At the top of each effect report is a pop-up menu icon to the left of the name of the effect. For example, the main effect C has the pop-up icon shown here at the left. Click on this submenu icon to select the command **Add Contrast** to bring up a new non-modal window for specifying contrasts with respect to that effect. This command is enabled only for pure classification effects.

The contrast dialog shown in **Figure 16.16** is for the effect called C. It shows the name of the effect and the names of the levels in the effect. Beside the levels is an area enclosed in a rectangle that has a column of numbers and some boxes of + and – signs. To construct a contrast, click on the + and – signs beside the levels that you want to increment or decrement. If possible, the dialog normalizes each time to make the sum for a column zero and the absolute sum equal to two. Each time you click, it adds to the plus or minus score proportionately.

Figure 16.16 The Contrast Dialog

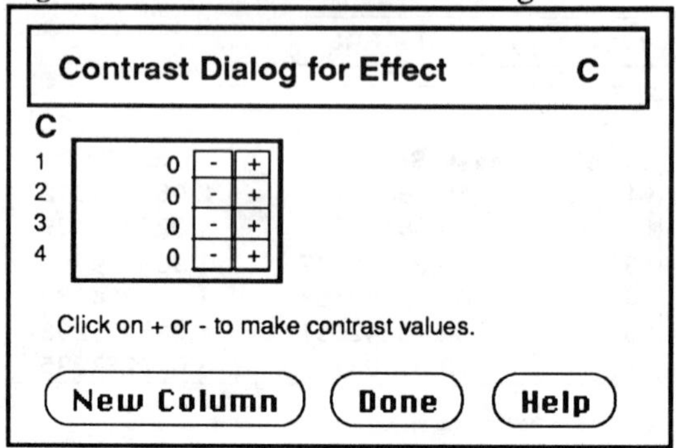

For example, to form a contrast that compares each of the first three levels with the others in turn, click on + for the C = 1 effect, and click on – for each of the other levels. The result is the example shown to the left in **Figure 16.17** Then click the **New Column** button to start a new column. After completing this kind of pattern for two more columns you see the completed contrast dialog shown on the right below.

Figure 16.17 Examples of a Contrast Specification

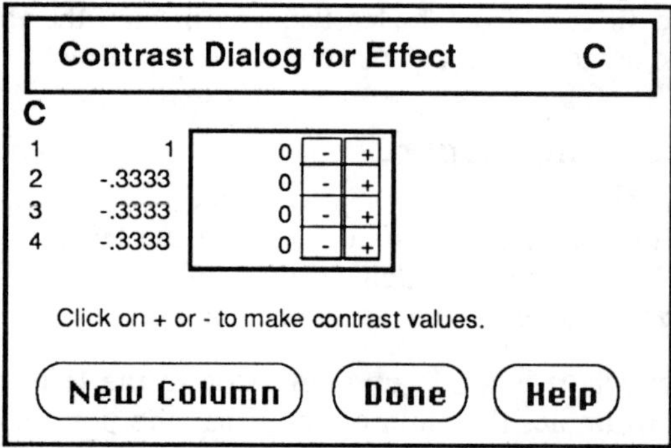

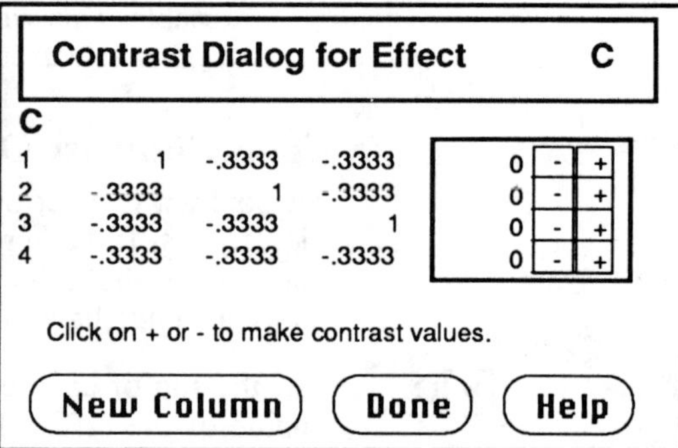

After you are through defining contrasts, click on **Done**. The contrast is estimated and the Contrast table shown to the left in **Figure 16.18** is appended below the other tables for that effect. If you click the Parameter Function reveal button you see the report shown to the right.

Figure 16.18 The Contrast and Parameter Function Tables

Contrast

1	1	-.3333	-.3333
2	-.3333	1	-.3333
3	-.3333	-.3333	1
4	-.3333	-.3333	-.3333
Estimate	-1.808	-.4961	.78166
Std Error	.48614	.48614	.48614
t Ratio	-3.719	-1.020	1.6078
Prob >|t|	.00067	0.3143	.11659

Sum of Squares	43.510841666
Numerator DF	3
F Ratio	6.8187366671
Prob > F	.00093340466

Parameter Function

Parameter Function

Parameter			
Intercept	0	0	0
A[1-3]	0	0	0
A[2-3]	0	0	0
C[1-4]	1.3333	0	0
C[2-4]	0	1.3333	0
C[3-4]	0	0	1.3333
A[1-3]*C[1-4]	0	0	0
A[1-3]*C[2-4]	0	0	0
A[1-3]*C[3-4]	0	0	0
A[2-3]*C[1-4]	0	0	0
A[2-3]*C[2-4]	0	0	0
A[2-3]*C[3-4]	0	0	0

The Contrast table contains the following items.

- the contrast specified by the dialog as a function of the least squares means.
- the estimates and standard errors of the contrast for the least squares means, and t-tests for each column of the contrast.
- the F-test for all columns of the contrast tested jointly.
- a reveal button containing the contrast expressed in terms of the parameters.

Examples with Statistical Details

One-Way Analysis of Variance with Contrasts

In a one-way analysis of variance, a different mean is fit to each of the different sample (response) groups, as identified by a nominal X variable. To specify the model for JMP, select the interval Y and a nominal X variable. The method translates this specification into a linear model as follows: The nominal variables are used to determine a sequence of *dummy* variables, which have only values 1, 0, and -1. For example, suppose the response Y is fit to a nominal variable group with three levels. The linear model is written

$$y_i = \beta_0 + \beta_1 x_{1i} + \beta_2 x_{2i} + \varepsilon_i$$

The first dummy variable is labeled group(1-3) in the report, which denotes that group=1 contributes a value 1 and group=3 contributes a value -1 to the dummy variable

$$x_{1i} = \begin{cases} 1 & \text{if group=1} \\ 0 & \text{if group=2} \\ -1 & \text{if group=3} \end{cases}$$

The second dummy variable is labeled group(2-3) in the report and is given values

$$x_{2i} = \begin{cases} 0 & \text{if group=1} \\ 1 & \text{if group=2} \\ -1 & \text{if group=3} \end{cases}.$$

No variable is needed for the last level because the model is constructed so that the last level is found by subtracting all the other parameters. Therefore, the coefficients sum to one across all the levels.

The estimates of the means for the three levels in terms of this parameterization are

$$\mu_1 = \beta_0 + \beta_1$$

$$\mu_2 = \beta_0 + \beta_2$$

$$\mu_3 = \beta_0 - \beta_1 - \beta_2$$

The following one-way analysis of variance example uses the data from Snedecor and Cochran (1967, p. 422). A variable called drug with values labeled a,d, and f is studied to see if there is any difference in the mean response for each drug. The drug effect has an observed significance probability (**Prob>F**) of .03, just barely significant compared to .05.

Figure 16.19 Summary Tables for the Drug Experiment

Summary of Fit

Rsquare	.2278264
Root Mean Square Error	6.070877
Mean of Response	7.9
Observations (or Sum Wgts)	30

Analysis of Variance

Source	DF	Sum of Squares	Mean Square	F Ratio
Model	2	293.6000	146.800	3.9831
Error	27	995.1000	36.856	**Prob > F**
C Total	29	1288.7000		0.0305

Parameter Estimates

Term	Estimate	Std Error	t Ratio	Prob>|t|
Intercept	7.9	1.10838	7.13	0.0000
drug[a-f]	-2.6	1.56749	-1.66	0.1088
drug[d-f]	-1.8	1.56749	-1.15	0.2609

Because the drug effect involves two parameters, a joint test on both parameters is needed. This joint test appears in the Effect Tests table. Because there is only one effect in the model, it is the same test as the whole-model F test in the Analysis of Variance table.

Figure 16.20 The Effect Tests Table for the Drug Experiment

Effect Tests

Source	Nparm	DF	Sum of Squares	F Ratio	Prob > F
Drug	2	2	293.60000	3.9831	0.0305

Whenever there are nominal effects, one interesting question is how to compare how well the levels predict the response. Rather than use the parameter estimates directly, it is usually more meaningful to compare the predicted values at the levels of the nominal values. These predicted values are the least squares means, and in this simple case they are the same as the ordinary means. They become different when other distorting effects are introduced into the model, as seen later.

Figure 16.21 The Least Squares Means Table for the Drug Experiment

Least Squares Means

Drug

Level	Least Sq Mean	Std Error	Mean
a	5.30000000	1.919780080	5.3000
d	6.10000000	1.919780080	6.1000
f	12.30000000	1.919780080	12.3000

Since there is only one effect, the leverage plot for drug shows a regression to the means of the actual values. This plot shown on the right in **Figure 16.22** can be interpreted as a regression, in the manner described by Levin, Serlin and Webne-Behrman (1989). The drugs a,d and f are labeled x, y and z respectively in the plots below. The only difference between the whole-model plot and the drug leverage plot is in the scaling of the X axis.

Figure 16.22 Whole Model Leverage Plot and Drug Effect Leverage Plot

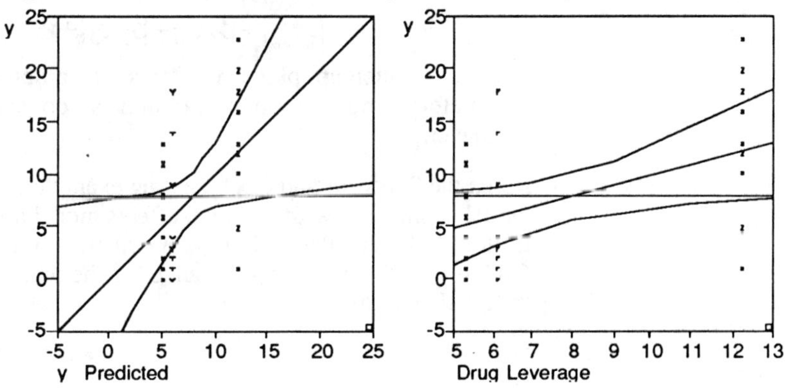

The drug effect can be studied in more detail by using a contrast. Click on the pop-up icon above the drug plot and issue the **Add Contrast** command to obtain a contrast specification dialog. Click the + keys for drugs a and d, and the – key for drug f to obtain the contrast shown to the left in **Figure 16.23**. Then click the **Done** button to obtain the contrast report shown to the right. The report shows that the effect looks more significant if the test is reduced to this one-degree-of-freedom comparison.

Figure 16.23 Contrast Example for the Drug Experiment

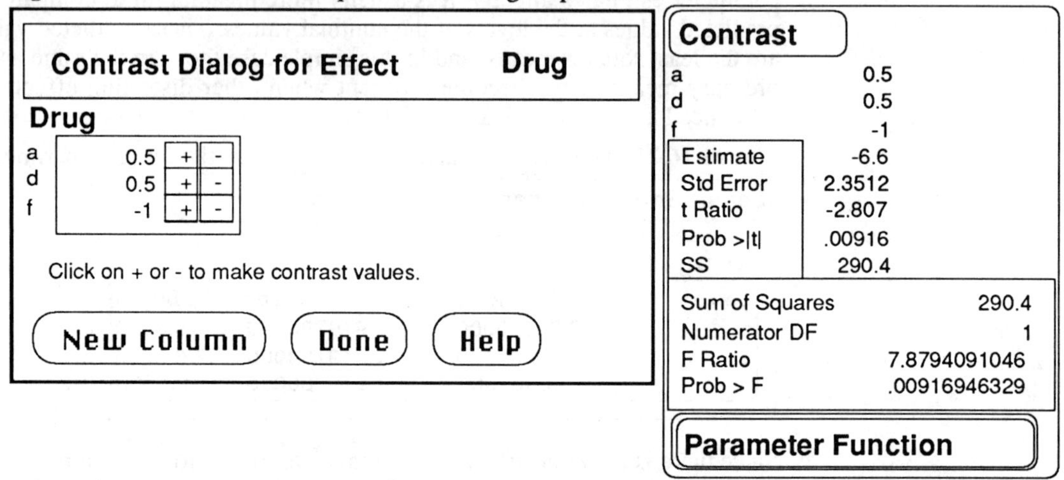

Analysis of Covariance

An analysis of variance model with an added regressor term is called an *analysis of covariance*. Suppose the data are the same as above, but with one additional term, X_3 as the new regressor, whereas X_1 and X_2 continue to be dummy variables to index over the three levels of the nominal effect.

$$y_i = \beta_0+\beta_1 x_{1i}+\beta_2 x_{2i}+\beta_3 x_{3i}+\varepsilon_i$$

There is an intercept plus two effects, one a nominal main effect using two parameters, and the other an interval covariate regressor using one parameter.

Rerun the Snedecor and Cochran drug example but add the covariate X to the model. Compared with the main effects model above, the Rsquare increases from 22.7% to 67.6% and the standard error of the residual is reduced from 6.07 to 4.0. The F test significance for the whole model has gone from .03 to less than .0001.

Figure 16.24 Summary Tables for the Drug Experiment with Covariate

Summary of Fit

Rsquare	.6762608
Root Mean Square Error	4.005777
Mean of Response	7.9
Observations (or Sum Wgts)	30

Analysis of Variance

Source	DF	Sum of Squares	Mean Square	F Ratio
Model	3	871.4974	290.499	18.1039
Error	26	417.2026	16.046	**Prob > F**
C Total	29	1288.7000		0.0000

In some models, you can investigate the precise functional contribution of a covariate. For example, the covariate might fit better if some transformation to it were done. If you happen to have data where there are exact duplicate observations for the regressor effects, then it is possible to partition the total error into two components. One component estimates error from the data where all the X values are the same. The other estimates error that might contain effects for unspecified functional forms of covariates, or interactions of nominal effects. This provides the basis for a lack of fit test. If the lack of fit error is significant, then you are warned that there is some effect in your data that is not in your model. Note that there is no lack of fit error in this example, as seen by the large probability value 0.7507.

Figure 16.25 Lack of Fit and Parameter Estimates Tables for the Drug Experiment with Covariate

Lack Of Fit

Source	DF	Sum of Squares	Mean Square	F Ratio
Lack Of Fit	18	254.86926	14.1594	0.6978
Pure Error	8	162.33333	20.2917	**Prob > F**
Total Error	26	417.20260		0.7507

Parameter Estimates

Term	Estimate	Std Error	t Ratio	Prob>|t|
Intercept	-2.695772	1.91108	-1.41	0.1702
Drug[a-f]	-1.185036	1.06082	-1.12	0.2742
Drug[d-f]	-1.076065	1.04129	-1.03	0.3109
x	.98718381	.164497	6.00	0.0000

The covariate X has a substitution effect with respect to drug. It accounts for much of the variation in the response previously accounted for by the drug variable. Thus, even though the model is fit with much less error, the drug effect no longer looks significant. The effect that was previously observed in the main effects model now appears explainable to some extent in terms of the values of the covariate.

Figure 16.26 The Effect Tests Table for drug with Covariate

Effect Tests

Source	Nparm	DF	Sum of Squares	F Ratio	Prob > F
Drug	2	2	68.55371	2.1361	0.1384
x	1	1	577.89740	36.0145	0.0000

The least squares means are now different from the ordinary means, because they are adjusted for the effect of X, the covariate, on the response. Now the least squares means are the predicted values that you expect for each of the three values of drug given that the covariate X is held at some constant value. The constant value is chosen for convenience to be the mean of the covariate, which is 10.7333. So, the prediction equation gives the least squares means as follows.

fit equation: -2.695 –1.185 drug(a-f) –1.0760 drug(b-f) + .98718 X

for a: -2.695 –1.185 (1) - 1.0760 (0) + .98718 (10.7333) = 6.71

for b: -2.695 –.185 (0) - 1.0760 (1) + .98718 (10.7333) = 6.82

for f: -2.695 –1.185 (-1) - 1.0760 (-1) + .98718 (10.7333) = 10.16

Figure 16.27 The Least Squares Means Table for Drug with Covariate

Least Squares Means

Drug

Level	Least Sq Mean	Std Error	Mean
a	6.71496346	1.288494280	5.3000
d	6.82393479	1.272468995	6.1000
f	10.16110174	1.315923424	12.3000

With two effects there is now a leverage plot for the whole model, and a leverage plot for each effect. If the covariate made no difference, the leverage plot for drug wouldn't be much different than above, but because the covariate makes a big difference, the leverage values for drug are dispersed a bit from their least-squares means.

Figure 16.28 Comparison of Leverage Plots for Drug Test Data

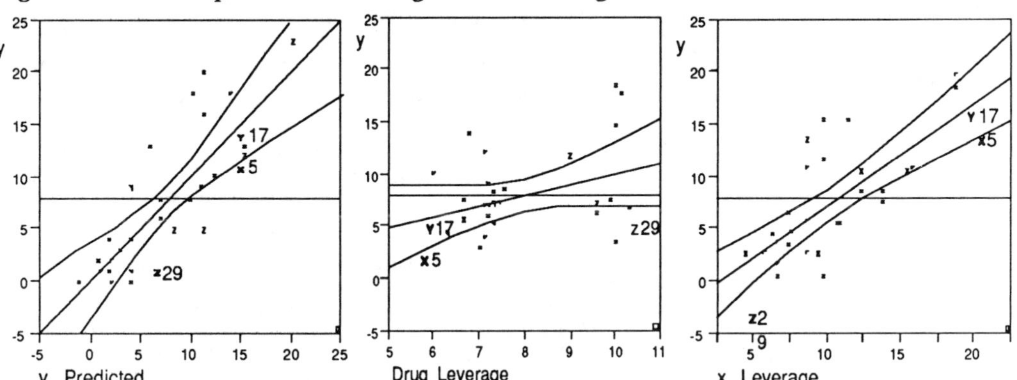

In the leverage plots above the three points 5, 17, and 29 are highlighted to illustrate the following situations.

- Point 29 makes drug appear less significant and X more significant.
- Point 5 makes drug appear more significant, but X less significant.
- Point 17 is important to the significance test for both drug and X. It has a small actual residual but is far out on the X axis so that it would have had a large residual if either effect or both were removed from the model.

If you use the Save pop-up border menu and choose **Save Prediction Formula** you can create a new column in the data table called Pred Formula Y with the prediction formula definition,

$$-2.6957729+\left(\begin{cases}2.26110174, & \text{if } Drug = \text{“f”}\\ -1.0760652, & \text{if } Drug = \text{“d”}\\ -1.1850365, & \text{if } Drug = \text{“a”}\\ \bullet, & \text{otherwise}\end{cases}\right)+.987183811x$$

Nominal Response Models

If the response variable is marked with the nominal level of measurement, then the fitting process and reports are very different from those of interval responses. The nominal and ordinal reports are not as richly developed with testing and descriptive and output machinery, and they have no supporting graphical displays. Yet they are similar in approach to the interval reports.

When the response is nominal the **Fit Y by X's** fits a linear model to a multi-level logistic response function using maximum likelihood. Likelihood Ratio test statistics are provided for the whole model. Lack of Fit tests and Wald test statistics are provided for each effect in the model.

The details of the model are in discussed in Chapter 13, *Understanding JMP Analyses* and Appendix A, *Statistical Details.*

Example

An experiment was performed on on metal ingots, which are prepared with different heating and soaking times and tested for readiness to roll (Cox 1970, 67-68).

Figure 16.29 The INGOTS Data Table

4 Cols / 38 Rows	heat (Int)	soak (Int)	ready (Nom)	count (Int)
18	14	2.8	0	31
19	14	4	1	0
20	14	4	0	19
21	27	1	1	1
22	27	1	0	55

The categorical variable called ready has values 1 and 0 for readiness and not readiness to roll, respectively. The **Fit Y by X's** platform fits the probability of the not readiness (0) response to a logistic cumulative distribution function applied to the linear model with regressors heat and soak.

$$\text{Probabililty (not ready to roll)} = \frac{1}{1 - \exp(\beta_0 + \beta_1 \text{heat} + \beta_2 \text{soak})}$$

The parameters are estimated by minimizing the sum of the negative logs of the probabilities attributed to the observations by the model (maximum likelihood).

The INGOTS data table is attributed as shown in **Table 16.1** for the **Fit Y by X's** command.

Table 16.1 Attribution of the INGOTS data table

Column	Measurement Level	Role
heat	Interval	X
soak	Interval	X
ready	Nominal	Y
count	Interval	Freq

Iterations

After launching **Fit Y by X's**, an iterative estimation process begins, and is reported iteration-by-iteration. If this takes too long, you can cancel by pressing the command-period key combination. Otherwise the iteration process stops when either the log-likelihood doesn't change by more than a very small relative amount (Obj-Critertion), or the parameter estimates don't change by a small relative amount (Delta-Criterion), or 15 iterations have been performed.

Figure 16.30 The Log Likelihood Iteration History Table

Iteration History

Iter	LogLikelihood	Step	Delta-Criterion	Obj-Criterion
0	-268.2479588	Initial	215940.610	•
1	-76.29480707	Newton	2.21362485	2.51561021
2	-53.38032919	Newton	1.50013773	.429187799
3	-48.34608552	Newton	0.66532357	.104107758
4	-47.69191264	Newton	.132895659	.013713766
5	-47.67282965	Newton	.005870740	.000400206
6	-47.67280662	Newton	.000009553	.000000482

Whole-Model Fit

The first test concerns the fit of the model as a whole, which is compared to the model that omits all the regressor effects and just fits intercept parameters. The test is analogous to the Analysis of Variance table for interval responses. The negative log likelihood corresponds to the Sums of Squares, and Chi-square tests correspond to F tests.

Figure 16.31 Summary Tables for the INGOT Data

Summary of Fit

Rsquare (U)	.1088231
Observations (or Sum Wgts)	387

Analysis of LogLikelihood

Source	DF	-LogLikelihood	ChiSquare
Model	2	5.821410	11.64282
Error	384	47.672807	**Prob>ChiSq**
C. Total	386	53.494217	0.002963

- The C Total line describes the negative log likelihood that results from a reduced model with only intercept parameters. For the ingot experiment, the –loglikelihood for the model that includes only the intercepts is 53.49.
- The Error line describes the negative log likelihood for the complete model.
- The Model line is the difference between the C Total and Error. It measures the significance of the regressors as a whole to the fit. Twice this difference, is a likelihood ratio chi-square test for the hypothesis that all regressor parameters are zero.

- The ratio of the Model to the C Total negative log likelihood is the Rsquare(U) seen in the Summary of Fit table and is sometimes referred to as U, the uncertainty coefficient.

After fitting the full model with two regressors, the –loglikelihood on the Model line shows a reduction to 5.82 from the C Total –loglikelihood of 53.49. The ratio of the reduction to the C Total (the proportion of the uncertainty that has been attributed to the fit), is 10% and is reported as the Rsquare (U). Nominal models rarely have high Rsquares, and have Rsquares of 1 only when all the probabilities of the events that occur are 1.

To test that the regressors as a whole are significant (the whole model test), a chi-square statistic is computed by taking twice the difference in –loglikelihoods between the fitted model and the reduced model that has only intercepts. In the INGOT example, this chi-square value is

$$2* 5.82 =11.64$$

which is significant at .002 as shown as **Prob>ChiSq** in **Figure 16.31.**

Lack of Fit

The next question is whether there is enough information using the variables in the current model, or whether more complex terms need be added. The lack of fit test, sometimes called a goodness of fit test, provides this information.It calculates a pure error negative log likelihood by constructing categories for every combination of the regressor values in the data, and tests whether this log likelihood is significantly better than the model fit. The degrees of freedom for Pure Error and Total Error are counted negatively from the Model line in the LogLikelihood table. This is convenient and avoids the arbitrariness of how to count total degrees of freedom. The error terms are summarized as follows.

- Total Error is the –loglikelihood for the fitted model
- Pure Error is the –loglikelihood for a model saturated in the regressor levels
- Lack of Fit is the difference between Total and Pure Error.

The Likelihood Ratio Chi-square Statistic for Lack of Fit =
2 * (Lack of Fit –loglikelihood)

Figure 16.32 Lack of Fit table for the INGOT Data

Lack Of Fit

Source	DF	-LogLikelihood	ChiSquare
Lack Of Fit	16	6.876314	13.75262
Pure Error	-18	40.796493	**Prob>ChiSq**
Total Error	-2	47.672807	0.617137

The Lack of Fit chi-square here is 13.75 with 16 degrees of freedom, which is not significant. This supports the conclusion that there is little to be gained by introducing variations on the regressor variables, such as using polynomials or cross terms, or reattributing the measurement level to nominal.

Wald Tests for Effects

To test an effect, fit the model both with and without the effect, and see if the difference in the -loglikelihood is significant. That is the Likelihood Ratio test for an effect. Unfortunately, this test involves refitting the whole model using another series of iterations. However a much cheaper test that serves the same function is provided. The Wald chi-square is a one-step linear approximation to the Likelihood ratio tests that is almost a by-product of the calculations. Though Wald tests are considered less trustworthy, they do provide a cheap significance indicator for screening effects. Each parameter estimate and effect is shown with a Wald test.

Figure 16.33 Parameter Estimates and Effect Tests for the INGOT Data

Parameter Estimates

Term	Estimate	Std Error	ChiSquare	Prob>ChiSq
Intercept	5.55915958	1.1196933	24.65	0.0000
Heat	-.08203067	.02373446	11.95	0.0005
Soak	-.05677077	.33121286	0.03	0.8639

Effect Test

Source	Nparm	DF	Wald ChiSquare	Prob>ChiSq
heat	1	1	11.945205	0.0005
soak	1	1	0.029379	0.8639

Likelihood Ratio Tests for Effects

It is recommended that you use the Wald tests only for casual screening, and obtain Likelihood Ratio tests for any effect you consider important. Likelihood Ratio chi-squares are calculated as twice the difference of the log likelihoods between the full model, and the model reduced by the hypothesis to be tested. This value can be conveniently computed by taking the differences in whole-model Chi-squares, since they are already scaled by two. The difference in them cancels out the C Total portion of the values.

Likelihood Ratio Chisquare(Effect A)

= 2*(–loglikelihood(full model) – –loglikelihood(model without A)

= Model Chisquare(full model) – Model Chisquare(model without A)

To do the Likelihood Ratio test on soak, refit the model using only heat. The resulting subset model has a Whole Model Chi-square of 11.61. The difference between this and the full model Chi-square of 11.64 yields a Likelihood Ratio chi-square of .03 for heat, which is the same as the Wald chi-square. With one parameter involved, the degrees of freedom is 1. The soak effect is not significant.

Figure 16.34 Analysis Omitting Soak but Including Heat

Analysis of LogLikelihood

Source	DF	-LogLikelihood	ChiSquare
Model	1	5.806938	11.61387
Error	385	47.687279	**Prob>ChiSq**
C Total	386	53.494217	0.000655

To do the Likelihood Ratio test on heat, refit without heat to obtain a subset Whole-Model chi-square of .59. The difference of this with the full model yields the Likelihood Ratio Chisquare of (11.64-.59) = 11.05, which is only a little different from the Wald chi-square of 11.94. The heat effect is thus very significant.

Figure 16.35 Analysis Omitting Heat but Including Soak.

Analysis of LogLikelihood

Source	DF	-LogLikelihood	ChiSquare
Model	1	0.296479	.5929579
Error	385	53.197738	**Prob>ChiSq**
C. Total	386	53.494217	0.441278

Ordinal Response Models

If the response variable has an ordinal level of measurement, then the platform fits the cumulative response probabilities to the logistic distribution function of a linear model using maximum likelihood. Likelihood Ratio test statistics are provided for the whole model and lack of fit. Wald test statistics are provided for each effect. The details of the model are in Chapter 13, *Understanding JMP Analyses* and Appendix A. The method is discussed in Walker and Duncan(1967), Nelson(1976), Harrell(1986), McCullaugh and Nelder(1983).

The ordinal model is much faster to fit and uses less memory than the nominal model if there are many response levels.

Example

McCullagh and Nelder (1983) report an experiment by Newell to test whether various cheese additives (A to D) had an effect on taste as measured by a tasting panel who recorded on an ordinal scale from 1 (strong dislike) to 9 (excellent taste).

Figure 16.36 The CHEESE Data Table

4 Cols / 36 Rows	Nom cheese	Ord response	Int count	Int score
8	A	8	8	-0.86
9	A	9	1	-0.86
10	B	1	6	2.48
11	B	2	9	2.48

The data table is attributed as shown in **Table 16.2.**

Table 16.2 Attribution of the CHEESE Data Table

Column	Measurement Level	Role
cheese	Nominal	X
response	Ordinal	Y
count	Interval	Freq

In this example the method required only 4 iterations, reducing the background –loglikelihood of 429.0 by 74 to 355.67. This reduction yields a likelihood ratio chi-square for the whole model of 148.4 with 3 degrees of freedom, showing the cheese treatment to be highly significant.

Figure 16.37 Summary Tables for Cheese Experiment

Iteration History

Iter	LogLikelihood	Step	Delta-Criterion	Obj-Criterion
0	-429.9009019	Initial	168555.124	4.1815E305
1	-373.9172814	Newton	2769.23639	.149717935
2	-356.6694121	Newton	.451006910	.048356784
3	-355.6806047	Newton	.129336796	.002779964
4	-355.6739714	Newton	.003081265	.000018649

Summary of Fit

Rsquare (U)	.1726605
Observations (or Sum Wgts)	208

Analysis of LogLikelihood

Source	DF	-LogLikelihood	ChiSquare
Model	3	74.22693	148.4538
Error	197	355.67397	**Prob>ChiSq**
C. Total	200	429.90090	0.000000

The lack of fit test happens to be testing the ordinal response model compared to the nominal model, since the model would be saturated if the response were treated nominal rather than ordinal, and there would be 21 more parameters if it were, which is the Lack of Fit degrees of freedom. The non-significance of Lack of Fit leads one to believe that the ordinal model is reasonable.

Figure 16.38 The Lack of Fit Table for CHEESE Experiment

Lack Of Fit

Source	DF	-LogLikelihood	ChiSquare
Lack Of Fit	21	10.15412	20.30823
Pure Error	-24	345.51986	**Prob>ChiSq**
Total Error	-3	355.67397	0.501820

There are 8 intercept parameters, since there are nine response categories. There are only three structural parameters. If this problem were treated nominally, it would have 8*3=24 structural parameters.

Figure 16.39 The CHEESE Experiment Parameter Estimates Table

Parameter Estimates

Term	Estimate	Std Error	ChiSquare	Prob>ChiSq
Intercept	-4.6041035	.42664542	116.45	0.0000
Intercept	-3.5491910	.30733329	133.36	0.0000
Intercept	-2.4503395	.23984446	104.37	0.0000
Intercept	-1.3818243	.20014145	47.67	0.0000
Intercept	-.04537241	.18030489	0.06	0.8013
Intercept	.906510315	.18864585	23.09	0.0000
Intercept	2.40797408	.23805266	102.32	0.0000
Intercept	3.96737905	.34662242	131.01	0.0000
Cheese[A-D]	-.86126990	.22890346	14.16	0.0002
Cheese[B-D]	2.48859627	.27026645	84.79	0.0000
Cheese[C-D]	.847561855	.22802510	13.82	0.0002

Because there is only one effect, the test for it is equivalent to the likelihood ratio test for the whole model. The likelihood ratio Chi-square was 148.45, which is different from the Wald chi-square of 115.12, and illustrates the point that Wald tests are to be regarded with some skepticism.

Figure 16.40 The CHEESE Experiment Effect Test Table

Effect Test

Source	Nparm	DF	Wald ChiSquare	Prob>ChiSq
Cheese	3	3	115.12417	0.0000

To see which cheese additive is preferred, look for the most negative values of the parameters. Cheese D's effect is the negative of the sum of the others.

A	-.8613	2nd most liked
B	2.4885	least liked
C	.8476	3rd place
D	-2.4748	most liked

You can also use the **Fit Y by X** platform for this model, which treats nominal and ordinal responses the same and shows a contingency table analysis. The log-likelihood that results agrees with the lack of fit log-likelihood shown above.

Figure 16.41 Analysis of log Likelihood for CHEESE Experiment

Analysis of LogLikelihood

Source	DF	-LogLikelihood
Model	24	84.38105
Error	4	345.51986
C. Total	28	429.90090

Test	ChiSquare	Prob>ChiSq
Likelihood Ratio	168.762	0.0000
Pearson	162.482	0.0000

If you want to see a graph of the response probabilities as a function of the scores for the four cheeses, you can create a new interval variable (score, as shown below) and use it as an interval effect in the **Fit Y by X** platform.

$$\text{score} = \begin{cases} -0.8613, & \text{if } \textit{Cheese}= \text{"A"} \\ 2.4885, & \text{if } \textit{Cheese}= \text{"B"} \\ 0.8476, & \text{if } \textit{Cheese}= \text{"C"} \\ -2.4748, & \text{otherwise} \end{cases}$$

The distance between each curve is the fitted response probability for the levels in the order shown for the levels on the left.

Figure 16.42 Cumulative Logistic Probability Pot for Cheese Scores

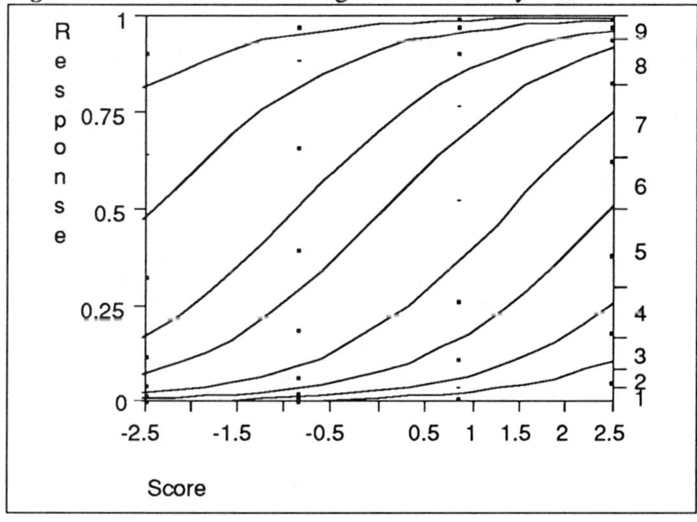

Chapter 17. Analyze: Specify Model

Using the **Specify Model** command lets you construct models that are more complex than those assumed by other JMP statistical platforms or by the **Fit Y by X's** command. You can specify crossed terms for factorial models, nested factors with specific error terms, polynomial models, response surface models, and add or remove terms from the model specification as needed.

Note that there is one fitting system and a single report platform that is launched from two commands:

- The **Specify Model** command, discussed in this chapter, displays a general dialog that defines complex models automatically, and lets you choose specific model effects and error terms.
- The **Fit Y by X's** command, discussed in the previous chapter, performs a fit of one Y variable by all the X variables selected. It is used for for multiple regression and main effects analysis of variance models.

  The graphical display of the **Specify Model** report is the same as that of the **Fit Y by X's**. Leverage plots are shown for each effect in the model.

This chapter focuses on specifying models for more complex experimental designs and interpreting the graphs and text reports. See Chapter 16, *Analyze: Fit Y by X's* for an introduction to linear models, a description of leverage plots, and definitions of the display options and save commands available for this platform.

Chapter Contents

The Specify Model Dialog

The **Specify Model** command uses the same fitting platform as the **Fit Y by X's** command. However, instead of producing a report based on simple X effects, this command displays the dialog shown in **Figure 17.1**. Using the **Specify Model** dialog, you can tailor the model for your data. The model can contain simple effects as well as crossed, nested, and polynomial effects.

The **Specify Model** dialog is *non-modal*, which means it persists until you close it explicitly. This is useful, because often an analyst wants to modify the model specification in order to fit several different models before closing the window.

Each part of the **Specify Model** dialog is described in the next section.

Figure 17.1 The **Specify Model** Dialog

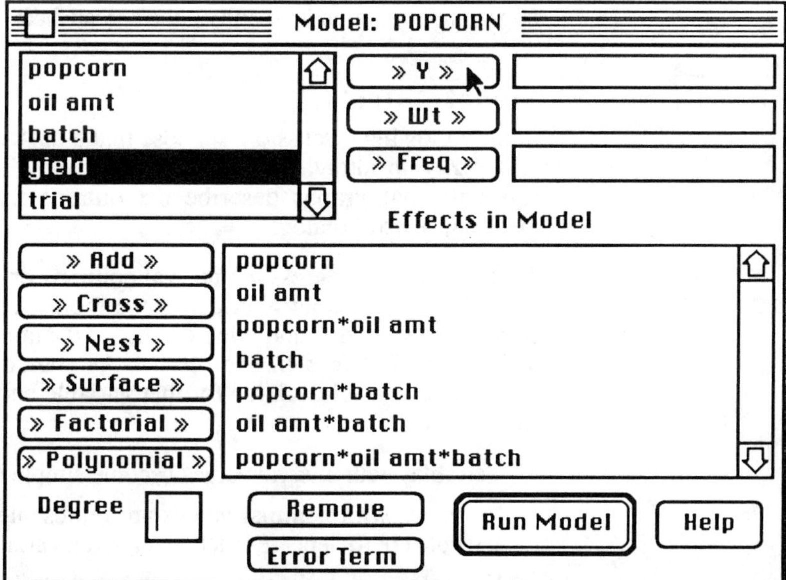

The Column Selector List

The column selector list on the upper left of the dialog lists all the columns in the current data table. When you click on a column name it highlights and responds to the action you choose using the other buttons on the dialog. If you want to change your column selection, click on a different column name. Use shift-click to extend a selection of column names and command-click to select names that are not contiguous in the list.

» Y »

» Wt »

» Freq »

Response Variable Roles

If you assigned Y, X, Weight, or Freq roles to columns on the spreadsheet, then they appear on the **Specify Model** dialog in the corresponding role box when you open the dialog. If you preassigned the Y role to more than one column, the leftmost column name on the spreadsheet is shown in the Y role box of the dialog. If you want to remove preassigned variables from roles, highlight them and click on the **Remove** button.

- **Y** identifies a single response variable (the dependent variable) for the model.
- **Wt** is an optional role that identifies one column whose values supply weights for the response. Weighting affects the importance of each row in the model.
- **Freq** is an optional role that identifies one column whose values assign a frequency to each row for the analysis and determines how degrees of freedom are counted.

If you want a column to have the **Label** role, you must assign it using the pop-up role assignment menu on the spreadsheet

Model Effects

In the following discussion, suppose that a data table contains the variables X and Z with interval values, and A, B, and C with nominal values. The following paragraphs describe the buttons that can be clicked in the **Specify Model** dialog.

Simple Effects

» Add »

To add a simple regressor (interval column) or a main effect (nominal or ordinal column) as an X effect to the model, select the column from the column selector list, and click on the **Add** button. The column names are added to the **Effects in Model** list. As you add effects, be aware of the measurement level declared for that variable and the consequences for the fitting model.

- Variables with interval values become simple regressors.
- Variables with nominal values are represented with *dummy* variables to fit separate coefficients for each level in the variable.
- Variables with ordinal values are handled like nominal variables, but with a slightly different coding.

See Appendix A, *Statistical Details* for details on coding of effects.

If you mistakenly add an effect, select it in the **Effects in Model** list and click the **Remove** button.

Crossed Effects and Interactions

» **Cross** »

To create a compound effect by crossing two or more variables, first select the columns in the column selector list. Next, click the **Cross** button to add the crossed effect to the effects list.

- Crossed nominal variables become interaction effects.
- Crossed interval variables become just multiplied regressors.
- Crossing a combination of nominal and interval variables produces special effects suitable for testing homogeneity of slopes.

If a variable is selected in the the column selector list and you click on the **Cross** button while an effect is highlighted in the **Effects in Model** list, the dialog crosses the variable with the effect.

See Appendix A, *Statistical Details* for for a discussion on how crossed effects are parameterized and coded.

Nested Effects

» **Nest** »

Nested effects are a special variety of crossed effects that absorb the main or outside effects too. The notation **B[A]** is read "B within A" and means that you want to fit a B effect within each level of **A**. Thus **B[A]** is like pooling B+A*B. To specify a nested effect, first specify the outside nesting effects with the **Add** or **Cross** buttons. Then add the nested variables with the **Nest** button.

For example, suppose that you want to specify **A*B[C]**. Use command-click to highlight the variables A and B in the column list. Then click **Cross** to obtain **A*B** in the **Effects in Model** list. Highlight the **A*B** term in the **Effects in Model** list and then select C in the column selector list. Now click **Nest** to obtain **A*B[C]** in the dialog effects list.

Warning➔

The current implementation is limited by the assumption that there is a matched crossing of levels across the terms in the nest. This will be ameliorated in a subsequent JMP release.

Response Surface Effects

» **Surface** »

A response surface effect is one where the object is to find the values of the terms that produce a maximum or minimum expected response. This is done by fitting a collection of terms in a quadratic model. The critical values for the surface are calculated from the parameter estimates, along with a report on the shape of the surface.

To specify a response surface effect, select two or more variables from the column list and click the **Surface** button. A term labeled **Surface(***term, term,...***)** appears on the **Effects in Model** list with each variable shown as a surface *term*. For example, if you have variables A and B,

Surface(A) fits $\beta_1 A + \beta_{11} A^2$

Surface(A,B) fits $\beta_1 A + \beta_{11} A^2 + \beta_2 B + \beta_{22} B^2 + \beta_{12} AB$

Factorial Models

» Factorial »

If you have many crossed factors, such as in a factorial design, you can specify your model most conveniently using the **Factorial** button. The factorial button adds the set of effects corresponding to all the crossings of all the variables selected in the column list.

For example, suppose that you select three variables A, B, and C. The **Factorial** button adds **A, B, A*B, C, A*C, B*C, A*B*C** to the effect list in the model dialog.

If you want to remove some of the higher-order interactions, highlight them in the effects list and click the **Remove** button.

Polynomial Effects

» Polynomial »

A polynomial effect is a series of terms which are powers of one variable. To specify a polynomial term,

- click on the variable in the column list
- enter the degree of the polynomial in the **Poly Degree** box
- click on the **Polynomial** button.

For example, suppose you select a variable X. A degree 2 polynomial, denoted **Poly(X,2)** in the model dialog, fits parameters for X and X^2. A degree 3 polynomial, denoted **Poly(X,3)** in the model dialog, fits parameters for X, X^2, and X^3.

Dispatching Buttons

Help

Help

The **Help** Button brings up a help window containing information on the **Specify Model** Dialog, which in turn contains buttons for help about specific models.

Run Model

Run Model

The **Run Model** button submits the model to the fitting platform. Because the **Specify Model** dialog is non-modal, the **Run Model** button does not close the dialog window. You can continue to use the window and make changes to the model for additional fits. You can also make changes to the data and refit a model using the same dialog window.

Validity Checks

The model is checked for several kinds of errors such as duplicate effects or missing effects in a hierarchy. If you get an alert message box, you can either continue the fitting despite the situation, or **Cancel** it.

Display Options and Save Commands

The display options and save commands are the same as those for the **Fit Y by X's** platform. For a complete discussion of these options, see Chapter 16, *Analyze: Fit Y by X's*.

Types of Complex Models

Table 17.1 presents a catalog of model examples that can be specified using the **Specify Model** dialog, where X and Z have interval values and A, B, and C have nominal values. This list is not meant to be exhaustive.

When you correctly specify the type of model you see the model effects shown to the right in **Table 17.1** listed as the **Effects in Model** on the **Specify Model** dialog. Refer to **Table 17.2** to see the clicking sequences that produce each of these sets of model effects.

Table 17.1 Types of Models

Type of Model	Model Effects
simple regression	X
quadratic polynomial	X, X*X Poly(X,2)
cubic polynomial	X, X*X, X*X*X or Poly(X,3)
multiple regression	X, Z, ... , *other interval columns*
one-way analysis of variance	A
analysis of covariance	A, X
two-way main effects anova	A, B
two-way anova with interaction	A, B, A*B
three-way full factorial	A, B, A*B, C, A*C, B*C, A*B*C
simple nested model	A, B[A]
compound nested model	A, B[A], C[A B]
simple split-plot or repeated measure	A, B[A]{error}, C, A*C
2-way response surface	X, Z, X*X, Z*Z, X*Z or Surface(X,Z)

Table 17.2 lists the sequence of clicking actions to specify each of the models in **Table 17.1**. The following convention is used in this table to specify clicking: If the column name is in plain typeface,click on the name in the column selector list. If the column or effect is shown as it appears in the **Effect in Model** list (**like these words**), highlight it in that list. Note that an effect remains highlighted when it appears in the effects list.

Note➜ Unless the effect itself is to be highlighted for the clicking sequence, be sure to deselect it in the effects list (by clicking anywhere in the list area) before you proceed with the sequence shown in the table.

The name of the button to click is shown in the same typeface as it appears in the dialog.

Table 17.2 Clicking Sequences for Models

Type of Model	Clicking Sequence
simple regression	X, **Add**
quadratic polynomial	X, **Add**, X, **X**, **Cross** or X, **Degree** 2, **Polynomial**
cubic polynomial	X, **Add**, X, **X**, **Cross**, X, **X*X**, **Cross** or X, **Degree** 3, **Polynomial**
multiple regression	X, **Add**, Z, **Add**, ... or X, Z, ... , **Add**
one-way anova	A, **Add**
analysis of covariance	A, **Add**, X, **Add** or A, X, **Add**
two-way main effects analysis of variance	A, **Add**, B, **Add** or A, B, **Add**
two-way analysis of variance with interaction	A, B, **Add**, A, B, **Cross** or A, B, **Factorial**
three-way full factorial	A, B, C, **Factorial**
simple nested model	A, **Add**, B, **Add**, A, **B**, **Nest** or A, B, **Add**, A, **B**, **Nest**
compound nested model	A, B, **Add**, A, **B**, **Nest** C, **Add** , A, B, **C**, **Nest**
simple split-plot repeated measure	A, **Add**, B, **Add**, A, **B**, **Nest**, **B[A]**, **Error** , A, C, **Cross**
2-way response surface	X, Z, **Surface**

Examples with Statistical Details

This section contains examples of typical models that either cannot be handled with other JMP analyze commands or are extended by using the **Specify Model** Dialog. The discussions describe how to specify the model in the dialog, how to interpret the results, and they provide the statistical details for each model.

Analysis of Covariance with Separate Slopes

This example continues with the DRUG example in Chapter 16, *Analyze: Fit Y by X's* that illustrates analysis of covariance. The example uses data from Snedecor and Cochran (1967, p. 422). A one-way analysis of variance for a variable called **drug** with values a,d, and f showed a difference in the mean response between the drug main effect with a significance probability of 0.03. A partial listing of the DRUG data table is shown in **Figure 17.2.** The lack of fit test for the model with main effect **drug** and covariate **x** was not significant. However, for the sake of illustration, the **Drug*X** term is now added to the model to test whether the regression on the covariate has separate slopes for different drug levels.

Figure 17.2 Partial Listing of the DRUG Data Table

3 Cols / 30 Rows	drug (Nom)	x (Int)	y (Int)
10	a	3	0
11	d	6	0
12	d	6	2

The **Specify Model** dialog includes the two simple effects, and the drug*x crossed effect as shown below.

Figure 17.3 Partial Listing of the DRUG Data Table

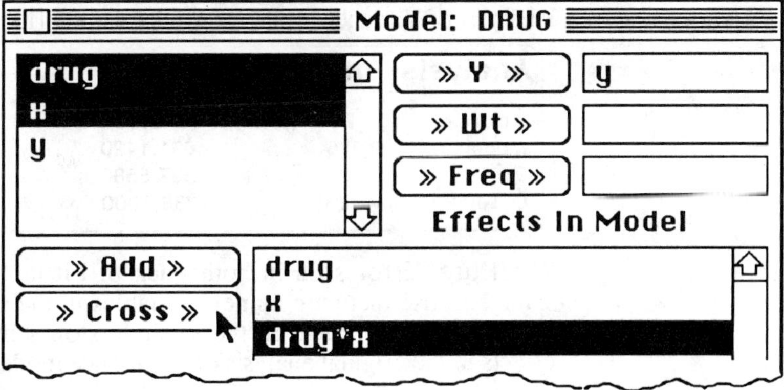

This specification adds two columns to the linear model (call them x_4 and x_5) which allow the slopes for the covariate to be different for each of the drug levels. The new variables are formed by multiplying the dummy variables for drug by the covariate values giving

$$y_i = \beta_0+\beta_1x_{1i}+\beta_2x_{2i}+\beta_3x_{3i}+\beta_4x_{4i}+\beta_5x_{5i}+\varepsilon_i$$

Table 17.4 summarizes the values assigned to the columns in the linear model.

Table 17.4 Coding of Analysis of Covariance with Separate Slopes

Regressor	Effect	Values
x1	Drug[a-f]	+1 if a, 0 if d, –1 if f
x2	Drug[d-f]	0 if a, +1 if d, –1 if f
x3	X	the values of X
x4	Drug[a-f]*X	+X if a, 0 if d, –X if f
x5	Drug[d-f]*X	0 if a, +X if d, –X if f

The comparison of the Summary tables shows the model having separate slopes increases the Rsquare from 67.7% to 69.1%. The separate slopes model shifts two degrees of freedom from Lack of Fit error (see **Figure 17.6** below) to the model, increasing the SS for Model from 3 to 5 as shown in **Figure 17.5.**

Figure 17.5 Comparison of Analysis of Variance Tables

Analysis of Variance Table for Model with Same Slopes

Analysis of Variance

Source	DF	Sum of Squares	Mean Square	F Ratio
Model	3	871.4974	290.499	18.1039
Error	26	417.2026	16.046	Prob > F
C Total	29	1288.7000		0.0000

Analysis of Variance Table for Model with Different Slopes

Analysis of Variance

Source	DF	Sum of Squares	Mean Square	F Ratio
Model	5	891.1420	178.228	10.7594
Error	24	397.5580	16.565	Prob > F
C Total	29	1288.7000		0.0000

The Pure Error seen in both Lack of Fit tables in **Figure 17.6** is the same because there are no new variables in the separate slopes covariance model. The new effect in the separate slopes model is constructed from effects in the original analysis of covariance model.

Figure 17.6 Comparison of Lack of Fit Tables

Lack of Fit Table for Model with Same Slopes

Lack Of Fit

Source	DF	Sum of Squares	Mean Square	F Ratio
Lack Of Fit	18	254.86926	14.1594	0.6978
Pure Error	8	162.33333	20.2917	Prob > F
Total Error	26	417.20260		0.7507

Lack of Fit Table for Model with Different Slopes

Lack Of Fit

Source	DF	Sum of Squares	Mean Square	F Ratio
Lack Of Fit	16	235.22462	14.7015	0.7245
Pure Error	8	162.33333	20.2917	Prob > F
Total Error	24	397.55795		0.7231

The remaining reports show further details. In the Effect Test reports (not shown), neither the **drug** nor **drug*x** effect is significant. The least squares means for each drug are adjusted due to fitting different slopes and have slightly greater standard errors.

Figure 17.7 Comparison of Least Squares Means Tables

Least Squares Means Table for Model with Same Slopes

Least Squares Means

Level	Least Sq Mean	Std Error	Mean
a	6.71496346	1.288494280	5.3000
d	6.82393479	1.272468995	6.1000
f	10.16110174	1.315923424	12.3000

Least Squares Means Table for Model with Different Slopes

Least Squares Means

Level	Least Sq Mean	Std Error	Mean
a	6.368152866	1.350271000	5.3000
d	6.883602151	1.300927285	6.1000
f	9.711994322	1.486066631	12.3000

Leverage plots in **Figure 17.8** illustrate the significance of the model and the **x** effect. The **drug** main effect is less significant than it was, with the separate slopes covariate absorbing some of its effect. You can highlight points 5 and 25, which are the only ones that have much influence on the separate slopes effect.

Figure 17.8 Leverage Plots for the Model with Separate Slopes

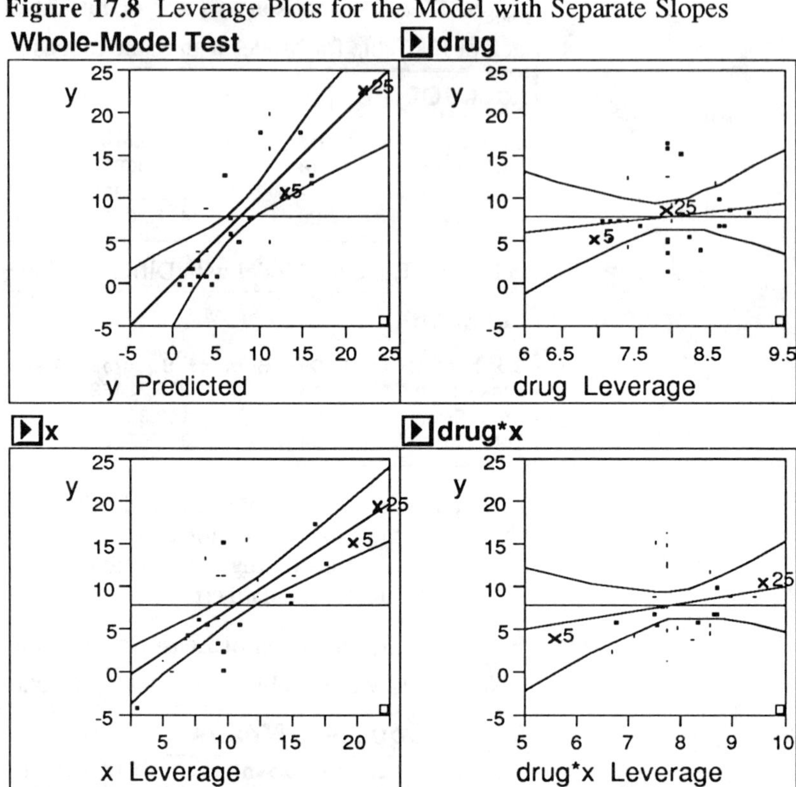

To summarize the fits made on the DRUG data, the following three graphs illustrate

- the simple main effects model (**Figure 17.9**)
- the model with a covariate, but with a common slope (**Figure 17.10**)
- the model that allows separate slopes to the covariate for each **drug** level (**Figure 17.11**)

The construction of the least squares means is also shown. The points are labeled x for drug a, y for drug d, and z for drug f. As an exercise, you can construct these plots with judicious use of features like the **Exclude** command in the **Rows** menu and the **Stack** table operation found in the **Transform** submenu of the **Edit** menu.

Figure 17.9 shows a scatterplot which illustrates the model **y** = **drug** with lines drawn at the mean of each drug group.

Figure 17.9 Illustration of the Simple Mean Model

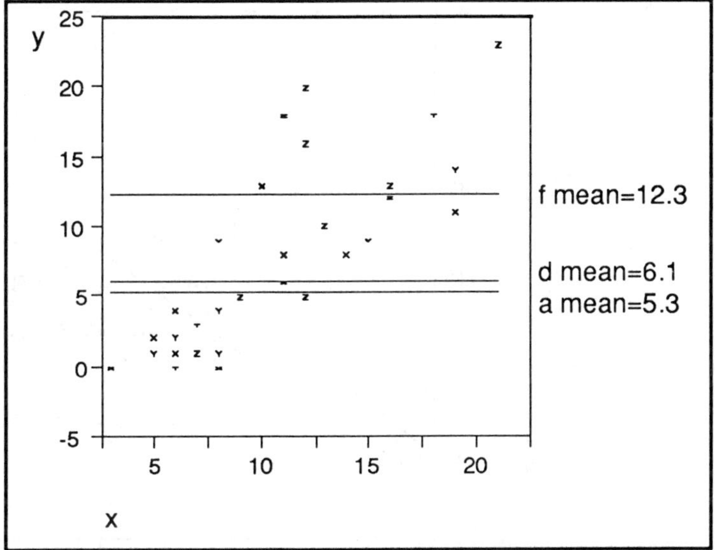

Figure 17.10 shows a scatterplot which illustrates the model **y** = **drug x** with lines drawn having the same slope for each drug group.

Figure 17.10 Illustration of the Analysis of Covariance Model

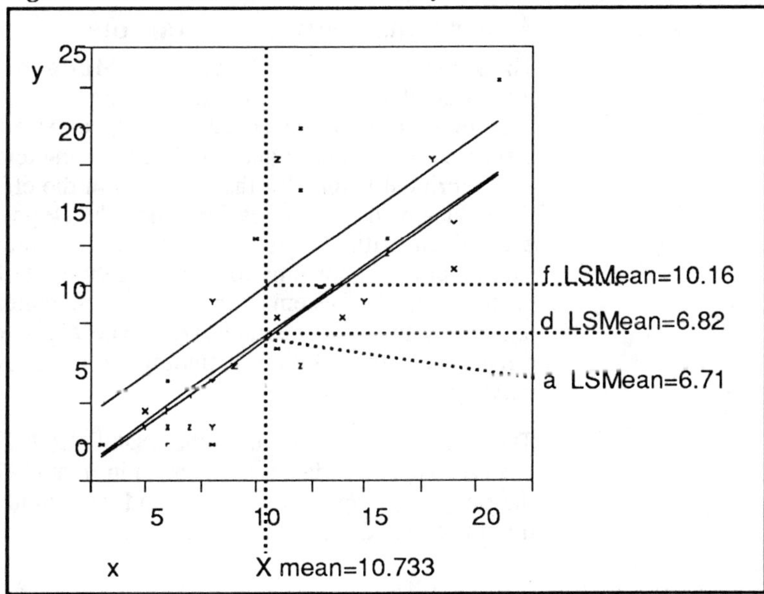

Figure 17.11 shows a scatterplot with lines drawn having different slopes for each drug group which illustrates the model **y = drug x drug*x.**

Figure 17.11 Illustration of the Covariance Model with Different Slopes

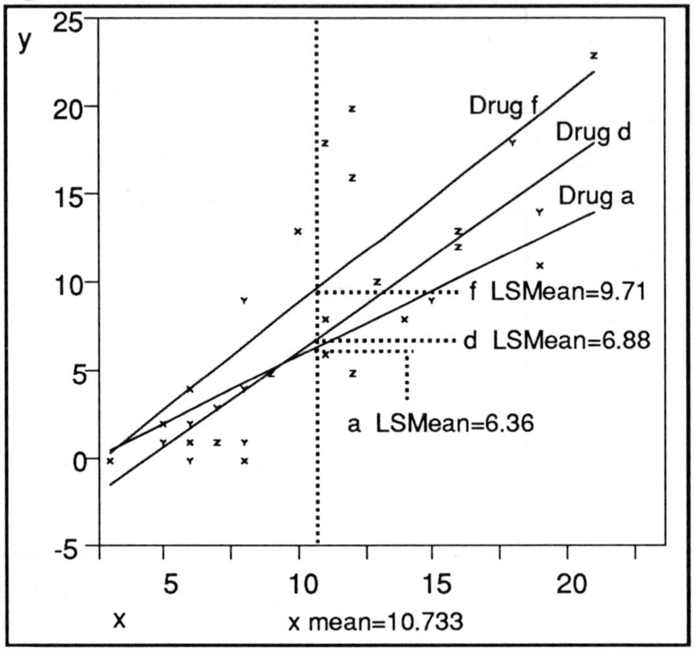

A Response Surface Example

The **Surface** button on the **Specify Model** dialog produces estimates for the parameters when a quadratic response surface is needed to find the maximum or minimum expected response. An equation using the parameter estimates to the linear and quadratic parameters in the effect is solved to obtain critical values for the variables in the effect. The critical values are the values of the variables for which the slope of the estimated surface is zero. Each critical value corresponds to a maximum, a minimum, or a saddle point according to the curvature at that point. By looking at eigenvalues of a system of the quadratic parameters, the curvature is found and the solution is characterized. The eigenvectors show the principal directions of the solution, including the directions of greatest and smallest curvature.

The following example uses the ODOR CONTROL file in the SAMPLE DATA folder. The data are discussed in John (1971). The objective is to find the range of temperature (temp) and gas-liquid ratio (gl ratio) values that minimize the odor of a chemical production process.

Figure 17.12 The ODOR CONTROL data table

3 cols / 15 rows	odor (Int)	temp (Int)	gl ratio (Int)
1	66	-1	-1
2	58	-1	0
3	65	0	-1
4	-31	0	0

To run a response surface model, select the response variable, **odor**, from the variable selector list and click on the **Y** button. Then use shift-click to select the factors, **temp** and **gl ratio**, and click the **Surface** button. The effect is shown in the **Effects in Model** window as **Surface(temp, gl ratio)**.

Figure 17.13 Specify a Response Surface Model

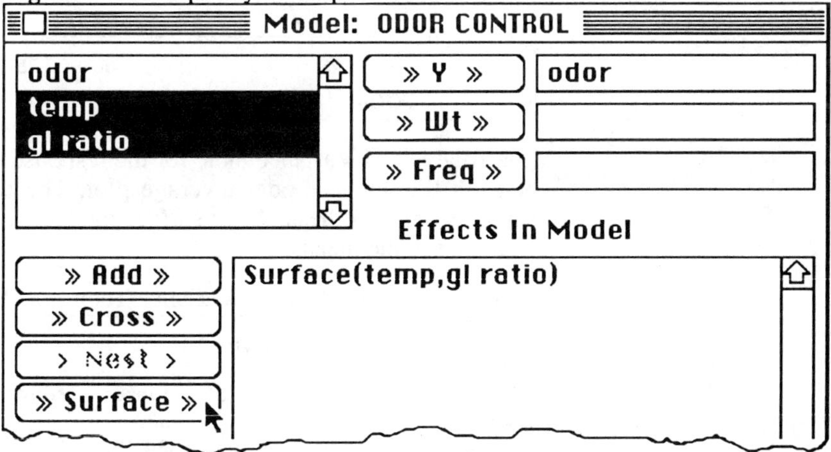

Because the whole model contains only the surface factor, both the whole model leverage plot and the factor (surface) plot are the same. The 95% confidence curves cross the mean line and indicate that the surface model accounts for a significant proportion of the total variation.

Figure 17.14 Leverage Plot for Response Surface Model

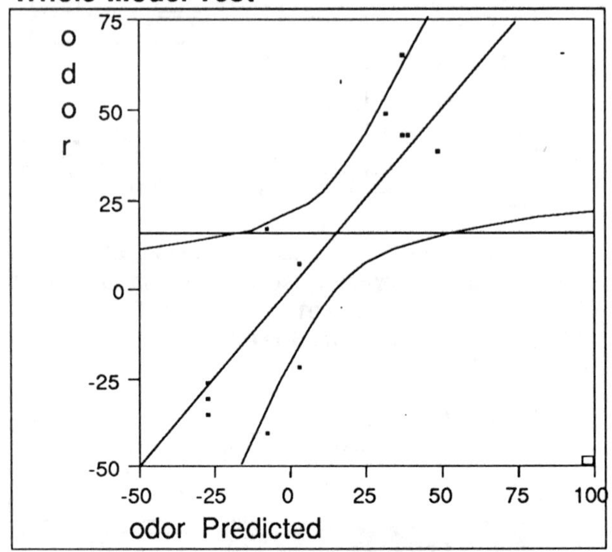

The Analysis of Variance table for the response surface model is displayed beneath the whole model leverage plot. The probability value of .0286 supports the hypothesis that a surface model is a significantly better model than the sample mean.

Figure 17.15 Analysis of Variance Table for Response Surface Model

Analysis of Variance

Source	DF	Sum of Squares	Mean Square	F Ratio
Model	5	15068.967	3013.79	4.2786
Error	9	6339.433	704.38	Prob > F
C Total	14	21408.400		0.0286

The Response Surface reveal button shown beneath the surface factor leverage plot displays two reports as shown in **Figure 17.16** The Solution report shows the critical values for the surface variables and indicates that the surface solution point is a minimum for this example.

The Eigenstructure table is useful for identifying the shape and orientation of the curvature and results from the eigenvalue decomposition of the matrix of second-order parameter estimates. The eigenvalues are negative if the response surface shape curves back from a unique maximum. The eigenvalues are positive if the surface shape curves up from a unique minimum. If the eigenvalues are mixed, the surface is saddle-shaped, curving up in one direction and down in another direction.

The eigenvectors show the orientations of the principal axes. In this example the eigenvalues (48.38 and 30.60) are positive so the curvature bends up from a minimum. The direction where the curvature is the greatest corresponds to the largest eigenvalue and is oriented mostly on gl ratio. The direction at right angles to this is loaded more on temp and has nearly as much curvature as that for gl ratio.

Sometimes a zero eigenvalue occurs. This means that along the direction described by the corresponding eigenvector, the fitted surface is flat.

Figure 17.16 Basic Reports for Response Surface Model

Response Surface

Solution

Variable	Critical Value
temp	.17027834
gl ratio	.16462659

Solution is a Minimum

EigenStructure

EigenValues and EigenVectors

Variable	48.3803	30.6004
temp	0.23892	0.97104
gl ratio	0.97104	-0.23892

You can use the results of the response surface fit to see the predicted surface in three dimensions. First, select the Save Predicted Formula option from the save ($) pop-up menu. This option adds column called Pred Formula odor to the data table as seen in **Figure 17.17**. Because the predicted formula is saved with this column, any modification you make to the temp and gl ratio values is automatically computed for the Pred Formula odor column.

Figure 17.17 Save Predicted Value and Predicted Formula

4 cols / 15 rows	Int odor	Int temp	Int gl ratio	Int Pred Formula odor
1	66	-1	-1	89.4326923
2	58	-1	0	16.8173076
3	65	0	-1	37.4423076
4	-31	0	0	-26.923076
5	39	1	-1	48.6826923

Next, modify the odor control data table by

- adding rows to form a square grid. In this example 385 rows are added to give a total of 400 rows
- using the technique described in Chapter 11, *The Calculator* to compute grid values for the temp and gl ratio columns.

A partial listing of the new data table is shown below.

Figure 17.18 Response Surface Grid and Predicted Values

4 cols / 400 rows	Int temp	Int gl ratio	Int Pred Formula odor
1	-1.5	-1.5	213.033653
2	-1.5	-1.3	187.140115
3	-1.5	-1.1	163.608286
4	-1.5	-1.0	142.438165
5	-1.5	-.86	123.629754

Now use the **Spin** command on the **Analyze** Menu to display this expanded table. **Figure 17.19** shows the temp, and gl ratio columns on the X and Y axes, and the Pred Formula odor column on the Z axis. You can see the minimum clearly and highlight the minimum odor points on the spinning plot to identify the range of temp and gl ratio on the spreadsheet that optimizes odor control in the chemical production process.

Figure 17.19 Predicted ODOR CONTROL Data Response Surface

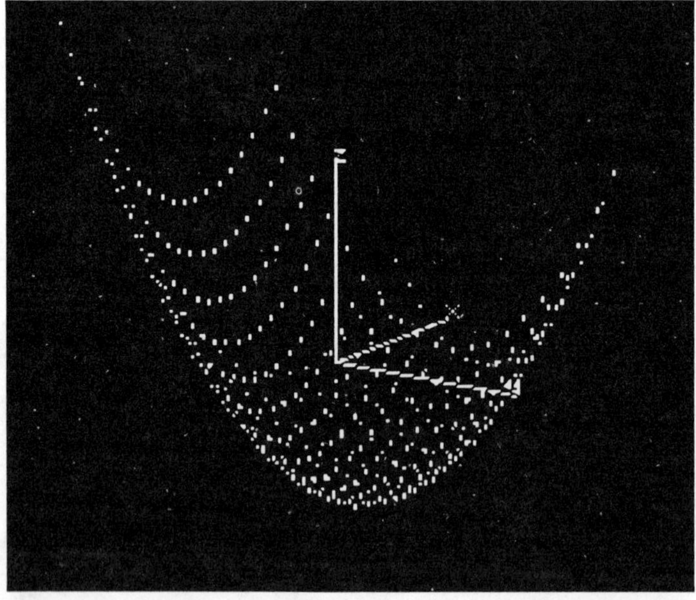

A Repeated Measures Example

A study was done to gain information about the difference in seasonal hunting habits of foxes and coyotes. Three foxes and three coyotes were marked and observed periodically for a year The average number of miles (rounded to the nearest mile) they wandered from their dens during different seasons of the year was recorded. The ANIMALS data table is shown in **Figure 17.20.** [the data is fictional.]

The model is defined by

- the interval response variable called **miles**
- the effect called **animal** with values fox or coyote
- the effect called **season** with values fall, winter, spring, and summer.
- an animal identification code called **subject**, with nominal values '1', '2', and '3' for both foxes and coyotes.

Figure 17.20 Data Table for Repeated Measure Experiment

4 cols / 24 rows	animal (Nom)	subject (Nom)	miles (Int)	season (Nom)
1	FOX	1	0	fall
2	FOX	1	0	winter
3	FOX	1	5	spring
4	FOX	1	3	summer
5	FOX	2	3	fall
6				
7				
8				
9				
10				
11				
12				
13	COYOTE	1	4	fall
14	COYOTE	1	2	winter
15	COYOTE	1	7	spring
16	COYOTE	1	8	summer
17	COYOTE	2	5	fall
18	COYOTE	2	4	winter
19	COYOTE	2	6	spring
20	COYOTE	2	6	summer
21	COYOTE	3	7	fall
22	COYOTE	3	5	winter
23	COYOTE	3	8	spring
24	COYOTE	3	9	summer

The design for this experiment is called a *repeated measures design* because each individual contributes several response measurements under different conditions. Although JMP does not specialize in handling this kind of design, it can be analyzed correctly if you accept certain assumptions and construct the model appropriately.

There are two layers to the model. The top layer is the across-subject layer in which the effect of being a fox or coyote (**animal** effect) is tested with respect to the variation from subject-to-subject. The bottom layer is the between-subject layer in which the repeated-measures factor for the four seasons is tested with respect to the variation from season-to-season within a subject.

The within-subject variability is reflected in the residual error. The **season** and **animal*season** effects can use residual error for the denominator of their F statistics.

However, the across-subject variability is not measured by residual error, and must be captured with an effect in the model, **subject[animal]**. The F statistic for the across-subject effect **animal** must use this subject within animal effect instead of residual error for its F ratio denominator.

The **Specify Model** command lets you construct model terms and specify error terms. To run this nested model with the correct F statistics, add terms to the **Effects in Model** list as follows to see the completed dialog in **Figure 17.21**.

- Select **animal** from the column selector list and click on **Add.**
- Select **subject** from the column selector list and click on **Add.**
- Select **subject** from the column selector list again and also select **subject** in the **Effects in Model** list. Whenever variables are selected in both lists, the **Cross** and the **Nest** buttons become active. Click on the **Nest** button to add the **subject** within **animal** effect to the model.
- Select **season** from the column selector list and click on **Add**
- Use command-click to select both animal and season from the column selector list. Whenever more than one variable is selected in this list the **Cross** button becomes active. Click on the **Cross** button to add the animal by season interaction effect to the model.
- Select the **subject** within animal effect and click on the **Error Term** button .

Each effect uses as its F-ratio denominator the nearest effect below it that is marked as an error effect. If no effects below it are error terms, then the residual error becomes the error term.

Figure 17.21 Specify Model Dialog for Repeated Measures Experiment

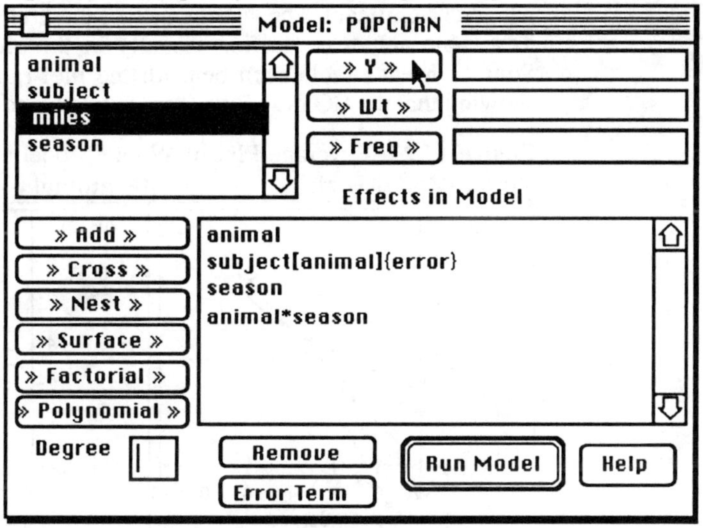

Click on the **Run Model** button when the dialog is complete. The fitting platform displays summary tables, leverage plots, and analysis of variance information as in other models. The summary information shown in **Figure 17.22** indicates that the model as a whole fits well (the Rsquare value is .89) and that both the animal and season effects are significant.

Figure 17.22 Summary Tables for Repeated Measure Experiment

Response: miles

Summary of Fit

Rsquare	.8924796
Root Mean Square Error	1.111805
Mean of Response	4.458333
Observations (or Sum Wgts)	24

Parameter Estimates

Effect Test

Source	Nparm	DF	Sum of Squares	F Ratio	Prob > F	
animal	1	1	51.041667	11.8932	0.0261	
subject[animal]	4	4	17.166667	3.4719	0.0419	{Error Effect}
season	3	3	47.458333	12.7978	0.0005	
animal*season	3	3	7.458333	2.0112	0.1662	

The leverage plot to the left in **Figure 17.23** shows that the model as a whole is significant. The significance of the animal effect seen in the Effect Test table above is verified visually in the plot to the right. The Least Squares Means table seen beneath the plot quantifies the animal effect, showing that the COYOTE wanders nearly twice as far as the FOX.

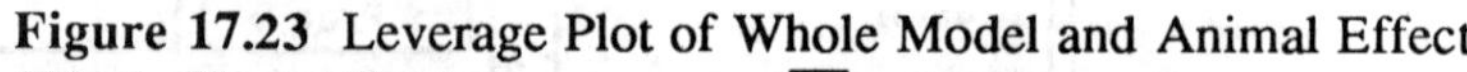

Figure 17.23 Leverage Plot of Whole Model and Animal Effect

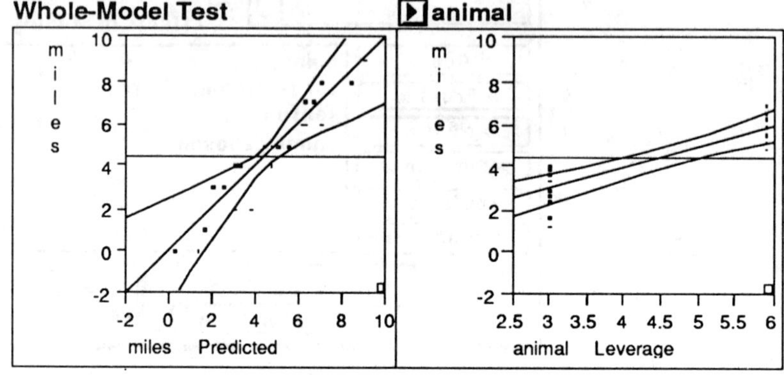

Least Squares Means

Level	Least Sq Mean	Std Error	Mean
COYOTE	5.916666667	0.3209505558	5.91667
FOX	3.000000000	0.3209505558	3.00000

The season effect is also highly significant as seen in the summary tables and in **Figure 17.24.** You can look at the Least Squares Means table beneath the season leverage plot to see that the animals wander further in the spring and summer than fall and winter, as might be expected.

Figure 17.24 Leverage Plot of the Season Effect

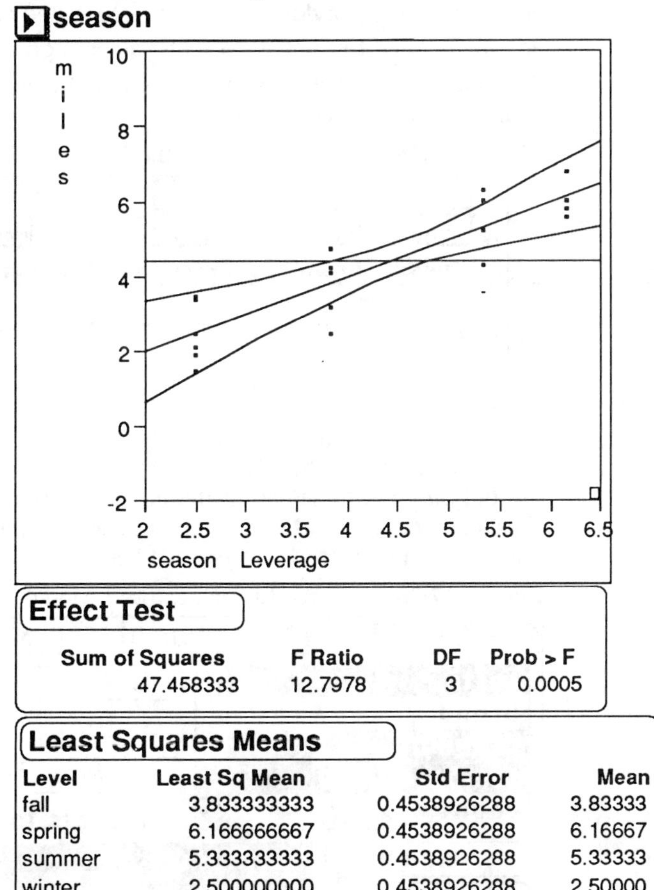

Effect Test

Sum of Squares	F Ratio	DF	Prob > F
47.458333	12.7978	3	0.0005

Least Squares Means

Level	Least Sq Mean	Std Error	Mean
fall	3.833333333	0.4538926288	3.83333
spring	6.166666667	0.4538926288	6.16667
summer	5.333333333	0.4538926288	5.33333
winter	2.500000000	0.4538926288	2.50000

Nominal Response Models

Plots are not available when the response is nominal or ordinal. For simple main effects you can use the **Fit Y by X** platform described in Chapter 15 to see a cumulative logistic probability plot for each effect. When there are complex effects such as interactions in the model, the **Specify Model** platform produces text reports only. The following example illustrates the text reports for a nominal response model.

A market research study was undertaken to evaluate preference for a brand of detergent. The model is defined by

- the response variable, **brand,** with values m and x
- an effect called **softness** (water softness) with values soft, medium, and hard
- an effect called **previous use** with values yes and no
- an effect called **temperature** with values high and low
- A count variable, **count**, that provides the frequency counts for each combination of effect categories

Figure 17.25 Partial Listing of the DETERGENT Data Table

	softness (Nom)	brand (Nom)	previous use (Nom)	temperature (Nom)	count (Int)
8	soft	m	no	low	53
9	med	x	yes	high	23
10	med	x	yes	low	47

The study begins by specifying the full three-factor factorial model as shown by the **Specify Model** dialog below. To specify a factorial model, use option-click to select all the main effects in the column selector list. Then click on the **Factorial** button to see all main effects and interactions are listed in the **Effects in Model** list.

Figure 17.26 A Three-factor Factorial Model with Nominal Response

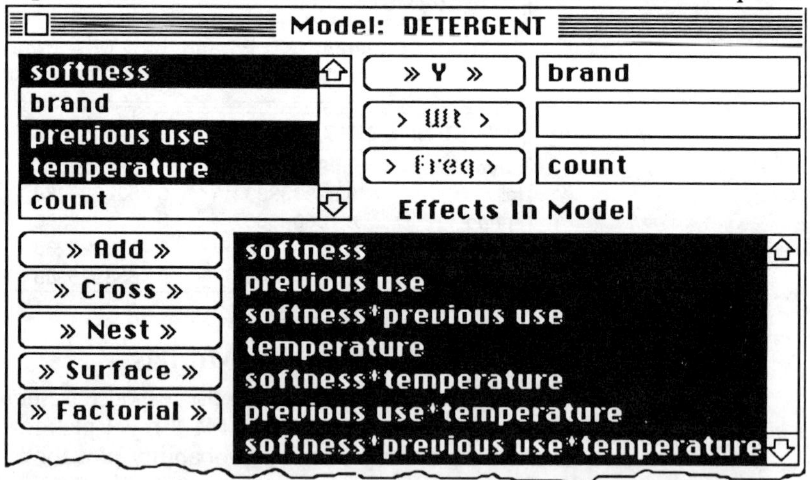

The tables in **Figure 17.23** selected from the text report show the three-factor model as a whole to be significant as indicated by the **ChiSquare** value of .000562 in the Analysis of LogLikelihood table. The Effect Test

table lists each effect in the model and indicates that all effects that include the **softness** effect do not contribute significantly to the model fit.

Figure 17.23 Tables for Nominal Response Three-factor Factorial

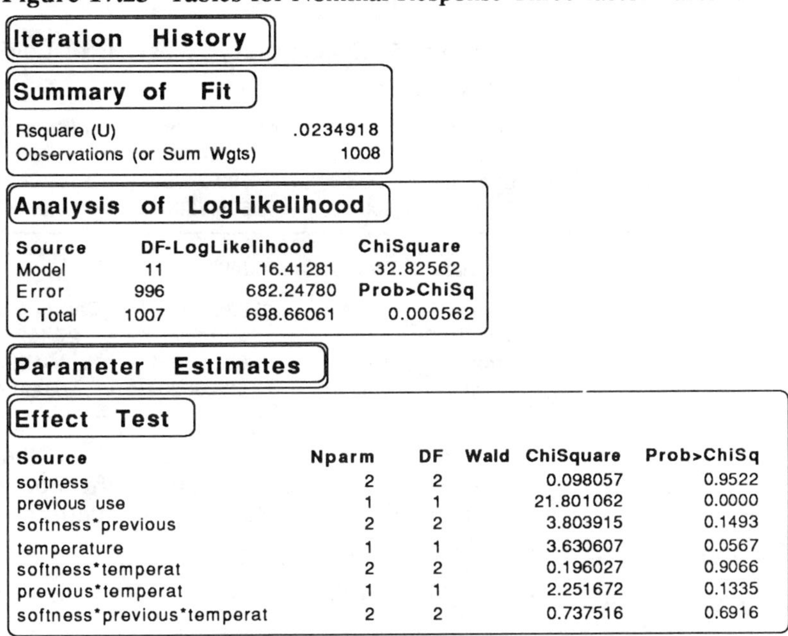

Iteration History

Summary of Fit

Rsquare (U)	.0234918
Observations (or Sum Wgts)	1008

Analysis of LogLikelihood

Source	DF	-LogLikelihood	ChiSquare
Model	11	16.41281	32.82562
Error	996	682.24780	Prob>ChiSq
C Total	1007	698.66061	0.000562

Parameter Estimates

Effect Test

Source	Nparm	DF	Wald ChiSquare	Prob>ChiSq
softness	2	2	0.098057	0.9522
previous use	1	1	21.801062	0.0000
softness*previous	2	2	3.803915	0.1493
temperature	1	1	3.630607	0.0567
softness*temperat	2	2	0.196027	0.9066
previous*temperat	1	1	2.251672	0.1335
softness*previous*temperat	2	2	0.737516	0.6916

The next step in the example removes the softness factor and its interactions which don't appear to be significant. This leaves the two-factor factorial model shown in the **Specify Model** dialog shown below.

Figure 17.24 A Two-factor Factorial Model with Nominal Response

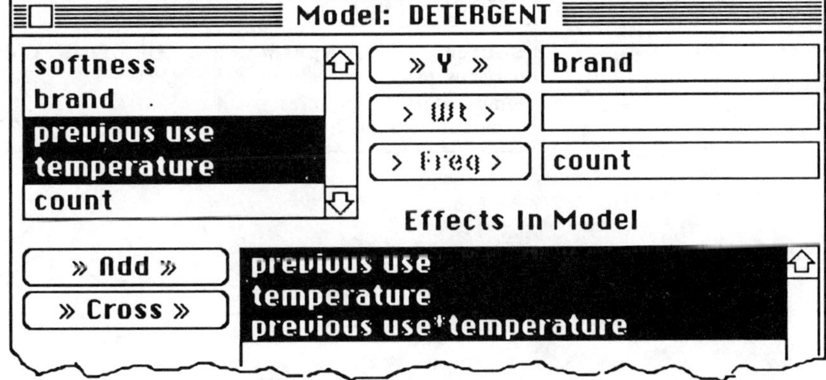

Selected tables from the analysis show that the two-factor model fits as well as the three-factor model. In fact, comparison of Analysis of LogLikelihood

tables for the two models seen in **Figure 17.25** show a smaller chi-square value for the two-factor model than the three-factor model. This results from the change in degrees of freedom used to compute the chi-square values and their probabilities.

Figure 17.25 Comparison of Analysis of LogLikelihood Tables

Summary of Three-factor Model

Source	DF	-LogLikelihood	ChiSquare
Model	11	16.41281	32.82562
Error	996	682.24780	Prob>ChiSq
C Total	1007	698.66061	0.000562

Summary of Two-factor Model

Source	DF	-LogLikelihood	ChiSquare
Model	3	13.58479	27.16957
Error	1004	685.07582	Prob>ChiSq
C Total	1007	698.66061	0.000005

The tables shown in **Figure 17.26** support the conclusion that previous use of a detergent brand has a strong effect on detergent preference, water temperature has a borderline-significant effect, but water softness does not seem to matter.

Figure 17.26 Tables for Nominal Response Two-factor Factorial

Parameter Estimates

Term		Estimate	Std Error	ChiSquare	Prob>ChiSq
Intercept	Unstable	.030423239	.06668262	0.21	0.6482
previous[no-yes]		-.31484007	.06668262	22.29	0.0000
temperat[high-low]		.133109960	.06668262	3.98	0.0459
previous[no-yes]*temperat[high-low]		-.11105724	.06668262	2.77	0.0958

Effect Test

Source	Nparm	DF	Wald ChiSquare	Prob>ChiSq
previous use	1	1	22.292291	0.0000
temperature	1	1	3.984701	0.0459
previous*temperat	1	1	2.773757	0.0958

Chapter 18.
Analyze: Spin

The **Spin** command produces a three-dimensional view of data and an approximation to higher dimensions through principal components. The plot is a spinable display of the values of numeric columns in the current data table. Spinning plots were introduced by Fisherkeller, Friedman, and Tukey, (1974). The spin platform displays three variables at a time from the columns you select.

To help capture and visualize variation in higher dimensions, the spinning platform displays a biplot representation of the points and variables when you request principal components.

Chapter Contents

The Display

The **Spin** platform, shown in **Figure 18.1**, is arranged with a spin control panel at the upper left and the plot frame on the right. Display options are available on a pop-up menu through the submenu triangle on the right of the spin control panel. You add variables to the components list and request principal component analysis with the pop-up menu accessed by the check-mark border button.

Figure 18.1 The Spinning Plot Platform

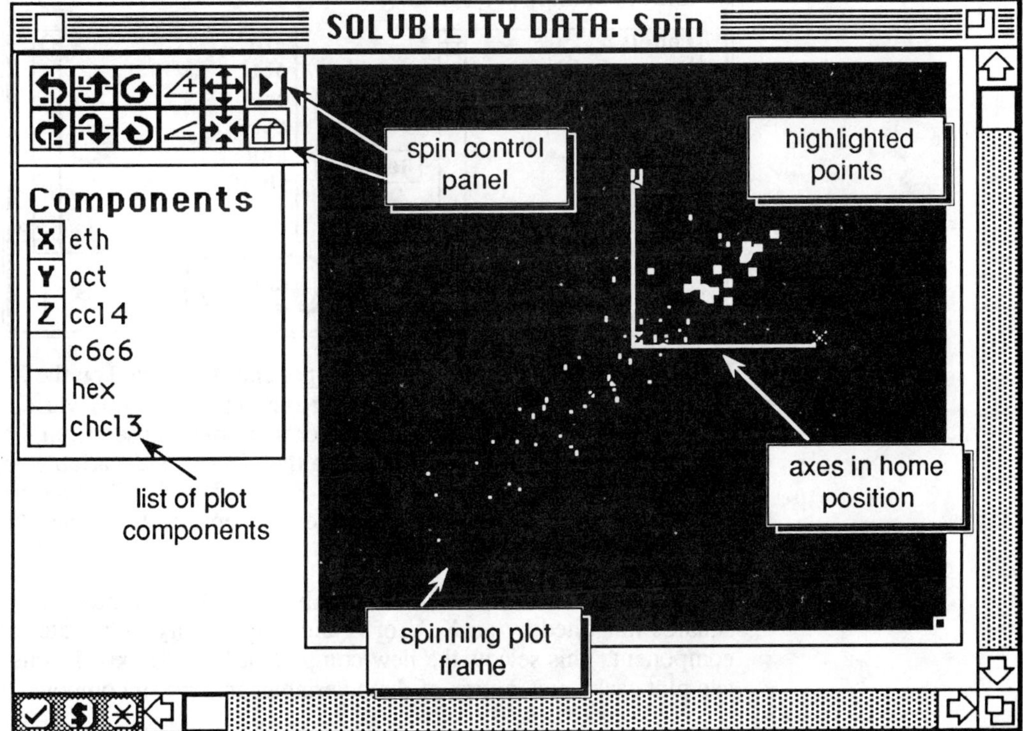

Selecting Variables to Plot

To begin a **Spin** session, columns must be selected as components for plotting. There are three ways that columns are available for plotting.

- When you select the **Spin** command, columns previously assigned either X or Y roles in the data table are listed as components on the platform.
- If no columns are previously assigned X or Y roles in the data table, the spin platform prompts you to select variables.

- You can add more variables at any time with the **Add Variables** command in the check-mark pop-up border menu.

The dialog for adding variables, shown in **Figure 18.2**, displays all the columns in the current data table. Use click or shift-click to select a range of columns, or command-click for a discontinuous selection from the list. and click **Add** to add them to the selection list. When you click **OK**, the columns are added to the list of plot components next to the spin platform.

Figure 18.2 The **Spin** Variable Selector Dialog

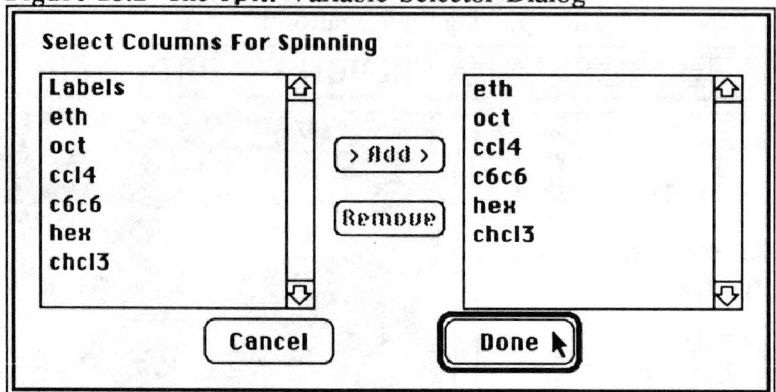

By default, the first three variable components are plotted on the X, Y and Z axes of the spinning plot (see **Figure 18.1**). If the variables have preassigned roles in the data table rather than roles chosen from the variable selector list, an X variable appears on the X axis, a Y variable appears on the Y axis, and the platform chooses a Z axis variable. The platform always assigns variables to three axes whenever there are three variables in the component list.

To plot a different set of three variables, click the mouse in one of the squares inhabited by an X, Y or Z and drag that axis character to another component. This selects the new component for that axis. In this way you can plot each combination of three variables in a list of components.

The Check and Save Border Pop-up Menus

These menus are brought up from the small icon buttons on the bottom left border of the window.

Figure 18.3 The Check Pop-up Menu

Add Variables
Principal Components
Remove Prin. Comp.
✓Show Prin. Rays

Each of these commands is described in the following list.

Add Variables brings up a selection dialog shown in **Figure 18.2** to add more variables to the spin components list.

Principal Components calculates principal components on the set of variables in the components list and adds them to the list. You can do this repeatedly with different current data values and variables.

Remove Prin. Comp removes the principal components and their report from the session.

Show Prin. Rays alternately shows or hides rays on the plot corresponding to the principal components on the variables, or the variables on the principal components.

Figure 18.4 The Save Pop-up Menu

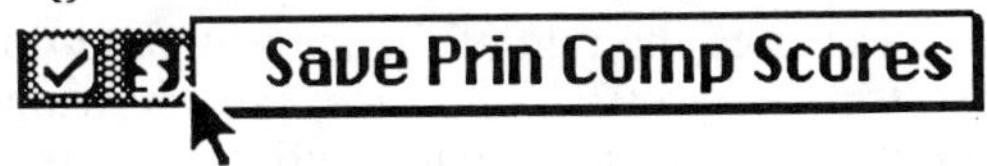

Save Prin Comp Scores saves the current principal component scores with their formula as a column in the current data table. For *n* variables in the components list, *n* principal component columns are created and named **Prin Comp 1**, **Prin Comp 2**, ... **Prin Comp *n***.

For details on Principal Components, see the section called **Principal Components** in this chapter.

Resizing the Plot

The size of the platform can be changed by clicking just inside the lower right corner of the plot frame, holding down the mouse button, and dragging out the rectangle to the desired size. If you click anywhere inside the plot frame, a small rectangle appears, which is the location of the resizing control. Size affects performance and memory needs, as explained in the last section of this chapter.

Spinning the Plot

After selecting three variables, you can spin the resulting point cloud in two ways:

- use the hand tool (the *grabber*) from the **Tools** menu and manually move the plot about the origin in any direction
- use the directional spin buttons on the control panel to spin the point cloud around each of the three axes of rotation: pitch, yaw, and roll.

In each case, the spin persists as long as you hold down the mouse button. Use shift-click to cause the spinning to continue after you release.

Manual Spinning

To spin the plot yourself, select the grabber (hand) from the **Tools** menu. Move it onto the display platform, grab the plot by holding down the mouse button, and move the hand about. While holding the mouse button down the rotation of the axes tracks the movement of the hand.

Alternatively, if you hold down the command key before you click the grabber tool , the spinning responds to the direction of the cursor from the center of the plot, and the spin speed is proportional to the distance from the center instead of tracking the mouse movement.

You can change the rate of the spin by hold down the option key and the mouse (option-grab). A circle appears showing the radius of the tracking leverage on the manual spinning action. Drag the circle smaller or larger and the plot spins faster or slower, respectively, for a given hand movement.

The plot can also spin by itself. Hold down the shift key and give the plot a push with the hand and release while you are still moving (the faster you push, the faster the plot spins). Clicking anywhere on the Macintosh screen stops the spinning.

You can spin a plot continuously in the following two modes:

- By default, the continuous spin dedicates the Macintosh to spinning. If you click anywhere on your desktop, the spinning stops.
- If you check **Background Coast** in the display options menu, then the spinning is a background task and continues during idle time even when you do other things. In this mode you have to click in the active spinning platform window to stop the spinning.

The grabber tool is special only on the spinning graph. When used elsewhere in the **Spin** platform, it behaves like the arrow pointer tool.

Table 18.1 Summary Table for Grabber Tool in the Spin Frame

Mouse Action	Effect
click	spins as mouse moves
shift-click (for coasting)	continuous spin when button is released
option-click	spin rate proportional to sizing circle
command-click	spin velocity controlled by mouse position
shift-command click	continuous command-click spin

Push-button Spinning

The spin control panel contains twelve small icons that let you direct the spinning axes, control the rate of spin, scale the points on the plot, and access a pop-up menu icon that provides display options.

Figure 18.5 The Spin Control Panel

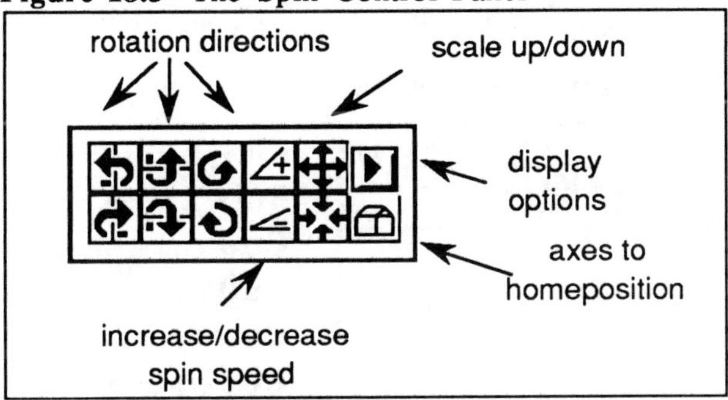

The six arrow icons provide three-dimensional rotation in six directions. Press each icon to see the following spin directions.

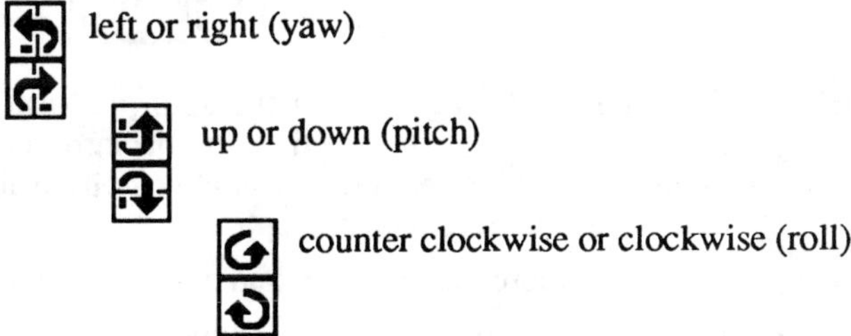

If you shift-click the spin buttons, the plot can continues the spin (the spin *coasts*). The **Background Coast** option also works here so that the coasting can be done as a background task as well as a dedicated task. Clicking anywhere on the Macintosh screen stops regular coasting, whereas clicking on the active spinning window stops background spinning.

The angle icons with + and – make the angle of each rotation step (and thus the speed) larger or smaller. The increment in step is small, multiple clicks may be needed to see a change in rotation speed. The angle is displayed in the plot frame for as long as you hold the button down. When you hold the option key down the angle increments in 10 degree quantities.

The scaling icons expand or shrink the scale of the points continuously for as long as you press them down.

The home icon brings the plot back to home position, with the X axis to the right, the Y axis to the top, and the Z axis on center.

The pop-up menu icon lists the spin platform display options.

Display Options

The options pop-up menu beside the spin control panel menu provides a variety of display options. All of the display options except scale down, scale up, and home are toggles that turn the feature on or off. The feature is in effect if there is a check mark next to it.

Figure 18.6 The Display Options

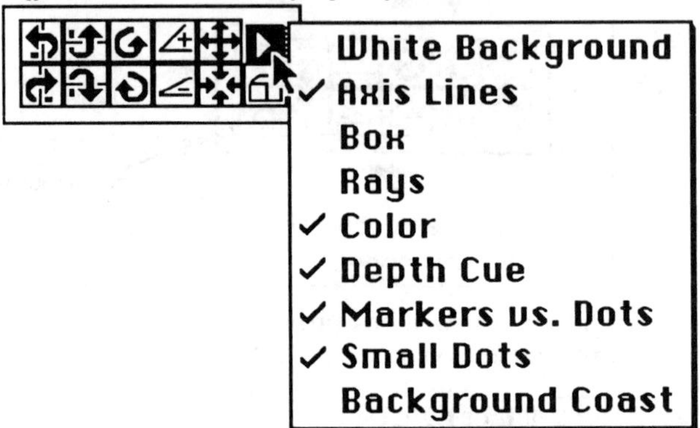

White Background alternates the background of the spin platform between black and white. This option is not connected to the background option in the Preferences window. A white background is always used in printing or cut-and-paste operations, regardless of this option.

Axis Lines shows or suppresses the x, y, and z axes in the spinning plot.

Box draws a cube around the point cloud as illustrated on the first page of this chapter.

Rays draws a line from each point to the origin of the plot.

Color specifies that the points be shown in the color from the current row state, rather than black or white only.

Depth Cue sets on or off the ability to display points in two depths (bright for near points and dim for distant points).

Markers vs. Dots specifies that the points use the markers assigned to rows in the data table, rather than uniform large dots.

Small Dots displays all the points as small dots and speeds up the spinning a bit. This option takes precedence over the **Markers vs. Dots** option.

Background Coast specifies whether coasting spins (continuous spin when you shift-click with either spin icons or the grabber tool) are dedicated tasks or background tasks.

Marking Features

Highlighting Points

Points on the spinning plot are highlighted with the arrow and brush tools found in the **Tools** menu, and are described in Chapter 12. *The Report Surface*. When you click near a point, it highlights on the spinning plot, on the data table, and all other plots that contain the point. Shift-click extends the highlighted selection.

Holding down the mouse button causes the row label to show. The example in **Figure 18.7** shows the SOLUBILITY data with the CAFFEINE point highlighted both in the spinning plot, and in the spreadsheet view of the data table.

Figure 18.7 Highlighting a labeled point

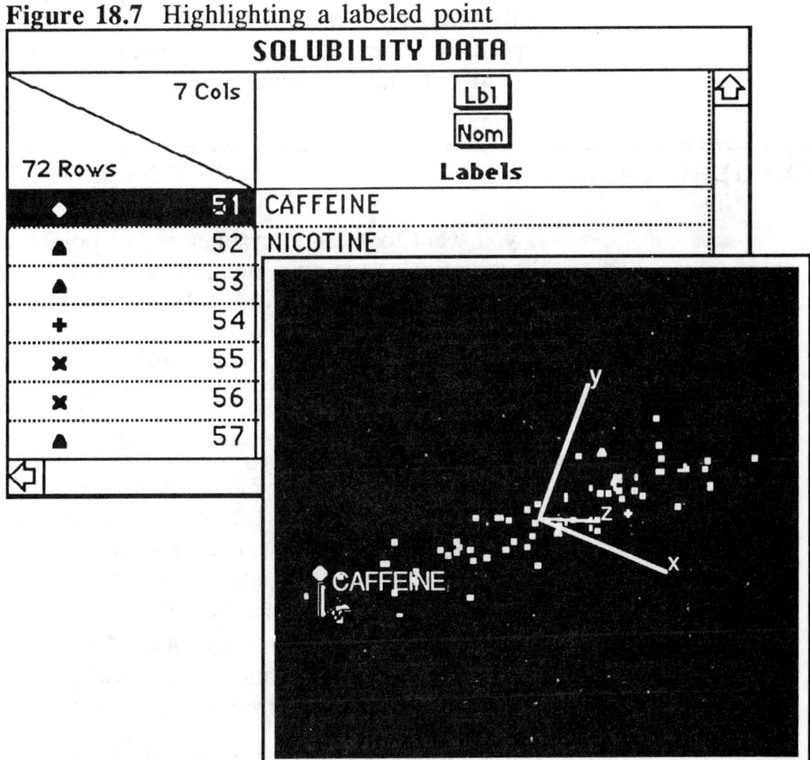

The brush tool in the **Tools** menu highlights all points within the brush rectangle and on the spreadsheet. Use option-click to change the brush shape and shift-click to extend a selection.

After points are highlighted you can apply various *row state* attributes to them using commands in the **Rows** menu: Row states are color, marker, hide, exclude, and label and are described in more detail in Chapter 10, *JMP Data Tables*.

Color and Marker Coding

To help visualize more dimensions on the spinning plot, you can code variables values with color and marker style. Often the easiest way to do this is to create a histogram for the variable with the **Distribution of Y** command on the **Analyze** menu. Then highlight the points bar by bar, assigning colors or markers in sequence as you work through the levels or ranges of the variable. You can also use the calculator that assigns row state characteristics to rows according to the formula you construct.

You can save these row states as a special row state variable in the data table through the **New Column** dialog by creating a new variable with the type Row State. This kind of column lets you copy the active row state information into the column as data.

Principal Components

If you want to see the arrangement of points across many variables and those variables are correlated, then you can use principal components to show the most prominent directions of the high-dimensional data. Using principal component analysis reduces the *dimensionality* of a set of data. This means finding a way to picture the structure of the data as completely as possible using as few variables as possible.

For *n* original variables, *n* principal components are formed as follows:

- The first principal component is the linear combination of the standardized original variables that has the greatest possible variance.
- Each subsequent principal component is the linear combination of the standardized original variables that has the greatest possible variance and is uncorrelated with all previously defined components.

Each principal component is calculated by taking a linear combination of an eigenvector of the correlation matrix with a standardized original variable. The eigenvalues show the variance of each component.

The set of *n* principal components has the same total variation and structure as the original variables. The restriction that each successive component be uncorrelated (orthogonal) with the previous components ensures that each new component will have a lower variance that its predecessor. It follows that most of the sample variation is often captured by the first few principal components.

Analyzing principal components can sometimes help to discriminate between groups and detect outlying observations using the first few components, particularly if the original variables are themselves correlated.

To illustrate, suppose that you have a scatterplot for two highly-correlated variables. If you compute a linear combination of the two variables that best captures the scatter of the points, then you have the combination indicated by the P1 line shown on the left in **Figure 18.8**. Rotating and reflecting the plot so that P1 is the variable on the main axis, as seen on the right in **Figure 18.8**, gives you the best one-dimensional approximation of a two-dimensional cloud of points. The remaining variation is described by the P2 line. This is the idea of a principal components decomposition for two variables.

Figure 18.8 Principal Components for Two Variables

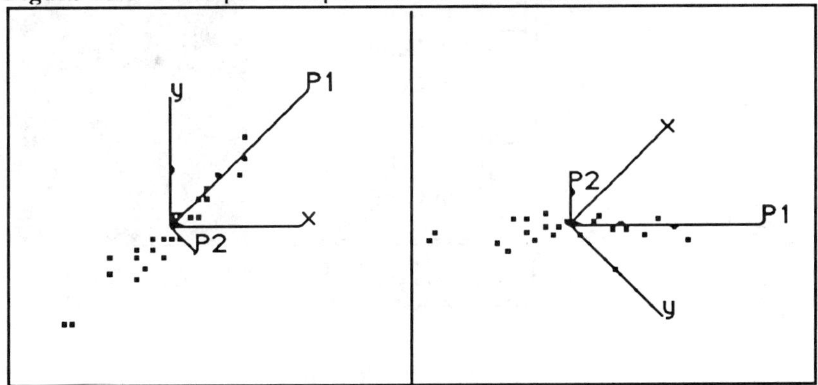

Similarly, for many dimensions, you can find the best three-dimensional view of the data by placing the first three principal components as the spinning axes and examining the plot.

Obtaining Principal Components

To obtain principal components on the variables in the plot selection list, use the **Principal Components** command from the check icon border menu. The principal components are listed as new variables in the components list. Principal components axes show on the plot, and a reveal button marked Principal Components makes a report available.

Figure 18.9 Display Principal Components

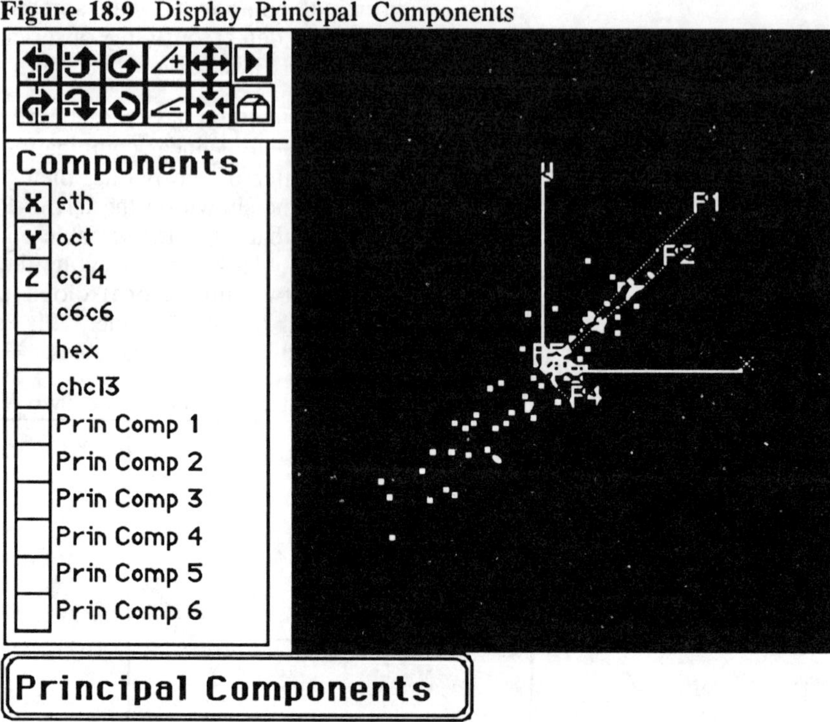

Principal Component Rays

If you load ordinary variables on two or three of your axes, and obtain principal components, then those components are shown as rays in the spinning plot. These rays, called *p-rays,* approximate the principal component as a function of the variables on the axes. If there are only two or three variables, then the rays represent the principal components exactly. The length of the ray corresponds to the eigenvalue or variance of the principal component. Because the first principal component captures the most variability possible, the P1 ray can be very long compared to the later components.

The principal components are orthogonal to each other in the high-dimensional space of all the variables, but they are not orthogonal in the plot if they are forced to approximate fewer dimensions. If you have just two or three variables, then you can see clearly how principal components work in the spinning plot.

Principal Components Text Report

The principal components text report summarizes the variation in a multivariate data table according to its principal components. The principal components are derived from an eigenvalue-decomposition of the correlation matrix of the variables. The details in the report reveal the breakdown in how the variation in the data is absorbed by the components, and in which direction the components lie in terms of the variables. The principal component points are derived from the eigenvector linear combination of the standardized variables.

Figure 18.10 Principal Components Text Report

Principal Components						
EigenValue:	4.7850	0.9408	0.1427	0.0630	0.0464	0.0221
Percent:	79.7503	15.6804	2.3779	1.0495	0.7738	0.3680
CumPercent:	79.7503	95.4307	97.8087	98.8582	99.6320	100.0000
Eigenvectors:						
eth	0.34887	0.64095	0.17653	0.63780	-0.12841	0.11419
oct	0.37314	0.56277	-0.17684	-0.67519	0.23816	0.01361
ccl4	0.43187	-0.29274	0.17821	-0.20373	-0.46785	0.66003
c6c6	0.44552	-0.14827	-0.21629	-0.01927	-0.52424	-0.67642
hex	0.42231	-0.27033	0.68072	-0.00820	0.47344	-0.24696
chcl3	0.41919	-0.30285	-0.62900	0.30884	0.45709	0.18056

Eigenvalues

The first row of the text report lists the eigenvalue corresponding to each principal component in the order Prin Comp 1, Prin Comp 2, and so on, from left to right. The eigenvalues represent a partition of the total variation in the multivariate sample. They sum to the number of variables.

Percent

Each eigenvalue is listed as a percent of the total eigenvalues. In the example above, the first principal component accounts for 79.7503 % of the variation in the sample.

CumPercent

The cumulative percent shows the cumulative percent of variation represented by the eigenvalues. In the example above, the first three principle components account for 97.8087% of the variation in the sample.

Eigenvectors

The columns of values are the eigenvectors for each of the principal components in the order Prin Comp 1, Prin Comp 2, and so on, from left to right. Using these coefficients to form a linear combination of the original variables produces the principal component variables.

Biplots

The biplot (Gabriel, 1971) is a technique for showing a an approximation to a set of both variables and points in one graph. Essentially, the contribution of the biplot to a plot of the principal component points is that you can show the variables as well as the points in one presentation.

If you select two or three principal components as the axes, then variable rays are shown, and the resulting presentation is what Gabriel terms the JK' principal component biplot.

These plots approximate the higher dimensional distance configuration between points in a two- or three-dimensional subspace. They also provide some information about the variables. Each ray corresponds to a hypothetical observation where all the values are at the mean except the value for one variable which is set to three standard deviations above the mean.

If you scale the data so that each principal component is standardized by its standard deviation, then the result is a GH' biplot, which represents the variable intercorrelations better than the JK' biplot, but is not as good at approximating interpoint distances.

As the higher dimensions are reduced to two or three dimensions, the variables that were originally thought of as orthogonal become closer to each other, and the variables that are the most correlated become closest. Often there are sets of variables that clump together in the two or three dimensional subspace. The combination corresponding to these clusters of variables is what factor analysts hope to find when they do factor analysis, with the number of factors corresponding to the number of dimensions represented.

Figure 18.11 shows the variable rays from the first three principal components of the SOLUBILITY data. The data are spread more over the X axis than the Y or Z axis as you expect because of the nature of the principal components. Note also that the variables tend to form into two groups, **eth** and **oct** in one group, and **cc14**, **c6c6**, **hex**, and **chc13** in the other group. Within each group the variables are highly correlated, but variables are not correlated across the two groups. You can confirm this by checking the correlation structure for this data with the **Y's by Y's** analysis platform.

Figure 18.11 Principal Components and Variable Rays

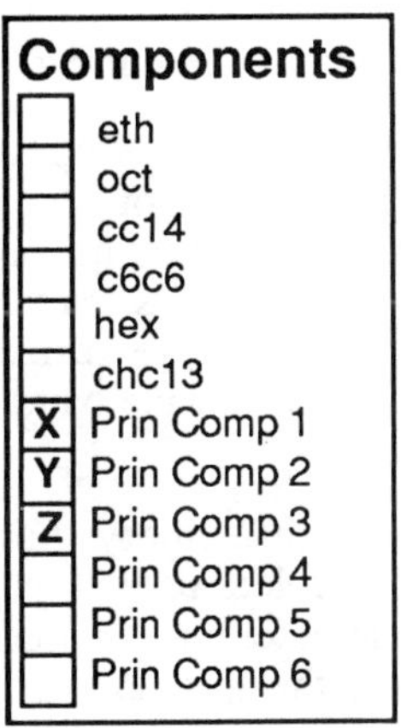

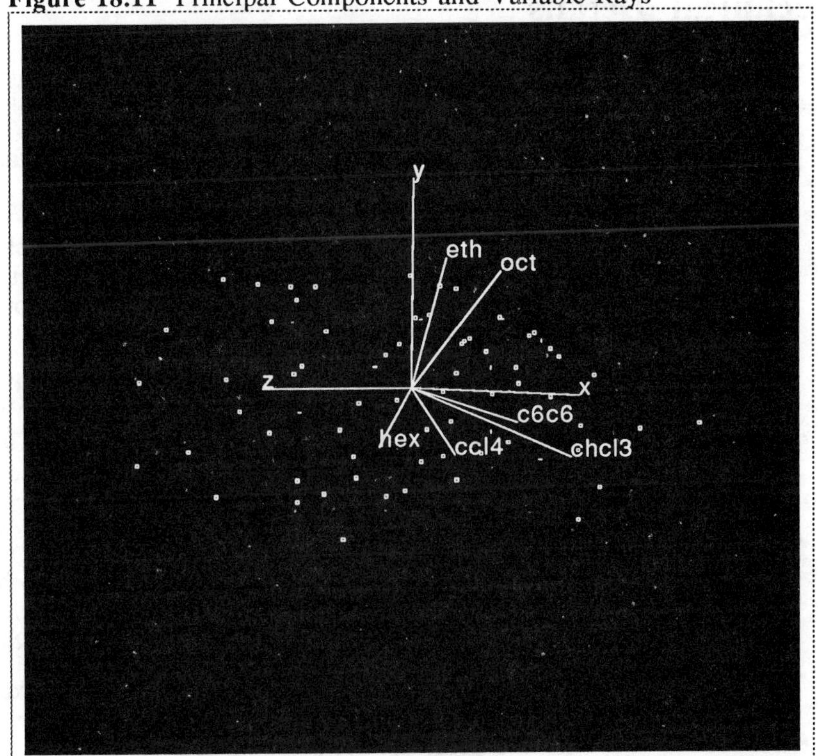

Searching for Outliers

A good way to look for outliers in multivariate data is to display the principal components on the plot axes, and then spin the three dimensional scatter cloud around and look for points away from the cloud. One approach is to look first at the first three principal components. This gives the best view of the shape of the data as a whole. To get details for each variable, you can sequence the Z axis through each variable while you keep the X and Y axes on the first two principal components. But this may still miss the outliers in the higher dimensions. These higher dimensional outliers are more likely to be found by looking at plots of the last principal components, since these last components show the least prominently supported directions of the space, and high-dimensional outliers are the distinguished inhabitants of this *unpopular* space.

You can verify the multivariate outliers you find with the Mahalanobis outlier plot given by **Y's by Y's** platform, highlight these points, and form a subset of them for reporting.

Technical Details

This is the fine print that you don't need to read unless you are interested, or are looking for some answers to questions or problems related to spinning.

Speed

The spinning plot for the Macintosh II gives fairly smooth motion up to about 500 points, after which the motion gets a bit jerky. Plots of up to 2500 points and more still give more than one frame per second on Macintosh II's. The speed can be improved by chosing small points, rather than markers. It also helps to reduce the picture size using the resize control. The plot spins slightly faster with the buttons than with the hand-tracked directions.

Memory Use

Large spinning plots may put a heavy demand on system memory. For example a 400 by 400 pixel frame (5.5 inches) for 256 colors uses about 160K of memory for an off-screen pixel map. Only one pixel map is used for possibly many simultaneous spinning plots, so only the spinning plot of the largest size costs memory.

If you want large pictures, we suggest that you reset your monitor options (through the control panel Desk Accessory) to black-and-white with two colors (1 bit), since this reduces the pixmap movement to one eighth as much as the 256-color version. We also support the other pixel depths.

Marker Notes

The spinning plot does not use the larger markers for highlighted points when doing fast rendering of the plot. However it does use larger markers when points are being highlighted.

Also the plot does not do a perfect job of unhighlighting large markers that run over neighboring points. This is all for the sake of performance and does not affect printed or copied versions of the plot. You can use the **Redraw** command in the **Window** menu to redraw the plot, if desired.

Distortion Control

After considerable rotation, the plots tend to shrink along the axis of rotation due to the accumulation of numerical error. JMP uses both low precision and shortcut arithmetic to improve spinning performance. This does not affect the accuracy of the plot, because the distortion is monitored, and when the plot detects distortion above some threshold, it reloads and retransforms the data. When this happens, you may see a brief Reorthonormalizing message in the plot itself.

Chapter 19.
Analyze: Y's by Y's

Y's by Y's is used for exploring how multiple variables relate to each other. The report produced by this platform is divided into two parts.

The first part of the report examines the relationship of each pair of response (Y) variables and includes

- a correlation matrix
- a matrix of scatter plots.

The second part of the report examines the multivariate relationship among all the response variables. Multivariate reports include

- the inverse correlation matrix
- a multivariate Mahalanobis distance outlier plot.

Chapter Contents

The Display

Correlation Matrix

When you choose the **Y's by Y's** platform a correlation matrix, a matrix of scatter plots, and a multivariate outlier plot are displayed. The inverse correlation matrix for the data is available through its reveal/conceal button.

The first report given by the **Y's by Y's** platform is a matrix of correlation coefficients to summarize the strength of the linear relationships between each pair of response (Y) variables.

Figure 19.1 Example of Correlation Matrix

Correlations

Variable	eth	oct	ccl4	hex
eth	1.0000	0.9167	0.5472	0.6110
oct	0.9167	1.0000	0.6445	0.6609
ccl4	0.5472	0.6445	1.0000	0.9395
hex	0.6110	0.6609	0.9395	1.0000

Scatter Plot Matrix

To help you visualize the correlations, scatter plots of each pair of response variables are displayed in a matrix arrangement as shown in **Figure 19.2.**

A 95% bivariate normal density ellipse is imposed on each scatter plot. If the variables are bivariate normally distributed, this ellipse encloses apporximately 95% of the points (when the number of points is large). The correlation of the variables is seen by the collapsing of the ellipse along the diagonal axis. If the ellipse is fairly round and is not diagonally oriented, the variables are uncorrelated.

To examine the relationship of two variables or see density ellipses other than 95% ellipse, you can use the **Distribution of Y by X** command on the **Analyze** menu.

Figure 19.2 Example of a Scatter Plot Matrix

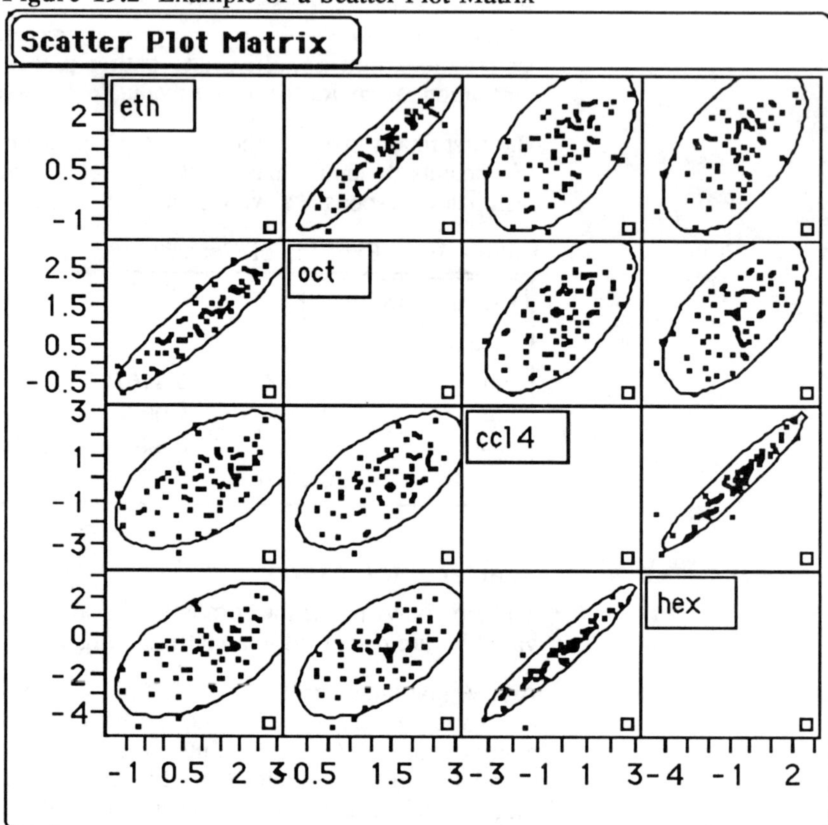

Inverse Correlation Matrix

The inverse correlation matrix provides useful multivariate information. The diagonal elements of the inverse correlation matrix are a function of how closely the variable is a linear function of the other variables.

Figure 19.3 Example of a Inverse Correlation Matrix

Inverse Corr

Inverse Corr

Variable	eth	oct	ccl4	hex
eth	7.1719	-6.6558	2.9835	-2.7868
oct	-6.6558	7.9673	-3.1288	1.7413
ccl4	2.9835	-3.1288	9.8409	-9.0012
hex	-2.7868	1.7413	-9.0012	10.0091

Multivariate Distance Outlier Plot

The outlier distance plot in **Figure 19.4** shows the the Mahalanobis distance of each point from the multivariate mean (centroid). The distance is plotted for each observation number. The extreme multivariate outliers can be identified by highlighting the points with the largest distance values.

Figure 19.4 Example of a Mahalanobis Outlier Distance Plot

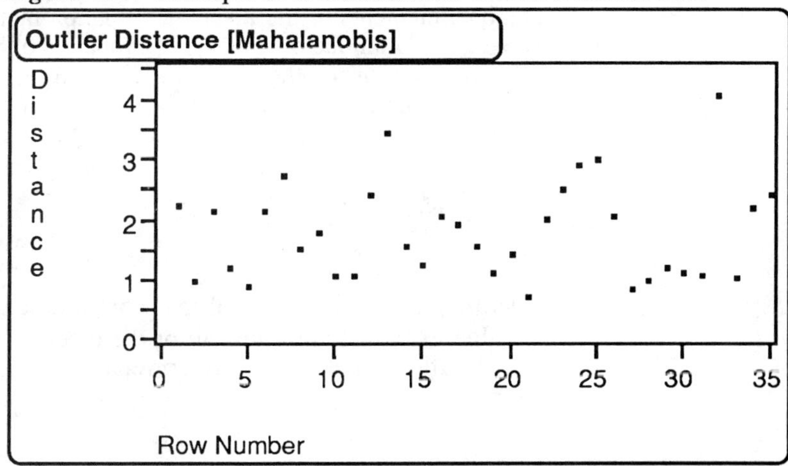

Statistical Details

Correlation Matrix

The correlation coefficient, referred to as the Pearson product-moment correlation, measures the strength fo the linear relationship between two variables. For response variables y_i and y_j it is denoted r_{ij} and computed as

$$r_{ij} = \frac{s_{ij}^2}{\sqrt{s_{ii}^2 s_{jj}^2}}$$

where

$$s_{ij}^2 = \frac{\Sigma(y_i - \bar{y}_i)(y_j - \bar{y}_j)}{N-1}$$

If there is an exact linear relationship between two variables then the correlation is 1 or –1 depending on whether the variables are positively or negatively related. If there is no linear relationship the correlation tends toward zero.

Inverse Correlation Matrix

The inverse correlation matrix provides useful multivariate information. The diagonal elements of the inverse correlation matrix, sometimes called the variance inflation factors (VIF), are a function of how closely the variable is a linear function of the other variables. Specifically, if the correlation matrix is denoted **R** and the inverse correlation matrix is denoted $\mathbf{R}^{-1}$, then the diagonal element is denoted r^{ii} and computed as

$$r^{ii} = \text{VIF}_i = \frac{1}{1-\text{R}_i^2}$$

where R_i^2 is the squared multiple correlation that results when the ith variable is regressed against all of the others. Thus, a large r^{ii} indicates that the ith variable is highly correlated with any number of the other variables.

Multivariate Distance Outlier Plot

The outlier distance plot shows the the Mahalanobis distance of each point from the multivariate mean (centroid). The Mahalanobis distance takes into account the correlation structure of the data as well as the individual scales. For each value the distance is denoted d_i and is computed as

$$d_i = \sqrt{(\mathbf{y}_i - \overline{\mathbf{y}})' \mathbf{S}^{-1} (\mathbf{y}_i - \overline{\mathbf{y}})}$$

where

$\mathbf{y}_i$ is the data for the *ith* row,

$\overline{\mathbf{y}}$ is the row of means,

$\mathbf{S}^{-1}$ is the estimated covariance matrix for the data.

If a variable is an exact linear combination of other variables, then the correlation matrix is singular and that row and column are zeroed out. The generalized inverse that results is still valid for forming the distances.

Multivariate distances are useful for spotting outliers in many dimensions. However, if the variables are highly correlated in a multivariate sense then a point can be seen as an outlier in multivariate space without looking unusual along any subset of dimensions Said another way, when data are correlated it is possible for a point to be unremarkable when seen along one or two axes, but still be an outlier in a multivariate sense.

Statistical Warning ➔ This outlier distance is not particularly robust in the sense that outlying points themselves can distort the estimate of the covariances and means in such a way that outliers are disguised.

Display and Save Options

The active border menus are the check menu for display options and the save ($) menu for saving computations.

Figure 19.5 Display Options Pop-up Menu

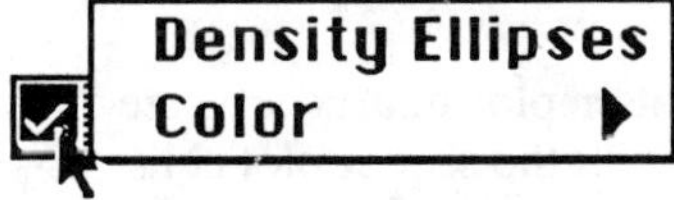

Density Ellipses
is a toggle that displays 95% ellipses on the scatter plots.

Color
lets you select from a palette of colurs to change the color of the ellipses if you are working on a color Macintosh.

The Save pop-up menu contains only the save distances command.

Figure 19.6 Save Option Pop-up Menu

Save Distances

saves the Mahalanobis distance computations as a new column in the current spreadsheet, named Mahal.Distance as shown in **Figure 19.7.**

Figure 19.7 Data Table with Saved Distances

SOLUBILITY SUBSET

5 Cols / 35 Rows	eth (Int)	oct (Int)	ccl4 (Int)	hex (Int)	Mahal. Distance (Int)
1	0.380	0.590	-3.300	-4.240	5.58714674
2	-0.720	0.050	-1.570	-4.600	5.56286475
3	1.100	1.350	-1.700	-0.950	4.30503732
4	0.880	2.020	2.060	1.350	4.03904813

The distance column is not saved as a locked formula-based computation linked to the columns from which it was created. This means the distance will not be recomputed if you modify the data table. If you add or delete columns, or change values in the data table you should select **Y's by Y's** again to compute new distances.

Dynamic Display Features

All plots and the current data table are linked. You can highlight points on any scatter plot in the scatter plot matrix, or the outlier distance plot. The points are highlighted on all other plots and are selected in the data table.

If a column is designated as a label variable, highlighting a point displays the label value.

The cells of the scatter plot matrix are size-linked so that the stretch box from any cell resizes all the scatter plot cells.

Appendix A.
Statistical Details

This appendix discusses the different types of response models, their factors, their design coding, and parameterization.

The JMP system fits linear models to three different types of response models labeled interval, ordinal, and nominal. Many of the details on the factor side are the same between the different response models, although JMP supports graphics and marginal profiles on interval responses, but not on ordinal and nominal.

Different computer programs use different design-matrix codings, and thus parameterizations, to fit effects and construct hypothesis tests. JMP uses a different coding than the GLM procedure in the SAS system, though in most cases JMP and SAS GLM procedure produce the same results. The following sections describe the details of JMP coding and highlight those cases when it differs from that of the SAS GLM procedure, which is frequently cited as the industry standard.

Contents

Interval Responses

Interval responses are modeled in the usual way through fitting a linear model by least squares. Interval responses are much more richly supported than the other response types. Interval responses are reported with leverage plots, least squares means, contrasts, and output formulas; none of these features has a corresponding feature for nominal or ordinal responses. Computationally, interval responses are the easiest to fit.

Nominal Responses

Nominal responses are analyzed with a straightforward extension of the logit model. For a binary (two-level) response, a logit response model is

$$\log\left(\frac{P(y=1)}{P(y=2)}\right)=\mathbf{X}\beta$$

which can be written

$$P(y=1) = F(\mathbf{X}\beta)$$

where F( ) is the cumulative distribution function of the standard logistic distribution:

$$F(x)=\frac{1}{1+e^{-x}}=\frac{e^{x}}{1+e^{x}}\ .$$

The extension for r responses is to relate each response probability to the rth probability, and fit a separate set of design parameters to these $r-1$ models.

$$\log\left(\frac{P(y=j)}{P(y=r)}\right) = \mathbf{X}\beta_{(j)} \qquad \text{for } j=1,\ldots,r-1$$

The fit is done with maximum likelihood using a modified Newton Raphson iterative algorithm.

Ordinal Responses

Ordinal responses are modeled by fitting a series of parallel logistic curves to the cumulative probabilities. Each curve has the same design parameters but a different intercept and is written

$$P(y\le j) = F(\alpha_j+\mathbf{X}\beta) \qquad \text{for } j=1,\ldots,r$$

where F( ) is the standard logistic cumulative distribution function,

$$F(x)=\frac{1}{1+e^{-x}}=\frac{e^{x}}{1+e^{x}}\ .$$

Another way to write this is in terms of an unobserved interval variable, z, that causes the ordinal response to change as it crosses various thresholds:

$$y = \begin{cases} r & \text{if } \alpha_{r-1} \le z \\ j & \text{if } \alpha_{j-1} \le z < \alpha_j \\ 1 & \text{if } z \le \alpha_1 \end{cases}$$

where z is an unobservable function of the linear model and error,

$$z = \mathbf{X}\beta + \varepsilon$$

and ε has the logistic distribution.

These models are attractive in that they recognize the ordinal character of the response, they need far fewer parameters than nominal models, and the computations are fast even though they involve iterative maximum likelihood calculation.

Interval Factors

Interval factors are placed directly into the design matrix as regressors. If a column is a linear function of other columns, then the parameter for this column is marked *zeroed* or *nonestimable*.

Nominal Factors

Nominal factors are transformed into indicator variables for the design matrix. SAS GLM constructs an indicator column for each nominal level. JMP constructs the same indicator columns for each nominal level except the last level. When the last nominal level occurs, a one is subtracted from all the other columns of the factor. For example, consider a nominal factor A with three levels coded for GLM and for JMP as shown below.

	GLM			JMP	
A	A1	A2	A3	A13	A23
A1	1	0	0	1	0
A2	0	1	0	0	1
A3	0	0	1	-1	-1

In GLM, the linear model design matrix has linear dependencies among the columns, and the least squares solution employs a generalized inverse. The solution chosen happens to be such that the A3 parameter is set to zero.

In JMP, the linear model design matrix is coded so that it achieves full rank unless there are missing cells or other incidental colinearities. The parameter for the A effect for the last level is the negative sum of the other levels. This makes the parameters sum to zero over all the effect levels. The parameters can be interpreted as the differences from the average over all the levels.

Interactions and Crossed Effects

Interaction effects with both GLM and JMP are constructed by taking a direct product over the rows of the design columns of the factors being crossed. For example, the GLM code

```
PROC GLM;
   CLASS A B;
   MODEL A B A*B;
```

yields this design matrix:

		A			B			AB								
A	B	1	2	3	1	2	3	11	12	13	21	22	23	31	32	33
A1	B1	1	0	0	1	0	0	1	0	0	0	0	0	0	0	0
A1	B2	1	0	0	0	1	0	0	1	0	0	0	0	0	0	0
A1	B3	1	0	0	0	0	1	0	0	1	0	0	0	0	0	0
A2	B1	0	1	0	1	0	0	0	0	0	1	0	0	0	0	0
A2	B2	0	1	0	0	1	0	0	0	0	0	1	0	0	0	0
A2	B3	0	1	0	0	0	1	0	0	0	0	0	1	0	0	0
A3	B1	0	0	1	1	0	0	0	0	0	0	0	0	1	0	0
A3	B2	0	0	1	0	1	0	0	0	0	0	0	0	0	1	0
A3	B3	0	0	1	0	0	1	0	0	0	0	0	0	0	0	1

Using the JMP **Specify Model** platform and requesting a factorial model for columns A and B produces the following design matrix. Note that A13 in this matrix is A1–A3 in the matrix above. However, A13B13 is A13*B13 in the current matrix.

		A		B		A13	A13	A23	A23
A	B	13	23	13	23	B13	B23	B13	B23
A1	B1	1	0	1	0	1	0	0	0
A1	B2	1	0	0	1	0	1	0	0
A1	B3	1	0	-1	-1	-1	-1	0	0
A2	B1	0	1	1	0	0	0	1	0
A2	B2	0	1	0	1	0	0	0	1
A2	B3	0	1	-1	-1	0	0	-1	-1
A3	B1	-1	-1	1	0	-1	0	-1	0
A3	B2	-1	-1	0	1	0	-1	0	-1
A3	B3	-1	-1	-1	-1	1	1	1	1

The JMP coding saves memory and some computing time for problems with interactions of factors with few levels.

Nested Effects

Nested effects in GLM are coded the same as interaction effects, since GLM determines the right test by what isn't in the model. Any effect not included in the model can have its effect soaked up by a containing interaction (or, equivalently, nested) effect.

Nested effects in JMP are coded differently. Essentially the terms inside the parentheses are expanded to all their levels, whereas the terms outside the parentheses are reduced by subtracting off the last level. The nested effect is composed by the horizontal direct product of all these indicator columns.

				B(A)					
	A		A1	A1	A2	A2	A3	A3
A B	13	23	B13	B23	B13	B23	B13	B23
A1 B1	1	0	1	0	0	0	0	0
A1 B2	1	0	0	1	0	0	0	0
A1 B3	1	0	-1	-1	0	0	0	0
A2 B1	0	1	0	0	1	0	0	0
A2 B2	0	1	0	0	0	1	0	0
A2 B3	0	1	0	0	-1	-1	0	0
A3 B1	-1	-1	0	0	0	0	1	0
A3 B2	-1	-1	0	0	0	0	0	1
A3 B3	-1	-1	0	0	0	0	-1	-1

Nested effects in JMP are parameterized as if the levels are crossed rather than nested. For example, suppose B has only the values B1, B2, and B3 for the A1 level of A, and B4, B5, and B6 for the A2 level of A. The coding allows for combinations that do not occur such as B1 with A2 and B4 with A1. Thus, JMP makes more parameters than is needed in these situations. We plan to produce a future version of JMP that eliminates the parameters that are not needed. In the meantime, we recommend that you set up the levels for nested effects so that they match across their nesting factor to the degree possible. This improves memory utilization and speed.

The coding of a nested level is refined in the cases of non-aligned levels so that the solution drops the parameters out for non-occuring combinations. The level that is the last occuring level for each nesting value combination is the one that is coded –1 across the other parameters representing values that occur, so that the correct test is produced.

Least Squares Means across Nominal-Factors

Least squares means are the predicted values corresponding to some combination of levels, after setting all the other factors to some *neutral* value. The neutral value for direct interval regressors is defined as the sample mean. The neutral value for an effect with uninvolved nominal factors is defined as the average effect taken over the levels (which happens to result in all zeroes in our coding). Ordinal factors use a different neutral value (explained below). The least squares means might not be estimable, and if not, they are marked nonestimable. JMP's least squares means agree with GLM's (Goodnight and Harvey, 1978) in all cases except when a weight is used, whereupon JMP uses a weighted mean and GLM uses an unweighted mean for its *neutral* values.

The expected values of the cells in terms of the parameters for a three-by-three crossed model are

	B1	B2	B3
A1	$\mu+\alpha_1+\beta_1+\alpha\beta_{11}$	$\mu+\alpha_1+\beta_2+\alpha\beta_{12}$	$\mu+\alpha_1-\beta_1-\beta_2-\alpha\beta_{11}-\alpha\beta_{12}$
A2	$\mu+\alpha_2+\beta_1+\alpha\beta_{21}$	$\mu+\alpha_2+\beta_2+\alpha\beta_{22}$	$\mu+\alpha_2-\beta_1-\beta_2-\alpha\beta_{21}-\alpha\beta_{22}$
A3	$\mu-\alpha_1-\alpha_2+\beta_1-\alpha\beta_{11}-\alpha\beta_{21}$	$\mu-\alpha_1-\alpha_2+\beta_2-\alpha\beta_{12}-\alpha\beta_{22}$	$\mu-\alpha_1-\alpha_2-\beta_1-\beta_2+\alpha\beta_{11}+$ $+\alpha\beta_{12}+\alpha\beta_{21}+\alpha\beta_{22}$

Hypothesis Tests

In the SAS GLM procedure, the hypothesis tests for Types III and IV are constructed by looking at the general form of estimable functions and finding functions that involve only the effects of interest and effects contained by the effects of interest (Goodnight 1978).

In JMP, the same tests are constructed, but since there is a different parameterization, an effect can be tested (assuming full rank for now) by doing a joint test on all the parameters for that effect. The tests do not involve containing interaction parameters because the coding has made them uninvolved with the tests on their contained effects.

If there are missing cells or other singularities, the JMP tests are different than GLM tests. Essentially, JMP produces tests which are equivalent to testing that the least squares means are different. If the least squares means are non-estimable, then the test cannot include some comparisons and therefore loses degrees of freedom.

Singularities and Missing Cells in Nominal Effects

Warning: This section gets into fine detail.

Consider the case of linear dependencies among the design columns. With JMP coding this does not occur unless there is insufficient data to fill out the combinations that need estimating, or unless there is some kind of confounding or colinearity of the effects.

With linear dependencies, a least squares solution for the parameters might not be unique, and some tests of hypotheses cannot be tested. The strategy chosen for JMP is to set parameter estimates to zero in sequence as their design columns are found to be linearly dependent on previous effects in the model. A special column in the report shows what parameter estimates are zeroed, and which parameter estimates are estimable. A separate *singularities* report shows what the linear dependencies are.

In cases of singularities the hypotheses tested by JMP can differ from those selected by GLM. Generally, JMP finds fewer degrees of freedom to test than GLM because it holds its tests to a higher standard of marginality. In other words, JMP tests always correspond to tests across Least Squares Means for that effect, but GLM tests do not always have this property.

Consider, for example, a two-way model with interaction and one missing cell: A with 3 levels, B with 2 levels, and the A3B2 cell missing.

A	B	A1	A2	B1	A1B1	A2B1	
A1	B1	1	0	1	1	0	
A2	B1	0	1	1	0	1	
A3	B1	-1	-1	1	-1	-1	
A1	B2	1	0	-1	-1	0	
A2	B2	0	1	-1	0	-1	
A3	B2	-1	-1	-1	1	1	<-missing

The expected values for each cell are

	B1	B2
A1	$\mu + \alpha_1 + \beta_1 + \alpha\beta_{11}$	$\mu + \alpha_1 - \beta_1 - \alpha\beta_{11}$
A2	$\mu + \alpha_2 + \beta_1 + \alpha\beta_{21}$	$\mu + \alpha_2 - \beta_1 - \alpha\beta_{21}$
A3	$\mu - \alpha_1 - \alpha_2 + \beta_1 - \alpha\beta_{11} - \alpha\beta_{21}$	$\mu - \alpha_1 - \alpha_2 - \beta_1 + \alpha\beta_{11} + \alpha\beta_{21}$

Obviously, any cell with data has an expectation that is estimable. The cell that is missing has an expectation that is non-estimable. In fact, its expectation is precisely that linear combination of the design columns that is in the singularity report,

$$\mu - \alpha_1 - \alpha_2 - \beta_1 + \alpha\beta_{11} + \alpha\beta_{21} \; .$$

Suppose that you want to construct a test that compares the least squares means of B1 and B2. In this example, we can average the rows in the above table to get these least squares means.

$$\begin{aligned} \text{LSM(B1)} &= 1/3\,(\mu + \alpha_1 + \beta_1 + \alpha\beta_{11} + \\ &\qquad \mu + \alpha_2 + \beta_1 + \alpha\beta_{21} + \\ &\qquad \mu - \alpha_1 - \alpha_2 + \beta_1 - \alpha\beta_{11} - \alpha\beta_{21}) \\ &= \mu + \beta_1 \end{aligned}$$

$$\begin{aligned} \text{LSM(B2)} &= 1/3\,(\mu + \alpha_1 - \beta_1 - \alpha\beta_{11} + \\ &\qquad \mu + \alpha_2 - \beta_1 - \alpha\beta_{21} + \\ &\qquad \mu - \alpha_1 - \alpha_2 - \beta_1 + \alpha\beta_{11} + \alpha\beta_{21}) \\ &= \mu - \beta_1 \end{aligned}$$

$$\text{LSM(B1)} - \text{LSM(B2)} = 2\beta_1$$

Note that this shows that a test on the β_1 parameter is equivalent to testing that the least squares means are the same. But since β_1 is not estimable, the test is not testable, meaning there are no degrees of freedom for it.

Now, construct the test for the least squares means across the A levels.

$$\text{LSM(A1)} = 1/2(\mu + \alpha_1 + \beta_1 + \alpha\beta_{11} + \mu + \alpha_1 - \beta_1 - \alpha\beta_{11}) = \mu + \alpha_1$$

$$\text{LSM(A2)} = 1/2(\mu + \alpha_2 + \beta_1 + \alpha\beta_{21} + \mu + \alpha_2 - \beta_1 - \alpha\beta_{21}) = \mu + \alpha_2$$

$$\text{LSM(A3)} = 1/2(\mu - \alpha_1 - \alpha_2 + \beta_1 - \alpha\beta_{11} - \alpha\beta_{21} + \mu - \alpha_1 - \alpha_2 - \beta_1 + \alpha\beta_{11} + \alpha\beta_{21}) = \mu - \alpha_1 - \alpha_2$$

$$\text{LSM(A1)} - \text{LSM(A3)} = 2\alpha_1 + \alpha_2$$

$$\text{LSM(A2)} - \text{LSM(A3)} = 2\alpha_2 + \alpha_1$$

These both turn out to be nonestimable, but there is another comparison that is estimable, namely comparing the two A columns that have no missing cells.

$$\text{LSM(A1)} - \text{LSM(A2)} = \alpha_1 - \alpha_2$$

This combination is indeed tested by JMP using a test with 1 degree of freedom, though there are two parameters in the effect.

The estimability can be verified by taking its inner product with the singularity combination, and checking that it is zero:

parameters	singularity combination	combination to be tested
μ	1	0
α_1	-1	1
α_2	-1	-1
β_1	-1	0
$\alpha\beta_{11}$	1	0
$\alpha\beta_{21}$	1	0

It turns out that the design columns for missing cells for any interaction will always *knock out* degrees of freedom for the main effect (for nominal factors). Thus there is a direct relation between the non-estimability of least squares means and the loss of degrees of freedom for testing the effect corresponding to these least squares means.

How does this compare with what GLM does? GLM and JMP do the same test when there are no missing cells. That is, they effectively test that the least squares means are equal. But when GLM encounters singularities, it focuses out these cells in different ways depending on whether it is Type III or Type IV. For Type IV it just looks for estimable combinations that it can find. These may not be unique and if you reorder the levels you may get a different result. For Type III, it does some orthogonalization of the estimable functions to obtain a unique test. But the test may not be very interpretable in terms of the cell means.

The JMP approach has several points in its favor, though at first it might seem distressing that you may lose more degrees of freedom than with GLM.

1. The tests are philosophically linked to LSM's.
2. The tests are easy computationally, using reduction sum of squares for reparameterized models.
3. The tests are whole marginal tests, meaning they always go completely across other effects in interactions.

The last point needs some elaboration: Consider a graph of the expected values of the cell means in the previous example with a missing cell for A3B2.

Figure A.1 Expected Cell Means with a Missing Cell

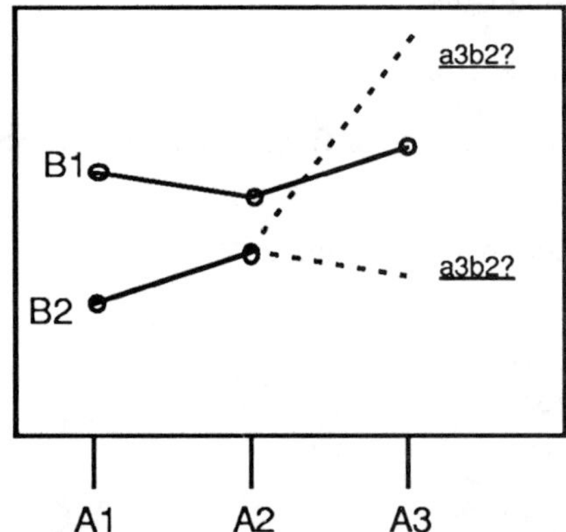

The graph shows the means of the A1 and A2 cells profiled across the B levels. What the JMP approach is saying is that you can't test the A main effect with a missing A2B3 cell, because the mean of the missing cell could be anything, as allowed by the interaction term. If the mean of the missing cell were the higher value shown, the A effect would test nonsignificant. If it were the lower value, it would test significant. The point is that you don't know. That is what the least squares means are saying when they are declared non-estimable. That is what the hypotheses for the effects should be saying too--that you don't know.

If you want to test hypotheses involving margins for subsets of cells, then that is what GLM Type IV does. In JMP you would have to construct these tests yourself by partitioning the effects with a lot of calculations or by using contrasts.

Ordinal Factors

Factors marked with the ordinal level of measurement are coded differently than nominal factors. The parameters estimates are interpreted differently, the tests are different, and the least squares means are different.

The theme for ordinal factors is that the first level of the factor is a control or baseline level, and the parameters measure the effect on the response as the ordinal factor is set to each succeeding level. The coding is appropriate for factors with levels representing various doses, where the first dose is zero.

JMP

A	a2	a3	
A1	0	0	control level, zero dose
A2	1	0	low dose
A3	1	1	higher dose

The pattern for the design is such that the lower triangle is 1's with zeros elsewhere.

For a simple main-effects model, this can be written:

$$y = \mu + \alpha_2 \mathbf{X}_{(a \le 2)} + \alpha_3 \mathbf{X}_{(a \le 3)} + \varepsilon$$

noting that μ is the expected response at $A = 1$, $\mu + \alpha_2$ is the expected response at $A = 2$ and $\mu + \alpha_2 + \alpha_3$ is the expected response at $A = 3$. Thus α_2 estimates the effect moving from $A = 1$ to $A = 2$, and α_3 estimates the effect moving from $A = 2$ to $A = 3$.

If all the parameters for an ordinal main effect have the same sign, then the response effect is monotonic across the ordinal levels. The parameterization is done with a future JMP feature in mind: the ability to impose and test monotonicity on the estimates (Barlow et al 1972, Robertson et al. 1988). We expect that monotonicity constraints could lead to tests of high power when compared to tests in unconstrained models.

Ordinal Interactions

The ordinal interactions, as with nominal effects, are produced with a horizontal direct product of the columns of the factors. Consider an example with two ordinal factors A and B, each with three levels.

JMP ordinal coding produces the design matrix:

		A		B		A*B			
A	B	A2	A3	B2	B3	A2B2	A2B3	A3B2	A3B3
A1	B1	0	0	0	0	0	0	0	0
A1	B2	0	0	1	0	0	0	0	0
A1	B3	0	0	1	1	0	0	0	0
A2	B1	1	0	0	0	0	0	0	0
A2	B2	1	0	1	0	1	0	0	0
A2	B3	1	0	1	1	1	1	0	0
A3	B1	1	1	0	0	0	0	0	0
A3	B2	1	1	1	0	1	0	1	0
A3	B3	1	1	1	1	1	1	1	1

The pattern for the interaction is a block lower-triangular matrix of lower-triangular matrices of 1's.

Hypothesis Tests for Ordinal Crossed Models

To see what the parameters mean, examine this table of the expected cell means in terms of the parameters, where μ is the intercept, α_2 is the parameter for level A2, and so forth.

	B1	B2	B3
A1	μ	$\mu+\beta_2$	$\mu+\beta_2+\beta_3$
A2	$\mu+\alpha_2$	$\mu+\alpha_2+\beta_2+\alpha\beta_{22}$	$\mu+\alpha_2+\beta_2+\beta_3+\alpha\beta_{22}+\alpha\beta_{23}$
A3	$\mu+\alpha_2+\alpha_3$	$\mu+\alpha_2+\alpha_3+\beta_2+\alpha\beta_{22}+\alpha\beta_{32}$	$\mu+\alpha_2+\alpha_3+\beta_2+\beta_3+\alpha\beta_{22}+$ $+\alpha\beta_{23}+\alpha\beta_{32}+\alpha\beta_{33}$

Note that the main effect test for A is really testing the A levels holding B at the first level. Similarly, the main effect test for B is testing across the top row for the various levels of B holding A at the first level. This is the approximate test for an experiment with doses of two different treatments. The main question is the efficacy of each treatment by itself, with fewer points devoted to looking for *drug interactions* when doses of both drugs are applied. In some cases it may even be dangerous to apply large doses of each drug.

Note that each cell's expectation can be obtained by adding all the parameters associated with each cell that is to the left and above it, inclusive of the current row and column. The expected value for the last cell is the sum of all the parameters.

Though the hypothesis tests for effects contained by other effects differs with ordinal and nominal codings, the test of effects not contained by other effects is the same. In the crossed design above, the test for the interaction would be the same no matter whether A and B were fit nominally or ordinally.

Ordinal Least Squares Means

As stated previously, least squares means are the predicted values corresponding to some combination of levels, after setting all the other factors to some *neutral* value. We define the neutral value for an effect with uninvolved ordinal factors as the effect at the first level, meaning the *control* or *baseline* level.

This definition of least squares means for ordinal factors maintains the idea that the hypothesis tests for contained effects are equivalent to tests that the least squares means are equal.

Singularities and Missing Cells in Ordinal Effects

With the ordinal coding, you are saying that the first level of the ordinal effect is the baseline. It is thus possible to get good tests on the main effects even when there are missing cells in the interactions---even if you have no data for the interaction.

Example with Missing Cell

The example is as above, with two observations per cell except that the A3B2 cell has no data.

Y	A	B
12	1	1
14	1	1
15	1	2
16	1	2
17	2	1
17	2	1
18	2	2
19	2	2
20	3	1
24	3	1

We now compare the results when the factors are coded nominally, and when they are coded ordinally. The model as a whole fits the same as seen in the tables shown in **Figure A.2.**

Figure A.2 Comparison of Summary Information for Nominal and Ordinal Fits

Nominal

Summary of Fit	
Rsquare	.8917322
Root Mean Square Error	1.483239
Mean of Response	17.2
Observations (or Sum Wgts)	10

Ordinal

Summary of Fit	
Rsquare	.8917322
Root Mean Sq ¨ Error	1.483239
Mean of Response	17.2
Observations (or Sum Wgts)	10

Nominal

Analysis of Variance

Source	**DF**	**Sum of Squares**	**Mean Square**	**F Ratio**
Model	4	90.6	22.65	10.2955
Error	5	11	2.2	**Prob > F**
C. Total	9	101.6		0.0125

Ordinal

Analysis of Variance

Source	**DF**	**Sum of Squares**	**Mean Square**	**F Ratio**
Model	4	90.6	22.65	10.2955
Error	5	11	2.2	**Prob > F**
C. Total	9	101.6		0.0125

The parameter estimates are very different because of the different coding. Note that the missing cell affects estimability for all the nominal parameters, but for none of the other ordinal parameters.

Figure A.3 Comparison of Parameter Estimates for Nominal and Ordinal Fits

Nominal

Parameter Estimates

Term		Estimate	Std Error	t Ratio	Prob>|t|
Intercept	Biased	18.083333	.741619	24.38	0.0000
A[1-3]	Biased	-3.833333	.856348	-4.48	0.0065
A[2-3]	Biased	-.3333333	.856348	-0.39	0.7131
B[1-2]	Biased	-0.75	.741619	-1.01	0.3583
A[1-3]*B[1-2]	Biased	-0.5	1.04880	-0.48	0.6537
A[2-3]*B[1-2]	Zeroed	0	0	•	•

Ordinal

Parameter Estimates

Term		Estimate	Std Error	t Ratio	Prob>|t|
Intercept		13	1.04880	12.40	0.0001
A[2]		4	1.48324	2.70	0.0429
A[3]		5	1.48324	3.37	0.0199
B[2]		2.5	1.48324	1.69	0.1527
A[2]*B[2]		-1	2.09761	-0.48	0.6537
A[3]*B[2]	Zeroed	0	0	•	•

The singularity details show the linear dependencies (and also identify the missing cell by examining the values).

Figure A.4 Comparison of Singularity Details for Nominal and Ordinal Fits

Nominal

Singularity Details

Term	A[2-3]*B[1-2]
Intercept	-1
A[1-3]	1
A[2-3]	1
B[1-2]	1
A[1-3]*B[1-2]	-1
A[2-3]*B[1-2]	-1

Ordinal

Singularity Details

Term	A[3]*B[2]
Intercept	0
A[2]	0
A[3]	0
B[2]	0
A[2]*B[2]	0
A[3]*B[2]	-1

The effect tests lose degrees of freedom for nominal. In the case of B there is no test. For ordinal, there is no loss, since there is no missing cell for the *base* first level.

Figure A.5 Comparison of Effects Tests for Nominal and Ordinal Fits

Nominal

Effect Tests

Source	Nparm	DF	Sum of Squares	F Ratio	Prob > F
A	2	1	24.5	11.1364	0.0206
B	1	0	0	•	•
A*B	2	1	0.5	0.2273	0.6537

Ordinal

Effect Tests

Source	Nparm	DF	Sum of Squares	F Ratio	Prob > F
A	2	2	81.3333333	18.4848	0.0049
B	1	1	6.25	2.8409	0.1527
A*B	2	1	0.5	0.2273	0.6537

The least squares means are also different. The nominal LSM's are not all estimable, but the ordinal LSM's are. You can verify the values by looking at the cell means. Note that the A*B LSM's are the same for the two.

Figure A.6 Comparison of Least Squares Means for Nominal and Ordinal Fits

Nominal

Least Squares Means

A

Level	Least Sq Mean		Std Error	Mean
1	14.25		.74161984871	14.2500
2	17.75		.74161984871	17.7500
3	22.25	NonEstimable	1.9621416870	22.0000

B

Level	Least Sq Mean		Std Error	Mean
1	17.333333333		.60553007081	17.3333
2	18.833333333	NonEstimable	1.3540064007	17.0000

A*B

Level	Least Sq Mean		Std Error
1,1	13		1.0488088481
1,2	15.5		1.0488088481
2,1	17		1.0488088481
2,2	18.5		1.0488088481
3,1	22		1.0488088481
3,2	22.5	NonEstimable	3.4785054261

Figure A.6 (continued)
Ordinal

Least Squares Means

A

Level	Least Sq Mean	Std Error	Mean
1	13	1.0488088481	17.9000
2	17	1.0488088481	18.1250
3	22	1.0488088481	6.5000

B

Level	Least Sq Mean	Std Error	Mean
1	13	1.0488088481	17.1429
2	15.5	1.0488088481	17.3333

A*B

Level	Least Sq Mean		Std Error
1,1	13		1.0488088481
1,2	15.5		1.0488088481
2,1	17		1.0488088481
2,2	18.5		1.0488088481
3,1	22		1.0488088481
3,2	23.5	NonEstimable	1.8165902124

Leverage Plot Details

Leverage plots generalize the partial residual leverage plots of Belsley, Kuh, and Welsch, 1980) to apply to any linear hypothesis. Suppose that the estimable hypothesis of interest is:

$$L\beta = 0$$

The leverage plot characterizes this test by plotting points so that the distance of each point to the sloped regression line displays the unconstrained residual, and the distance to the x-axis displays the residual when the fit is constrained by the hypothesis.

Of course the difference between the sums of squares of these two groups of residuals is the sum of squares due to the hypothesis, which becomes the main ingredient of the F test.

The parameter estimates constrained by the hypothesis can be written

$$\mathbf{b}_0 = \mathbf{b} - (\mathbf{X'X})^{-1}\mathbf{L}'\lambda$$

where **b** is the least squares estimate

$$\mathbf{b} = (\mathbf{X'X})^{-1}\mathbf{X'y}$$

and where lambda is the lagrangian multiplier for the hypothesis constraint, which is calculated

$$\lambda = (\mathbf{L}(\mathbf{X'X})^{-1}\mathbf{L}')^{-1}\mathbf{Lb} \; .$$

Compare the residuals for the unconstrained and hypothesis-constrained residuals, respectively.

$$\mathbf{r} = \mathbf{y} - \mathbf{Xb}$$

$$\mathbf{r}_0 = \mathbf{r} + \mathbf{X}(\mathbf{X'X})^{-1}\mathbf{L}'\lambda \; .$$

Constructing a Test on the Leverage Plot

To get a leverage plot, the x-axis values $\mathbf{v}_x$ of the points are the differences in the residuals due to the hypothesis, so that the distance from the line of fit (with slope 1) to the x-axis is this difference. The y-axis values $\mathbf{v}_y$ are just the x-axis values plus the residuals under the full model as illustrated in **Figure A.7**. Thus the leverage plot is composed of the points

$$\mathbf{v}_x = \mathbf{X}(\mathbf{X'X})^{-1}\mathbf{L}'\lambda$$

$$\mathbf{v}_y = \mathbf{r} + \mathbf{v}_x \; .$$

Figure A.7 Construction of Leverage Plot

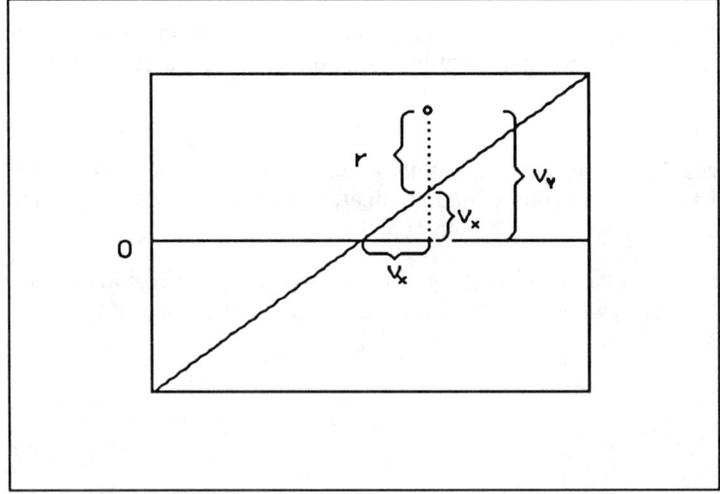

Superimposing a Test on the Leverage Plot

In simple linear regression, you can plot the confidence limits for the expected value as a smooth function of the regressor variable x.

$$\text{Upper}(x) = \mathbf{x}\mathbf{b} + t_{\alpha/2}\, s \sqrt{\mathbf{x}(\mathbf{X}'\mathbf{X})^{-1}\mathbf{x}'}$$

$$\text{Lower}(x) = \mathbf{x}\mathbf{b} - t_{\alpha/2}\, s \sqrt{\mathbf{x}(\mathbf{X}'\mathbf{X})^{-1}\mathbf{x}'}$$

Where $\mathbf{x} = [1\ x]$ is the 2-vector of regressors.

This hyperbola has the following nice property which makes it a useful significance-measuring instrument:

- If the slope parameter is significantly different from zero the confidence curve will cross the horizontal line of the response mean (left plot in **Figure A.8**).
- If the slope parameter is not significantly different from zero the confidence curve will not cross the horizontal line of the response mean (right plot in **Figure A.8**).
- If the t test for the slope parameter is sitting right on the margin of significance, the confidence curve will have the horizontal line of the response mean as an asymptote (middle plot in **Figure A.8**).

This property can be verified algebraically, or it can be understood by considering what the confidence region has to be when the regressor value goes off to plus or minus infinity.

Figure A.8 Cases of Significant, Borderline, and Non-significant Confidence Curves.

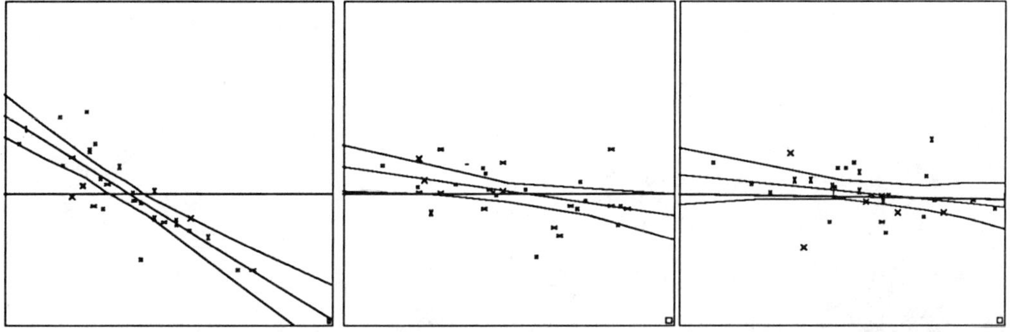

On leverage plots, we can use the same device by calculating a confidence function with respect to one variable while holding all other variables constant at their mean value in the sample.

For general hypothesis tests, we can display a curve with the same properties: we show the confidence interval at the mean value in the middle, and adjust the curvature so that it crosses the horizon if the F test is significant at some α level like .05.

Consider the following functions,

$$\mathbf{upper}(z) = z + \sqrt{s^2 t^2_{\alpha/2} \bar{h} + \frac{F_\alpha}{F} z^2}$$

and

$$\mathbf{lower}(z) = z - \sqrt{s^2 t^2_{\alpha/2} \bar{h} + \frac{F_\alpha}{F} z^2}$$

where F is the F statistic for the hypothesis, F_α is the reference value for significance α, and $\bar{h} = \bar{x}(X'X)^{-1}\bar{x}'$, where $\bar{x}$ is the regressors at a suitable middle value, such as the mean.

These functions serve the same properties listed above. If the F statistic is greater than the reference value, the confidence functions cross the x-axis. If the F statistic is less than the reference value, the confidence functions do not cross. If the F statistic is equal to the reference value, the confidence functions have the x-axis as an asymptote. And the range between the confidence functions at the middle value is a valid confidence region of the predicted value at the point.

Appendix B. Questions and Answers

Q? What is the difference between JMP and JMP IN ?

A: JMP IN is an introductory version of JMP that is best suited for student and demonstration use. The only difference between JMP IN and JMP is that JMP IN limits the following operations to 500 cells.

- **Save** and **Save As**
- **Import** and **Export**
- **Copy**, **Paste**, and **Paste at End**

There are many useful problems that fit within these limits. You can use JMP IN to analyze larger data tables, but if you enter or create data tables with more than 500 cells you can not save them directly. You can analyze JMP files in the SAMPLE DATA folder distributed with both JMP and JMP IN, or tables available to you in some other way, such as from an instructor in a statistics course or files sent to you by a colleague or coworker. You can reassemble a large data table whenever you want to use it with repeated copy-and-paste operations or concatenation but this becomes tedious. If you have data tables larger than 500 cells, an upgrade from JMP IN to JMP is the best solution.

JMP IN comes with the same documentation as JMP. Your JMP IN disk has trade-in value. You can return your copy of JMP IN and receive JMP at a slightly reduced price from SAS Institute.

Q? What do I do when I run out of memory?

A: If you are in the middle of a session, close any windows that you don't need. When you close windows, it frees memory both for the data and for the the program code used by the window.

Also, if you are running Multifinder, you can make available Multifinder memory to JMP by going to the preferences dialog and clicking Multifinder Memory if is not already checked. This option makes memory outside the region dedicated to the application available. However, using this method for the long term use does violate some guidelines from Apple, though it seems to work fine in practice.

If you get out of JMP, you can reset its *Application Memory Size*, which is the memory JMP tries to get when it is launched. Under Finder, select the JMP application file and issue the **Get Info** command under the **File** menu. This brings up a window of information about the JMP file. In the lower right corner you see a field labeled *Application Memory Size (K)* , in which you can type to the amount of memory you think you may need. We distribute JMP with that field set to 800K, but recommend you set it to several times that value if you have enough memory.

If you have only 1 meg of memory, we recommend that you install more, because JMP is fairly limited in what it can do in a 1-meg machine.

Q? I have a Version 5 SAS data set. How do I get it into a JMP data table?

A: JMP imports SAS transport data libraries directly. First, copy the SAS data set into a SAS transport data library. Under Version 5 on a mainframe operating system you use PROC XCOPY with the EXPORT option as follows.

```
PROC XCOPY  IN=libref  OUT=libref EXPORT;
RUN;
```

On a mini-computer system, you use the statements

```
PROC COPY IN=libref  OUT=libref EXPORT;
RUN;
```

In both examples above, IN=*libref* references a standard format SAS data library and the OUT=*libref* references a transport data library with a record format of FB, a logical record length of 80 and a blocksize of 8000. The procedures COPY and XCOPY procedures are documented in the *SAS User's Guide: Basics,Version 5 Edition.*

The transport data library can then be transferred to the Macintosh through any number of methods such as TCP/IP, FTP, asyncronous connection, and so on. Whichever method you use, you must specify a BINARY file transfer.

Q? I have a Version 6 SAS data set. How do I get it into a JMP data table?

A: The process to transfer a Version 6 SAS data set to JMP is a little simpler than the Version 5 transfer. The Version 6 SAS system uses a data engine which tells SAS what format a data library is and the translation takes place transparently. The engine that you need to use is called SASV5XPT. For example, you specify the engine on the LIBNAME statement with the statement

```
LIBNAME MYLIB SASV5XPT 'C:\subdir\filename.ext;
```

Now, any time that you reference the data library, MYLIB, you access a transport data library. Use PROC COPY to copy data from one library into the transport data library.

```
PROC COPY IN=libref OUT=MYLIB MEMTYPE=DATA;
RUN;
```

These statements copy all data sets from the IN=*libref* into the transport data library called MYLIB. The transport data library can then be copied onto the Macintosh using the Apple File Exchange with the Default Translation. Once on the Macintosh, the file can be opened using the **Import** command found in the JMP **File** menu. Select the **SAS Transport** option and and click **OK**.

Q? But how do I actually (physically) get the file from the 'other' system to the Macintosh.

A: The two approaches to transfer of information are diskette media and communications.

Diskette media is one simple way to transfer information. The most common is through a DOS-formatted diskette and the Apple File Exchange. The Apple File Exchange is documented in the Macintosh Utilities User's Guide from Apple. It reads files from DOS-formatted diskettes in either Apple's 5.25 PC Drive or the FDHD™ 3.5 inch drive.

Communications is the other approach, and here the opportunities are numerous. The simplest way is through a terminal emulator and file-transfer package, usually through the serial port. Kermit from Columbia University is popular, especially since it is usually available on the 'other side', whatever it may be. Many other packages are on the market. Many users hook on to other machines through a network connection, either through on-board hardware, or through an Appletalk gateway. It is possible to hook into IEEE 802.5 Token Ring, IBM 3270 SNA, or IEEE 802.3 Ethernet communicating through various upper layers. Your data might be available through file transfer, list capture, or in some cases as a server file. If on the mainframe, you may have to make sure that it is translated EBCDIC or ASCII, or if it is a SAS binary transport file, that it is not so translated. You may have to translate line ending codes: Apple uses Carriage Returns (CR) where as Vax uses LineFeeds(LF) and PCs use the combination (CR,LF). The communcations package you use should be able to handle all these differences.

Q? What is the difference between the **Paste** and **Paste at End** commands?

A: The **Paste** command requires rows or columns (or both) to be selected and pastes data from the clipboard into the region defined by the selection, writing over any values that are there.

The **Paste at End** command uses selected rows or selected columns *but not both.* If there are selected rows, **Paste at End** creates new columns and pastes the data into the new columns at the end of the selected rows. If there are selected columns, **Paste at End** creates new rows and pastes the data into the new rows at the end of the selected columns. The **Paste at End** command never writes over values in the spreadsheet.

Note: If there are no selected rows or columns, **Paste at End** copies the data from the clipboard to the end of the spreadsheet. It behaves as if all the columns are selected. To transfer data from another application into JMP, copy it from the application to the clipboard. Then use the **New** command from the **File** menu to create a new empty spreadsheet. When you select **Paste at End**, the contents of the clipboard are written to the new table with columns and rows created as needed.

Q? Different import options appear when I click the **SAS Transport** option in the JMP **Import** dialog. What do they mean?

A: A SAS transport file can contain multiple members of a SAS data library. JMP cannot tell the number of members nor the name and size of each member until the entire transport file is read. By default, JMP reads the entire SAS transport file and creates a spreadsheet window for each member. If the transport file is large and has many members, this can be very memory demanding! The **SAS Transport** import options let you save the members as JMP files rather than opening spreadsheet windows for them, or select only the one that you want to open as JMP spreadsheet windows.

Q? What is the header information used for that is exported by the **Save As** command when I click on **Text format** and the **Header** option?

A: These options write a JMP data table in text format with JMP header information preceding the data lines. The header contains information about the JMP data table, as well as information about the columns. The following is an example of a header extracted from a JMP export file. The header is followed by the data values arranged in rows and columns as they are organized in the JMP data table

```
Header Begin   TableName="big class"
   NRows=45
   NColumns=5
   Mode=ReadOnly
   TableNotes="this is a test"
   Column Begin
      Label="name"
      DataType=Character
      FieldWidth=0
      Measure=Nominal
      Notes="Student's Name"
   End
         .
         .     more column header information
         .
   Column Begin
      Label="height"
      DataType=Numeric
      Format=Best12.
      Measure=Interval
   End
End
ALFRED  M  14  69.0  112.5
ALICE   F  13  56.5   84.0
BARBARA F  13  65.3   98.0
      .
      .
      more data lines
```

Because the header is just a set of information written by JMP at the beginning of a standard text file, it can easily be created with a text editor. The format of JMP export header is fairly flexible. There can be multiple items on a line. The header can be as complete as the one shown above created by JMP, or as brief as

```
Header NRows = 5
```

The keyword `Header` is required to identify the beginning of header information. The keywords `Begin` and `End` are used both for the main header and for the column information to bracket more than one item in a list. Text strings with embedded blanks should be enclosed in double quotes ("). Keywords have to be spelled exactly as they appear in the JMP export header. The key words are stored in the STR# resource 21266.

Table B.1 and **Table B.2** summarize the keywords and the column information used as JMP header information.

Table B.1 Summary of JMP Header Keywords

Key Word	Information
`Header`	start of the JMP Export Header
`TableName=`	data table name
`NRows=`	maximum number of rows to be read
`NColumns=`	maximum number of columns
`Mode=`	the word `ReadOnly` if the data table is locked; omit this information otherwise
`TableNotes=`	data table notes which appear in the **Notes** field of the **Table Info** dialog
`Column`	start of column definition

Table B.2 Summary of Items inside a Column Definition List

Key Word	Information
`Label=`	a string of up to 31 characters for column name
`DataType=`	one of the words Numeric, Character, RowState, which define the data type of the column
`FieldWidth=`	maximum number of characters for a character column
`Format=`	format specification for a numeric column. usually in the form n.d, where n is the number of display position for the number, and d is the number of decimal places; the special key word **Best** can precede the format specification as in the case of the variable Height in the above example.
`Measure=`	the measurement level of the column
`Mode=`	the keyword Lock if the column is locked; omit this information otherwise
`Notes=`	column notes which appear in the Notes field of the **Col Info** dialog

This text header is very useful for importing data. When importing data that has no JMP header information (the usual situation), JMP must make some basic assumptions about the data which may or may not reflect the intended usage. For example, JMP interprets a column of integer data as numeric when importing. Should the data happen to be row state data, JMP still reads it as simple numeric data. If you define the data type of the column in the header of the text file as RowState, JMP interprets the column of data as row state data instead of numeric data.

Another useful example is to add the `NRows=` header information at the beginning of a data file and specify the maximum number of rows for JMP to import. This gives you the ability to limit the number of rows that JMP reads from the text file.

Sources and References

Agresti, Alan (1984), *Analysis of Ordinal Categorical Data*, New York: John Wiley & Sons, Inc.

Atkinson, A.C. (1985), *Plots, Transformations and Regression*, Oxford: Clarendon Press.

Barlow, R.E., D.J. Bartholomew, J.M. Bremner, and H.D. Brunk (1972), *Statistical Inference under Order Restrictions*, New York: John Wiley & Sons, Inc.

Becker, R.A., and Cleveland, W.S. (1987), "Brushing Scatterplots", *Technometrics*, 29:2.

Belsley, Kuh, and Welsch (1980), *Regression Diagnostics*, New York: John Wiley & Sons.

Box, G.E.P., Hunter,W.G., and Hunter, J.S. (1978), *Statistics for Experimenters*, New York: John Wiley & Sons, Inc.

Cook, R.D. and Weisberg, S. (1982), *Residuals and Influence in Regression*, New York: Chapman and Hall.

Cox, D.R. (1970), *The Analysis of Binary Data*, London: Metheun.

Daniel C. and Wood, F. (1980), *Fitting Equations to Data*, Revised Edition, New York: John Wiley & Sons, Inc.

Draper, N. and Smith, H. (1981), *Applied Regression Analysis*, 2nd Edition, New York: John Wiley & Sons, Inc.

Eppright, E.S., Fox, H.M., Fryer, B.A., Lamkin, G.H., Vivian, V.M., and Fuller, E.S. (1972), "Nutrition of infants and preschool children in the north central region of the United States of America," *World Review of Nutrition and Dietetics*, 14.

Eubank, R. L., (1988), *Spline Smoothing and Nonparametric Regression*, New York: Marcel Dekker.

Fisherkeller, M.A., Friedman, J.H., and Tukey, J.W. (1974), "PRIM-9: An Interactive Multidimensional Data Display and Analysis System", SLAC-PUB-1408, Stanford Linear Accelerator Center, Stanford, CA.

Gabriel, K.R. (1982), "Biplot", *Encyclopedia of Statistical Sciences*, Volume 1, Kotz and Johnson editors, New York: John Wiley & Sons, Inc.

Goodnight, J.H. (1978), "Tests of Hypotheses in Fixed Effects Linear Models", *SAS Technical Report R-101*, Cary: SAS Institite Inc, also in *Communications in Statistics* (1980) A9 167-180.

Goodnight, J.H., and W.R. Harvey (1978), "Least Square Means in the Fixed Effect General Linear Model", *SAS Technical Report R-103*, Cary NC: SAS Institute Inc.

Hauck, W.W., and Donner, A. (1977) "Wald's Test as Applied to Hypotheses in Logit Analysis" *Journal of the American Statistical Assocation*, 72, 851-863. (why Wald tests are undesirable)

Harrell, Frank (1986), "The Logist Procedure", *SUGI Supplemental Libary User's Guide*, Version 5 Edition, SAS Institute Inc, Cary NC.

Harris, R.J. (1975), *A Primer of Multivariate Statistics*, New York: Academic Press.

Hartigan J.A. and B. Kleiner (1981), "Mosaics for contingency tables", *Proceedings of the 13th Symposium on the Interface between Computer Science and Statistics,* W. F. Eddy editor, New York: Spinger.

Hawkins, D.M., (1974), "The Detection of Errors in Multivariate Data Using Principal Components", *Journal of the American Statistical Assocation*, 69.

John, P. M. (1971) Statistical Design and Analysis of Experiments, New Yord, Macmillan.

Koehler, Grigorus, Dunn (1988), "The Relatiohnship Between Chemical Structure and the Logarithm of the Partition", QSAR, 7.

Leven, Joel R., Ronald C. Serlin, and Lisa Webne-Behrman (1989), "Analysis of Variance Through Simple Correlation", *American Statistician*, 43.

Myers, Ray, "Response Surface Methodology", Virginia Polytechnic and State University.

Mosteller, F. and Tukey, J.W. (1977), *Data Analysis and Regression,* Reading Mass: Addison-Wesley.

McCullagh, P. and Nelder, J.A. (1983), *Generalized Linear Models*, London: Chapman and Hall Ltd.

Nelson, Forrest (1976), "On a General Computer Algorithm for the Analysis of Model with Limited Dependent Variables", *Annals of Economic and Social Measurement*, 5/4.

Rawlings, John O. (1988), *Applied Regression Analysis: A Research Tool*, Pacific Grove CA: Wadsworth & Books/Cole.

Reinsch, C.H. (1967), *Smoothing by Spline Functions*, Numerische Mathematik, 10, 177-183.

Robertson, T., Wright, F.T., and R.L. Dykstra, R.L., (1988), *Order Restricted Statistical Inference*, New York: John Wiley & Sons, Inc..

SAS Institute (1987), *SAS/Stat Guide for Personal Computers, Version 6 Edition*, Cary NC: SAS Institute Inc.

Snedecor and Cochran (1967), *Statistical Methods*, Ames Iowa, Iowa State University Press.

Walker, S.H. and Duncan, D.B. (1967), "Estimation of the Probability of an Event as a Function of Several Independent Variables", *Biometrika*, 54.

Winer, B.J. (1971), *Statistical Principals in Experimental Design*, 2nd Edition, New York: McGraw-Hill, Inc.

A

B

C

T

U

V

W

SOFTWARE AGREEMENT

The enclosed software is provided on the express condition that you consent to the terms of this Agreement. OPENING THE DISK PACKAGE INDICATES THAT YOU ACCEPT THIS LICENSE AGREEMENT AND WILL ABIDE BY IT. If you cannot agree to the terms of this Agreement, do not open the disk package, but return it and accompanying items to the place from which you obtained them for a full refund.

JMP IN® software is copyrighted. You may store the software on your personal computer and make backup copies of the software for archiving purposes only, provided that each copy of the software residing on transportable media contains a copyright notice identical to the one set forth in the original media. You agree not to reverse engineer the software or mask or suppress any copyright notice.

By paying the fee for a single copy of the software, one user may use the software on one workstation at a time. You may install the software on a file server in a network as long as there is a licensed copy of the software corresponding to each current user.

SAS INSTITUTE INC. DISCLAIMS ALL WARRANTIES, EXPRESS OR IMPLIED, INCLUDING BUT NOT LIMITED TO ANY IMPLIED WARRANTIES OF MERCHANTABILITY AND/OR FITNESS FOR A PARTICULAR PURPOSE. IN NO EVENT WILL THE INSTITUTE'S LIABILITY FOR ANY ACTION, INCLUDING LOST PROFITS, INCIDENTAL OR CONSEQUENTIAL DAMAGES ARISING OUT OF YOUR USE OF THE SOFTWARE, EXCEED THE FEE YOU PAID FOR THE SOFTWARE EVEN IF THE INSTITUTE OR AN AUTHORIZED INSTITUTE DEALER HAS BEEN ADVISED OF THE POSSIBILITY OF SUCH DAMAGES. SOME STATES DO NOT ALLOW THE LIMITATION OR EXCLUSION OF LIABILITY FOR INCIDENTAL OR CONSEQUENTIAL DAMAGES, SO THE ABOVE LIMITATION OR EXCLUSION MAY NOT APPLY TO YOU.

This Agreement constitutes the complete agreement between you and the Institute. Any additions or modifications must be approved in writing by both parties. The terms of the Agreement supersede the terms of any purchase order or other instrument you may issue for the software provided under this Agreement. This Agreement is governed by the laws of North Carolina.

U.S. GOVERNMENT RESTRICTED RIGHTS

For U.S. Federal Government Customers:

Notice: Notwithstanding any other purchasing documents, lease or license agreement that may pertain to, or accompany the delivery of, this computer software, the rights of the Government regarding its use, reproduction and disclosure are as set forth in this Agreement, or, if applicable in DFAR 252.227-7013(c)(1)(ii).

JMP IN is a registered trademark of SAS Institute Inc. A footnote should accompany the first use of each trademark or registered trademark and should state that the referenced trademark is used to identify products or services of SAS Institute Inc.